Plasma Astrophysics

Plasma Astrophysics

Nonthermal Processes in Diffuse Magnetized Plasmas

VOLUME 1 The Emission, Absorption and Transfer of Waves in Plasmas

D. B. MELROSE

Department of Theoretical Physics
Faculty of Science
Australian National University

GORDON AND BREACH SCIENCE PUBLISHERS
New York London Paris

Gordon and Breach, Science Publishers, Inc.
One Park Avenue
New York, NY10016

Gordon and Breach Science Publishers Ltd.
42 William IV Street
London WC2N 4DE

Gordon & Breach
7–9 rue Emile Dubois
Paris 75014

Library of Congress Cataloging in Publication Data

Melrose, D. B., 1940–
Plasma astrophysics.
Bibliography: v. 1, p. ; v. 2, p.
Includes index.
CONTENTS: v. 1. The emision, absorption, and transfer of waves in plasmas.—v. 2. Astrophysical application.
1. Astrophysics. 2. Plasma (Ionized gases)
I. Title.
QB461.M46 523.01 77–26360
ISBN 0–667–03490–3

Library of Congress catalog card number 77–26360. ISBN 0 677 02340 5. 2-vol. set ISBN 0 677 03490 3.
 Printed in Great Britain.

Contents of Volume 1

Preface

A little more than a decade ago "plasma astrophysics" became the accepted name for the particular sub-branch of astrophysics concerned with microscopic plasma processes in astrophysical contexts. The main topics included in plasma astrophysics are the emission, absorption and transfer of radiation, mainly at radio wavelengths, and the scattering and acceleration of fast particles. At least for the purpose of this book, other plasma-physical topics in "cosmic electrodynamics" such as the MHD theory (the macroscopic theory of plasmas) and orbit theory (the motion of charged particles) are regarded as of secondary interest.

The important emission processes in radio astronomy are synchrotron emission, plasma emission (in solar radio bursts), and bremsstrahlung. Although synchrotron emission and bremsstrahlung may be treated using conventional classical electrodynamics, plasma emission involves some quite sophisticated plasma concepts. The transfer problem, namely that of how emission and absorption within a source leads to escaping radiation, can be complicated; besides emission and absorption, scattering, Faraday rotation and mode-coupling can all affect the escaping radiation.

Astrophysical plasmas are usually permeated by magnetic fields and normally contain suprathermal particle populations. These fast particles are scattered on a suprisingly short time scale. The theory of "resonant scattering" by waves was developed in order to explain the rapid scattering of particles in the magnetosphere and in the interplanetary medium. In these contexts, resonant scattering has proved to be successful, both qualitatively and semi-quantitatively; it has been less successful in explaining the properties of galactic cosmic rays. Resonant scattering is also an important ingredient in some current ideas on the acceleration of fast particles.

There are interesting parallels and important differences in emphasis in the development of plasma ideas for laboratory plasmas and for naturally occurring plasmas. The two earliest plasma-physical theories were developed independently in the late 1920's in connection with Langmuir's laboratory experiments on arc discharges and with the propagation of radio waves in the ionosphere. Another early theory whose initial development was motivated by astrophysical problems in the MHD theory. There was a particularly productive exchange of ideas concerning synchrotron radiation, which became of interest in connection with laboratory machines (synchrotrons) just before it became recognized as the most important radiation process in radio astronomy. Another parallel development came in the theory of nonlinear processes which was developed initially in the late 1950's both in connection with the ionosphere, laboratory plasmas and solar radio bursts.

There are two important differences in emphasis between astrophysical and laboratory plasma physics. One concerns dimensions: in plasma astrophysics the only observable phenomena are those which occur over large volumes and long times; by contrast, laboratory plasmas are confined to relatively small volumes and short times. Consequently, many of the important phenomena in the laboratory are unimportant boundary or transient effects in astrophysical contexts. Further, because of the small length scales and short time scales of laboratory devices, many plasma astrophysical effects can be simulated only with difficulty, if at all. However, when such simulation is possible a great deal of additional information can be obtained from laboratory experiments due to the possibilities of studying the microscopic processes directly and of making controlled changes in parameters (the study of anomalous electrical conductivity is an example).

The other major difference in emphasis between astrophysical and laboratory plasma physics stems from the fact that in astrophysics the observer is outside the source and is concerned with how the observed radiation (in the form of waves or particles) is generated and how it is modified as it escapes from the source. In the laboratory one can probe the plasma internally and observe waves which could not escape from an astrophysical plasma. A more subtle point is that such waves can generate radiating currents in surface layers of the plasma and thereby allow external detection of these waves. Such a surface phenomenon is the intermediate mechanism which allows radio receivers on the surface of the Earth to detect whistler waves trapped in the magnetosphere. At least in this connection the magnetosphere is more analogous to a laboratory plasma than to astrophysical plasma: we are inside the source and we are concerned with trapped radiation which could not be observed directly in an astrophysical source.

In this book I have attempted, firstly, to give a thorough and systematic development of the basic plasma theory involved in radiation processes and in resonant scattering; and, secondly, to discuss the astrophysical applications as illustrations of the use of the theory and as motivation for its further extension. The book has been designed for two kinds of readers: graduate students in astrophysics or plasma physics, and research workers in these or related fields. It should be useful as a textbook for a graduate course on plasma astrophysics and, with this use in mind, problem sets have been included at the end of each of Chapters 1 through 6. Also, it should be useful as a reference book for research workers who frequently or occasionally encounter plasma physical problems. The book is written at a relatively advanced level and although no detailed knowledge of either plasma physics or astrophysics is assumed, a general background knowledge of both subjects is desirable.

The book is divided into two volumes, the first emphasizing the plasma theory and the second emphasizing the astrophysical applications. The systematic treatment of plasma astrophysical problems is based on a general theory developed in Volume 1 as follows. A general notation for waves in different wave modes is introduced (Section 2.3) and used in deriving a general formula for the emission of waves in an arbitrary mode by an arbitrary source (Section 3.1). The results are used (Section 3.2) to rederive standard formulae for emission by classical particles *in vacuo* and to treat the specific emission processes of interest in plasma astrophysics (Sections 3.3 to 4.4). Then the semi-classical (or "semi-quantum") theory is introduced (Section 5.1) simply by regarding the waves as a collection of wave quanta. The Einstein coefficients are used (Section 5.2) to relate absorption to emission, and finally the quasi-linear equations are derived (Section 5.3). Although the resulting theory has its limitations (it cannot be used to treat intrinsically growing waves or nonresonant diffusion, for example) the vast majority of plasma astrophysical problems can be treated using it.

Among those who have assisted me in various ways towards the writing of this book, foremost is Dr. A. G. W. Cameron who first gave me the opportunity and encouragement to develop my interest in plasma astrophysics. I have also profited from many discussions with Dr. D. G. Wentzel, the late Dr. D. F. Smerd, Dr. G. A. Dulk, and Prof. H. A. Buchdahl.

Department of Astro-Geophysics
University of Colorado

D. B. MELROSE

CHAPTER 1

Introduction

Plasma astrophysics is the application of microscopic plasma theory to diffuse (i.e. low-density) ionized gases of astrophysical interest. These diffuse plasmas include the solar corona and its extension into the interplanetary medium (the "solar wind"), the magnetosphere of Jupiter, the interstellar medium, (astrophysical) sources of synchrotron radiation and pulsar magnetospheres. Radiation from these diffuse plasmas is observed at radio wavelengths and an important part of plasma astrophysics is the theory of the associated radiation processes. There are very few radiation processes which are known to be important; indeed bremsstrahlung, synchrotron radiation, and plasma emission (from the solar corona) are the only well-established mechanisms, although it is clear that the decametric emission from Jupiter and pulsar emission are not due to any of these. Another important class of problems involves resonant particle–wave interactions in these plasmas. Resonant interactions can lead to very effective scattering of fast particles, whose presence seems to be a characteristic feature of astrophysical plasmas. Resonant scattering can also play an important role in the acceleration of fast particles.

Not surprisingly, the amount of information we have concerning the astrophysical plasmas mentioned above decreases roughly with increasing distance from the observer. We have most information on the geophysical plasmas (the ionosphere and magnetosphere) which can be sampled directly by detectors on space craft. Although these geophysical plasmas will be mentioned where appropriate below, no systematic discussion of them will be given. Magnetospheric plasma physics has a significantly different emphasis from plasma astrophysics. This is essentially because we are inside the magnetosphere and see trapped particles and radiation, whereas we see only escaping radiation and particles from astrophysical sources. Of the astrophysical sources, we have most information concerning the solar corona and the interplanetary medium (which is where astrophysical and geophysical interests overlap). Consequently, it is in connection with the solar corona that the most sophisticated plasma astrophysical ideas are required. For sources outside the solar system, detailed information on the plasma parameters is meagre, and relatively simple plasma astrophysical ideas are appropriate. The important types of process are those mentioned above, namely radiation processes and resonant wave–particle interactions.

Virtually all our information concerning astrophysical plasmas comes from radio astronomy, and it is appropriate to start with an outline of the development of radio astronomy (Section 1.1). In Section 1.2 the development of plasma physics is outlined with particular emphasis on the theory of waves, and some simple theories of wave motions are given in Section 1.3. The important plasma parameters are summarized in Section 1.4.

1.1 EMISSION PROCESSES IN RADIO ASTRONOMY

Radio astronomy began with the first successful observation of radio noise from our own galaxy, the Milky Way, by Jansky in 1932 (Jansky, 1932, 1935) at the relatively low frequency of 20.5 MHz. The galactic continuum was observed also by Reber (1940) at 160 MHz. In 1942 the Sun was identified as a source in the microwave range (3 to 10 GHz) by Southworth (1945). Earlier in the same year Hey recognized that periodic jamming of radar (at around 30 MHz) was due to the Sun and he pointed out that the jamming correlated with flaring activity in a large group of sunspots (Hey, 1946). After World War II three groups of radar physicists, two in England and one in Sydney, initiated programs in radio astronomy. The first successes were by Hey's group, who inferred the existence of the first descrete source (Hey, Parsons and Phillips, 1946) which was subsequently localized by Bolton and Stanley (1948), and by Pawsey's group who used an ingenious technique to pinpoint the source of solar radio noise as a specific group of sunspots. Subsequent notable developments in radio astronomy were the first identification of a radio source (Taurus A) with a known astronomical object (the Crab Nebula) by Bolton (1948), the first identification of an extragalactic source by Bolton, Stanley and Slee (1949), observation of the 21 cm line of neutral hydrogen in the galactic spectrum by Ewen and Purcell (1951) as had been predicted by van de Hulst (1945), the identification of Jupiter as the source of radio bursts in the decametric range by Burke and Franklin (1955), the gradual recognition of the spectacular properties of certain quasi-stellar objects, later called quasars, and the discovery of pulsars by Hewish, Bell *et al.* (1968). Solar radio astronomy is discussed separately below.

Nonthermal (Synchrotron) Emission

From the earliest days it was recognized that the observed radio emission must be nonthermal. The idea is that from a knowledge of the observed flux density, the angular size of the source and the distance to the source, one can estimate the (radio) surface brightness temperature. Let us indicate how this is done.

A radio antenna (i.e. a radio telescope), when pointed in a particular direction records the intensity of radio waves in the beam solid angle (or beam area), $\Delta\Omega_A$ say. (The beam pattern of specific antennas may be quite complicated, with a main beam and various side lobes.) A particular astronomical source is resolved by the antenna if it fills a solid angle $\Delta\Omega_s$ greater than $\Delta\Omega_A$. For $\Delta\Omega_s < \Delta\Omega_A$ the source is unresolved.

The observed quantity is the *flux density* $F(f)$, and it is usually expressed in flux units (f.u.),

$$1 \text{ f.u.} = 10^{-26} \text{ Wm}^{-2} \text{ Hz}^{-1} = 10^{-23} \text{ erg s}^{-1} \text{ cm}^{-2} \text{ Hz}^{-1}. \tag{1.1}$$

A flux unit is sometimes called a Jansky. The *brightness* or *specific intensity* of the source, $I(f)$ say, is the flux density per unit solid angle. The apparent brightness $I_A(f)$ is obtained by dividing the observed flux density by the beam solid angle,

$$I_A(f) = \frac{F(f)}{\Delta\Omega_A}. \tag{1.2}$$

If the source is resolved, $I_A(f)$ is the brightness of the region of the source under observation. If the source is unresolved the brightness at the source is

$$I_s(f) = \frac{\Delta\Omega_A}{\Delta\Omega_s} I_A(f) = \frac{F(f)}{\Delta\Omega_s}. \tag{1.3}$$

It is usually the brightness $I(f)$ which is to be compared with the corresponding quantity calculated for a model of the source.

It should be remarked that the specific intensity $I(f)$ is defined here per unit cyclic frequency f, whereas for formal purposes the specific intensity $I(\omega)$ per unit angular frequency is more convenient. The two are related by

$$I(f) = 2\pi I(\omega), \qquad \omega = 2\pi f. \tag{1.4}$$

The brightness temperature $T_b(f)$ may be defined in terms of $I(f)$ by writing the intensity for thermal radiation and equating the temperature to $T_b(f)$. The relevant thermal intensity is that in the classical (Rayleigh–Jeans) limit. Thus one has

$$I(f) =: \frac{2f^2}{c^2} T_b(f), \tag{1.5}$$

where units in which Boltzmann's constant is equal to unity are used. The symbols := and =: are used to imply that the written relation defines the quantity on the left and right respectively. Now, if the source were at a temperature T then its surface brightness temperature would be $T_b(f)=T$ if it were optically thick and $T_b(f)<T$ if it were optically thin. (A source if optically thin if the mean free path for absorption of a photon is much greater than the size of the source.) Besides implying unacceptably high temperatures (for thermal emission), the observed brightness temperatures depend on frequency and this is incompatible with thermal emission. Typically radio sources have power-law spectra, $F(f)\propto f^{-\alpha}$, with "spectral index" $\alpha>0$.

It is now widely accepted that synchrotron radiation is the nonthermal emission mechanism for our galaxy, for a class of discrete galactic sources and for extragalactic sources. This mechanism was first suggested by Alfvén and Herlofson (1950). Independently, Kiepenheuer (1950) suggested that the galactic radio background might be synchrotron radiation from the electron component of galactic cosmic rays. The theory of synchrotron radiation is developed in Sections 4.2, 4.3 and 6.3. Briefly, synchrotron radiation is emission by relativistic electrons due to their spiralling motion in a magnetic field. For ultrarelativistic electrons (energy $\varepsilon \gg m_e c^2$) with a power-law energy spectrum $N(\varepsilon)$ (the number of electrons per unit energy range) proportional to ε^{-a}, the predicted synchrotron spectrum is a power-law $I(f)\propto f^{-\alpha}$ with

$$\alpha=\tfrac{1}{2}(a-1). \tag{1.6}$$

A typical value of α is around unity (with $0.5\lesssim\alpha\lesssim 1.5$ for most sources) implying $a\approx 3$ as a typical value.

Solar Radio Astronomy

Compared with the interpretation of most galactic and extragalactic sources as synchrotron sources, the identification of the emission processes involved in solar radio astronomy proved formidable. The initial definitive classification of meter wavelength (30 to 300 MHz) solar radio bursts was due to Wild (Wild and McCready, 1950; Wild, 1950a, b). He classified the bursts into types I, II, and III. (Types IV and V were added subsequently, by Boischot (1957) and by Wild, Sheridan and Trent (1959) respectively.) In both types II and III, emission is observed at two harmonically related frequencies with steady drifts to lower frequencies (Wild, Murray and Rowe, 1954). It was inferred that the two frequencies were the plasma frequency and its second harmonic at the point of emission in the corona and the frequency drift was interpreted as due to localized exciting agency propagating outwards through the corona. (The plasma density, and hence the plasma frequency, decreases with height in the corona. Consequently, a source emitting at the plasma frequency or its second harmonic drifts from higher to lower frequencies as it propagates outwards through the corona.) These qualitative ideas are now well established and the quantitative theory of these so-called *plasma emission processes* remains a field of current interest.

The theory of the plasma emission processes is developed in Chapter 10, and applied to the interpretation of solar radio bursts in Chapter 11. The plasma emission processes involve several stages. The first stage is generation of Langmuir waves, i.e. longitudinal electron plasma waves. These waves have frequencies close to the plasma frequency. The other stages involve conversion of the energy in Langmuir waves into energy in escaping transverse waves near the plasma frequency and its second harmonic. Only transverse waves (slightly modified by the effect of the magnetic field due to the "magnetoionic" character of the medium) can escape from an astrophysical source.

The first attempt at a quantitative theory of plasma was by Ginzburg and Zheleznyakov (1958). They developed a theory for type III bursts, and incidently they also needed to develop the nonlinear plasma processes involved. Their theory remained the accepted theory for type III bursts for about a decade. The present-day theory of plasma emission includes essentially the same ideas as in Ginzburg and Zhelzynakov's theory, with the details of the various plasma processes updated to take account of further developments in theory.

Before leaving this brief outline of solar radio emissions, it is appropriate to point out that there is a rich variety of emissions. Even for meter-wave bursts the classification into five types (types I to V) is inadequate and further sub-classifications are required before most observed meter-wave emissions can be included. There are also bursts at centimeter and decimeter wavelengths. The centimeter-wave bursts are called microwave bursts and their study is a major aspect of solar radio astronomy. Also, besides the nonthermal emission from the solar corona, the "quiet Sun" can also be observed at (radio) quiet times. The "quiet Sun" emission is due to bremsstrahlung from the coronal gas. Consequently, the observed brightness temperature is a direct measure of the temperature of the corona (1 to 2×10^6 K).

Besides plasma emission, the only other emission processes thought to be relevant to solar radio bursts involve gyromagnetic emission. Gyromagnetic emission, or gyroemission for short, is due to the spiralling motion of particles in a magnetic field. Synchrotron emission is gyroemission from ultrarelativistic electrons. Certain types of solar radio emission (specifically microwave bursts and moving type IV bursts) are attributed to gyroemission from mildly relativistic electrons, e.g. electrons with energies $\approx 100\,\text{keV}$. (The rest energy of an electron is 511 keV.) It has also been suggested that type I emission is due to amplified gyroemission by nonrelativistic electrons, see Section 12.3.

"Coherent" Emission

When particles radiate independently of each other their emission is said to be "incoherent". Absorption is the simplest example of a non-incoherent or cooperative process. One expects that whenever incoherent emission would lead to a brightness temperature greater than about the kinetic energy of the radiating electrons (remember we use units in which Boltzmann's constant is unity), absorption should become important and limit the brightness temperature to about the kinetic energy of the radiating particles. However there are cases where the observed brightness temperature exceeds this limit. For example, type III solar radio bursts have brightness temperatures up to about 10^{11} K ($\approx 10^7$ eV), and are generated by electrons with kinetic energies around 10^5 eV. The decametric radiation from Jupiter is also thought to be generated by mildly relativistic electrons (e.g. energies $\approx 10^5$ eV) and yet brightness temperatures up to 10^{17} K ($\approx 10^{13}$ eV) have been estimated. Brightness temperatures for the radio emission from pulsars are much higher still.

The recognition that sources with such high brightness temperatures exist led to the concept of "coherent" emission processes. Unfortunately, the word "coherent" is used with a variety of meanings, and despite it being appropriate in this connection, its use should be discouraged. The word is used here to imply that the emission must be "coherent" in the sense that incoherent emission by individual particles is inadequate. In other words, a cooperative emission process is required. The most familiar cooperative emission process is amplification, as it occurs in a maser or laser for example. Such amplification may be regarded as negative absorption. Two other types of cooperative process could also occur in a plasma. One is the appearance of intrinsically growing waves, referred to as a "reactive instability." The other is a localized bunch of N particles radiating in phase, and hence radiating N^2 times the power from an individual particle.

In the plasma emission processes in the solar corona, the cooperative process occurs in the generation of the Langmuir waves. It is also possible for amplification to occur in the conversion of the Langmuir waves into transverse waves at the fundamental of the plasma frequency. There have been several ideas advanced for the cooperative process involved in the decametric emission from Jupiter. One specific mechanism is discussed in Section 13.2. The cooperative mechanism responsible for the radio emission from pulsars is not known, perhaps the most favorable candidates are coherent curvature radiation and amplified linear acceleration emission (Section 13.3).

Summary

Let us summarize the emission processes which are thought to be important in astrophysics:

1) Bremsstrahlung is emission due to electron–ion encounters in a thermal plasma. It is the basic thermal emission process (Section 3.3).

2) Synchrotron emission is gyromagnetic emission from ultrarelativistic electrons. It is the nonthermal emission process for supernova remnants, radio galaxies, quasars and a variety of other sources. These sources collectively are called "synchrotron sources" (Sections 4.2, 4.3, 6.3, 9.1 to 9.4.)

3) Plasma emission is emission of transverse waves at the plasma frequency and its second harmonic due to nonlinear plasma processes involving Langmuir waves. It is the emission process for type II, type III and certain other meter-wave solar radio bursts (Chapters 10 and 11).

4) Inverse Compton emission is scattering of soft photons into harder photons by relativistic electrons. It is of possible significance in the generation of the diffuse x-ray background, and is thought to play a subtle role in limiting the brightness temperature of synchrotron sources (Sections 3.4, 4.4, 6.4, 9.2). A related mechanism is "nonlinear inverse Compton emission" or "Rees' mechanism" (Sections 3.4, 4.4).

5) Amplified gyromagnetic emission is emission at near the electron gyrofrequency (in a plasma in which this is greater than the plasma frequency) amplified due to the form of the momentum distribution of the emitting electrons. It appears to be a favorable mechanism for the decametric radio emission from Jupiter (Sections 12.3, 12.4).

6) Coherent curvature emission is emission due to bunches of electrons streaming along curved magnetic field lines. It is one of the favored mechanisms for the radio emission from pulsars (Section 13.3).

7) Amplified linear acceleration emission is emission due to acceleration by an electric field parallel to the ambient magnetic field amplified due to the form of the energy spectrum of the relativistic electrons. It is a plausible alternative emission mechanism for pulsars (Section 13.3).

1.2 INTRODUCTION TO PLASMA PHYSICS

Modern plasma physics commenced its rapid development in the 1950's. (Some of the earlier work was initially classified in connection with fusion research.) However, many of the basic concepts had been developed before about 1950, and an outline of the history of this development provides a useful introduction to the subject.

Firstly, to avoid any confusion, let it be understood that (unless specified otherwise) a "plasma" means a partially or completely ionized charge-neutral gas in which the only negatively charged particles are electrons and the only positively charged particles are ions. Further, if the plasma is permeated by a magnetic field then it is said to be magnetized (or magneto-active).

Waves in Plasmas

One of the characteristic features of plasmas is the rich variety of wave motions which they can support, and the earliest contributions to what is now plasma physics involved theories for various wave motions. Langmuir, who was the first (in 1928) to use the name "plasma" to describe an ionized gas, experimented with arc discharges and identified a wave motion which he called "plasma–electron oscillations". In a classic paper, Tonks and Langmuir (1929) presented a theory for these waves and for "plasma–ion oscillations". Nowadays these waves are referred to as "Langmuir waves" and "ion sound waves" respectively. Along with the familiar transverse waves (with refractive index $n=(1-\omega_p^2/\omega^2)^{1/2}$ here), these are the wave modes of an isotropic plasma with thermal distributions of electrons and ions.

The term "mode" is used in plasma physics to describe a type of wave motion. The names given to various modes are not according to any system; some names are descriptive of the wave motion itself ("ion sound", "electron cyclotron"), some are of historical origin ("Langmuir", "Alfvén") and some are descriptive of the theory used to describe the wave motion ("cold plasma", "MHD", "magnetoionic").

About the same time as Langmuir's investigations, theoretical ideas on the propagation of radio waves in the ionosphere were being developed. These ideas culminated in the magnetoionic theory of Appleton (1932) and Hartree (1931). In the magnetoionic theory the plasma is regarded as a cold magnetized electron gas. In other words, the thermal motion of the electrons is neglected and the ions are assumed immobile. Unfortunately, although much of the terminology of the magnetoionic theory is incompatible with modern usage in plasma physics, it continues to be used, especially in ionospheric and magnetospheric physics. For example, "magnetoionic" itself is a misnomer, according to current usage, in that the ions play no active role whatsoever.

To digress briefly, it is interesting that several logical extensions of these two earliest theories were not made until many years later. One extension to the magnetoionic theory would be to allow for the effect of the ions, and this was done by Åström (1950). The resulting theory is called cold plasma theory. A logical extension of both cold plasma theory and the theory of waves in an isotropic thermal plasma is to allow the particles to have finite thermal motions in a magnetized plasma. This was done by Sitenko and Stepanov (1956).

In all these theories the plasma is assumed to be approximately collisionless. In other words, the effects of inter-particle collisions are either neglected entirely or taken into account as perturbations. Such a procedure is justified provided the wave frequency is much greater than the highest relevant collision frequency (and the wavelength much shorter than the relevant collisional mean free path). In the opposite limit, collisions cause all particles to move together, at least to a first approximation, and a fluid theory is appropriate. Put another way, a collision-dominated plasma may be described as a magnetized fluid.

Alfvén (1942) discussed waves in a magnetized conducting fluid. He showed that a type of shear wave can exist. In these waves, now called Alfvén waves, the magnetic field lines act like stretched strings which are plucked. Alfvén considered only an incompressible fluid, and the generalization to a compressible fluid was carried out by Herlofson (1950), de Hoffman and Teller (1950) and van de Hulst (1951). Three modes can exist in the compressible case. One is the Alfvén mode. The other two are modified versions of the familiar compressional sound waves (which exist in a neutral gas) and magnetoacoustic waves (which exist in the absence of thermal motions, as had been found by Åström). The fluid approach has developed into the macroscopic theory of plasmas, now called magnetohydrodynamics, or MHD theory for short.

The Kinetic Theory of Plasmas

Modern plasma physics is based on kinetic theory, i.e. on the application of statistical mechanics to a plasma. An important idea was introduced by Vlasov (1938, 1945). The idea is a subtle one and its significance can easily escape anyone unfamiliar with the basic difficulty encountered in a statistical theory of plasmas. Before introducing Vlasov's idea it is appropriate to outline this basic difficulty.

The difficulty in a statistical description of a system in which the particles interact via an inverse-square-law force is that any particle must be regarded as interacting simultaneously with all the other particles. This is because the net effect of the interactions on a specific particle is proportional to the integral over r of the number of particles between r and $r+\mathrm{d}r$ times the force. However, this integral diverges because the increase as r^2 of this number of particles balances the decrease as r^{-2} of the inverse-square-law force each exerts. This basic difficulty arises in a statistical description of a cluster of stars (which interact via the inverse-square-law force of gravity). The way the difficulty is overcome in connection with stellar clusters is analogous to the way in which Vlasov overcame it in plasma theory.

In developing a model for a stellar cluster one looks for a self-consistent solution for the "mean" gravitational field and the "smoothed-out" mass distribution of the stars. The "mean" gravitational potential is related to the "smoothed-out" mass density by Poisson's equation. Once a self-consistent solution for the mean field and the smoothed-out density has been found, one regards individual stars as both moving under the influence of the "mean" field, and experiencing binary interactions with their neighbours.

In a plasma the self-consistent solution consists of the "mean" electromagnetic field and the "smoothed-out" distributions of particles, with particles of each species described by a single-particle distribution function. Vlasov's idea was to suppose that the self-consistent solution could be found by solving the collisionless Boltzmann equation

$$\left\{\frac{\partial}{\partial t}+\mathbf{v}\cdot\frac{\partial}{\partial \mathbf{r}}+q\left(\mathbf{E}+\frac{\mathbf{v}\times\mathbf{B}}{c}\right)\cdot\frac{\partial}{\partial \mathbf{p}}\right\}f(\mathbf{p},\mathbf{r},t)=0 \tag{1.7}$$

for each species, with the fields $\mathbf{E}$ and $\mathbf{B}$ determined by Maxwell's equations in which the charge density ρ and current density $\mathbf{j}$ are to be identified as

$$\rho=\sum q\int \mathrm{d}^3\mathbf{p}f, \qquad \mathbf{j}=\sum q\int \mathrm{d}^3\mathbf{p}\mathbf{v}f, \tag{1.8a, b}$$

where the sums are over all species of particle. Now the point is that the collisionless Boltzmann equation (1.7) is usually derived under the assumption that **E** and **B** are *external* fields, and here (1.7) is being invoked where **E** and **B** are self-consistent fields. As a consequence (1.7) is not a linear differential equation for f (as it is under normal circumstances) but is a nonlinear integro-differential equation for f due to the functional dependence of **E** and **B** on f. The self-consistent equations, e.g. (1.7) with (1.8a, b) and Maxwell's equations, are sometimes called the Vlasov equations.

One consequence of the self-consistent-field approach is that the field attributed to an individual particle is not its usual field, but rather the self-consistent field. For a charge at rest *in vacuo*, the field is a Coulomb field. For a charge in a plasma the field which it would have *in vacuo* is often called its "bare" field and the self-consistent field is called the "shielded" or "dressed" field. The self-consistent field for a charge at rest in an isotropic plasma may be calculated in the following simplified manner.

The bare field for a charge q at rest at a point $\mathbf{r}=\mathbf{r}_0$ is the Coulomb field,

$$\phi=\frac{q}{|\mathbf{r}-\mathbf{r}_0|}, \qquad \mathbf{E}=-\text{grad}\,\phi, \tag{1.9a, b}$$

where ϕ is the electrostatic potential. Now suppose that a charge-neutral plasma composed of electrons and ions (charge $Z_i e$, mass m_i) is present. The potential energy of a particle of charge q is an electrostatic field is $q\phi$. Moreover in the presence of a potential field the number density of thermal particles varies as $\exp[-U/T]$, where U is the potential energy. Consequently the number density of electrons and ions depends on ϕ according to

$$n_e=n_0\exp\left[\frac{e\phi}{T_e}\right], \qquad n_i=\frac{n_0}{Z_i}\exp\left[-\frac{Z_i e\phi}{T_i}\right]. \tag{1.10a, b}$$

The electrostatic potential is determined by $\text{div}\,\mathbf{E}=4\pi\rho$, which gives

$$-\nabla^2\phi=-4\pi e(n_e-Z_i n_i). \tag{1.11}$$

Equations (1.10a, b) and (1.11) constitute an example of a set of self-consistent equations which we wish to solve for ϕ.

Let us assume $\mathbf{r}_0=0$ for simplicity, and then for $|e\phi|\ll T_e$, T_i, (1.11) with (1.10a, b) gives

$$\frac{1}{r^2}\frac{\partial}{\partial r}\left(r^2\frac{\partial\phi}{\partial r}\right)=\left(\frac{4\pi n_e e^2}{T_e}+\frac{4\pi n_i Z_i^2 e^2}{T_i}\right)\phi. \tag{1.12}$$

The only acceptable solution of (1.12) is

$$\phi=\frac{q}{r}\exp\left(-r/\lambda_D\right), \tag{1.13}$$

with

$$\lambda_D^{-2} := \frac{4\pi n_e e^2}{T_e} + \frac{4\pi n_i Z_i^2 e^2}{T_i}. \tag{1.14}$$

Thus the Coulomb field is shielded out over distances greater than a Debye length λ_D. An alternative derivation of (1.13) is given in Section 10.4.

The Debye length $\lambda_{D\alpha}$ for each species α may be defined by

$$\lambda_{D\alpha} := \frac{V_\alpha}{\omega_{p\alpha}}, \tag{1.15}$$

with V_α the thermal speed for that species,

$$T_\alpha =: m_\alpha V_\alpha^2, \tag{1.16}$$

and with $\omega_{p\alpha}$ the plasma frequency for that species,

$$\omega_{p\alpha} := \left(\frac{4\pi n_\alpha q_\alpha^2}{m_\alpha}\right)^{1/2}. \tag{1.17}$$

The definition (1.14) of the Debye length may be replaced by the more general definition

$$\lambda_D^{-2} := \sum_\alpha \lambda_{D\alpha}^{-2}. \tag{1.18}$$

The concept of a "smoothed-out" charge distribution and a "mean" electromagnetic field breaks down if the number of particles per cubic Debye length is small. In more formal theories one expands in inverse power of the number of particles per cubic Debye length. The kinetic theory of plasmas (the lowest order theory) is valid only when this number is large. Actually this number appears in connection with the Coulomb logarithm, which is effectively its natural logarithm, and with the collision frequency, which is roughly the plasma frequency times the Coulomb logarithm divided by this number.

1.3 SOME SIMPLE WAVES MODES

The theory of weakly damped waves is developed from a formal viewpoint in Chapter 2, and some relevant wave properties are summarized in Section 2.4. Here simplified treatments of Langmuir waves, transverse waves in an isotropic plasma, and MHD waves are given. The purpose is to illustrate the qualitative properties of the waves.

Langmuir Waves

Langmuir waves may be regarded as oscillations of the electrons relative to the ions. The natural frequency of the oscillations is the plasma frequency. Suppose we start with the following four equations. The first is one of Maxwell's equations

$$\operatorname{div} \mathbf{E} = 4\pi\rho, \tag{1.19}$$

where ρ is the charge density. For the second the electron gas is regarded as a fluid with number density n_e and fluid velocity $\mathbf{v}$, and the equation of motion is

$$m_e \frac{\mathrm{d}\mathbf{v}}{\mathrm{d}t} = -e\mathbf{E}. \tag{1.20}$$

The third equation is

$$\frac{\partial n_e}{\partial t} + \operatorname{div} n_e \mathbf{v} = 0, \tag{1.21}$$

which is the fluid equation for the continuity (i.e. conservation) of electrons. Finally, the charge density is related to the number density n_e by

$$\rho = -e(n_e - n_0), \tag{1.22}$$

where the ions are assumed to form a uniform background positive charge density en_0. (Charge neutrality requires that the average value of n_e be equal to n_0.) Now we may replace n_e in div $n_e\mathbf{v}$ by n_0 and $\mathrm{d}\mathbf{v}/\mathrm{d}t$ by $\partial\mathbf{v}/\partial t$ to lowest order in the small fluctuating quantities $\mathbf{v}$, $\mathbf{E}$ and $n_e - n_0$. Then multiplying (1.20) by n_0 and taking its divergence, div $n_0\mathbf{v}$ may be replaced using (1.21) and div $\mathbf{E}$ replaced using (1.19) with (1.22). Hence one finds

$$\left\{\frac{\partial^2}{\partial t^2} + \frac{4\pi n_0 e^2}{m_e}\right\}(n_e - n_0) = 0. \tag{1.23}$$

One concludes that oscillations occur at the (electron) plasma frequency

$$\omega_p := \left(\frac{4\pi n_e e^2}{m_e}\right)^{1/2}. \tag{1.24}$$

The oscillations occur in the number density of the electrons, in the electric field $\mathbf{E}$ and the velocity $\mathbf{v}$ of the electrons. The electric field is longitudinal in the sense that if one supposes the variation to be as for a plane wave, e.g. as $\cos(\mathbf{k}\cdot\mathbf{r} - \omega t)$, $\mathbf{E}$ (and also $\mathbf{v}$) is parallel to $\mathbf{k}$. The electric field is due to the local charge separation between electrons and ions as the electrons oscillate (in phase with each other) about their mean position.

Langmuir waves in this approximation do not propagate (their group velocity is zero). They are oscillations in the density of electrons in which the restoring force is the electric field caused by local charge separation between electrons and ions. One would expect there to be equipartition between the electric energy density and the kinetic energy density of the electrons, and this turns out to be the case, i.e.

$$\frac{E^2}{8\pi}=\tfrac{1}{2}n_e m_e v^2. \tag{1.25}$$

When thermal corrections are taken into account, the frequency of Langmuir waves is slightly greater than the plasma frequency, and their group velocity is non-zero, cf. Equation (2.109).

Transverse Waves

Transverse waves may be treated in a similar way. In this case the following two of Maxwell's equations are used,

$$\text{curl}\,\mathbf{E}=-\frac{1}{c}\frac{\partial\mathbf{B}}{\partial t},\quad \text{curl}\,\mathbf{B}=\frac{4\pi}{c}\mathbf{j}+\frac{1}{c}\frac{\partial\mathbf{E}}{\partial t}. \tag{1.26a, b}$$

These are combined in the form

$$\text{curl curl}\,\mathbf{E}+\frac{1}{c^2}\frac{\partial^2}{\partial t^2}\mathbf{E}=-\frac{4\pi}{c^2}\frac{\partial\mathbf{j}}{\partial t}. \tag{1.27}$$

With the electrons described as a fluid, the current density is

$$\mathbf{j}=-n_e e\mathbf{v}, \tag{1.28}$$

and (1.20) implies

$$\frac{4\pi}{c^2}\frac{\partial\mathbf{j}}{\partial t}=\frac{\omega_p^2}{c^2}\mathbf{E}. \tag{1.29}$$

Now suppose that $\mathbf{E}$ is a plane wave propagating in the z-direction and varying as $\cos(kz-\omega t)$, then (1.27) with (1.29) implies

$$\omega^2=\omega_p^2+k^2c^2, \tag{1.30}$$

which is the dispersion relation for transverse waves. The dispersion relation may also be written in terms of the refractive index n,

$$n^2:=\frac{k^2c^2}{\omega^2}=1-\frac{\omega_p^2}{\omega^2}. \tag{1.31}$$

The qualitative properties of transverse waves include the following. The electric vector is orthogonal to the direction of propagation (the z-direction here and the direction of $\mathbf{k}$ more generally) and the magnetic vector is orthogonal to both **E** and **k**. The electric, magnetic and kinetic energies are in the ratios

$$\frac{E^2}{8\pi}:\frac{B^2}{8\pi}:\frac{1}{2}n_e m_e v^2=\frac{1}{2}:\frac{1}{2}\left(1-\frac{\omega_p^2}{\omega^2}\right):\frac{1}{2}\frac{\omega_p^2}{\omega^2}, \tag{1.32}$$

which implies equipartition between the electric energy and the sum of the other two. For $\omega<\omega_p$ transverse waves are evanescent. This means that they do not propagate but decay in space. This decay involves no loss of energy. In fact, reflection of light from a mirror is an example of the response of a plasma-like medium to transverse waves with $\omega<\omega_p$. The "wave" decays spatially over a skin depth $c/(\omega_p^2-\omega^2)^{1/2}$ over which distance the wave energy is reflected.

Sound Waves

Sound waves as such do not exist in a collisionless plasma, and in a collision-dominated plasma their properties are analogous to their properties in a neutral gas. Sound waves are discussed here for two purposes. The first is as an introduction to the use of hydrodynamics. The other is for comparison with ion sound waves, which may be regarded as the counterpart of sound waves in the collisionless regime, cf. Equation (2.113).

The relevant equations for sound waves are the equation of continuity

$$\frac{\partial\eta}{\partial t}+\operatorname{div}\eta\mathbf{v}=0, \tag{1.33}$$

where η is the mass density, the equation of fluid motion

$$\eta\frac{\mathrm{d}\mathbf{v}}{\mathrm{d}t}=-\operatorname{grad}P, \tag{1.34}$$

where P is the pressure, and the equation of state. For adiabatic sound waves the equation of state is the adiabatic equation

$$P\eta^{-\gamma}=\text{constant}, \tag{1.35}$$

where γ is the adiabatic index. (For isothermal sound waves the equation of state is $T=$constant, i.e. $P\eta^{-1}=$constant for a perfect gas.) The sound speed is

$$c_s:=\left(\frac{\gamma P}{\eta}\right)^{1/2}. \tag{1.36}$$

Now η and P are assumed to have small fluctuating parts, $\delta\eta$ and δP say. Then (1.35) implies $\delta P=c_s^2\delta\eta$. Differentiating (1.33) with respect to time and linearizing in small quantities, one finds

$$\frac{\partial^2}{\partial t^2}\,\delta\eta = -\eta\,\frac{\partial}{\partial t}\,\mathrm{div}\,\mathbf{v} = c_s^2\,\mathrm{div}\,\mathrm{grad}\,\delta\eta, \tag{1.37}$$

where (1.34) and (1.35) have been used. Hence any plane wave solution has

$$\omega^2 = k^2 c_s^2, \tag{1.38}$$

which is the dispersion relation for sound waves. The energetics in the waves correspond to equipartition between kinetic and potential energies

$$\frac{1}{2}\,\eta v^2 = \frac{1}{2}\left(\frac{\delta\eta}{\eta}\right)^2 \eta c_s^2. \tag{1.39}$$

Alfvén Waves

Alfvén waves exist in a magnetized fluid, i.e. a fluid permeated by a magnetic field. An important assumption is that the fluid is perfectly (electrically) conducting. This assumption implies that the fluid and the magnetic field move together. The magnetic field is said to be "frozen-in" to the fluid, and the fluid said to be "tied" to the magnetic field lines.

The properties of Alfvén waves do not depend on the assumed equation of state. The simplest case is for an incompressible fluid for which the equation of state is

$$\mathrm{div}\,\mathbf{v} = 0. \tag{1.40}$$

The simplicity of this case is due to the fact that Alfvén waves are the only waves which can exist in a perfectly conducting, magnetized incompressible fluid. Apart from (1.40), the relevant equations are (1.26a), (1.26b) without the displacement current, (1.33), the equation of fluid motion

$$\eta\,\frac{\mathrm{d}\mathbf{v}}{\mathrm{d}t} = \frac{\mathbf{j}\times\mathbf{B}}{c} \tag{1.41}$$

and the equation

$$\mathbf{E} = -\frac{\mathbf{v}\times\mathbf{B}}{c}. \tag{1.42}$$

Equation (1.42) expresses the fact that the fluid is assumed to be infinitely conducting. For a finite conductivity σ, the current density $\mathbf{j}$ is given by Ohm's law $\mathbf{j} = \sigma(\mathbf{E} + \mathbf{v}\times\mathbf{B}/c)$ which implies (1.42) for $\sigma = \infty$ and arbitrary $\mathbf{j}$.

After linearizing in small quantities, the time-derivative of (1.41) gives

$$\eta\,\frac{\mathrm{d}^2\mathbf{v}}{\mathrm{d}t^2} = -\frac{1}{c}\,\mathbf{B}_0\times\frac{\partial\mathbf{j}}{\partial t} = \frac{c}{4\pi}\,\mathbf{B}_0\times\mathrm{curl}\,\mathrm{curl}\,\mathbf{E}$$

$$= -\frac{1}{4\pi}\,\mathbf{B}_0\times\mathrm{curl}\,\mathrm{curl}\,(\mathbf{v}\times\mathbf{B}_0), \tag{1.43}$$

where the background magnetic field $\mathbf{B}_0$ is assumed uniform, and where (1.27) and then (1.42) have been used. Further reduction of (1.43) involves some lengthy vector algebra. It is appropriate to anticipate a result derived in Section 2.1 to simplify the reduction. The result is that if (1.43) is Fourier transformed then the differential operators $\partial/\partial t$, div and curl are replaced by $-i\omega$, $i\mathbf{k}\cdot$ and $i\mathbf{k}\times$ respectively. Hence (1.43) reduces to

$$(-i\omega)^2\eta\mathbf{v} = -\frac{1}{4\pi}\,\mathbf{B}_0\times[(i\mathbf{k})\times\{(i\mathbf{k})\times(\mathbf{v}\times\mathbf{B}_0)\}], \tag{1.44}$$

which, in view of $\mathbf{k}\cdot\mathbf{v}=0$ from (1.40), implies the dispersion relation for Alfvén waves

$$\omega^2 = k^2 v_A^2 \cos^2\theta, \tag{1.45}$$

where

$$v_A := \left(\frac{B_0^2}{4\pi\eta}\right)^{1/2} \tag{1.46}$$

is the Alfvén speed and θ is the angle between $\mathbf{k}$ and $\mathbf{B}$. Both the fluid velocity $\mathbf{v}$ and the fluctuation in the magnetic field are in the direction $\mathbf{k}\times\mathbf{B}$.

The ratios of the electric, magnetic and kinetic energies in Alfvén waves may be derived as follows. Firstly, (1.42) implies

$$\frac{E^2}{8\pi} = \tfrac{1}{2}\eta v^2\left(\frac{v_A^2}{c^2}\right), \tag{1.47}$$

where we have used $\mathbf{B}_0\cdot\mathbf{v}=0$, which is implied by (1.41). Secondly, (1.26a) implies

$$\mathbf{k}\times\mathbf{E} = \frac{\omega}{c}\,\delta\mathbf{B}, \tag{1.48}$$

where $\delta\mathbf{B}$ is the magnetic fluctuation in the wave. Then, because $\mathbf{v}$ is orthogonal to both $\mathbf{k}$ and $\mathbf{B}_0$, (1.42) implies $|\mathbf{k}\times\mathbf{E}|^2 = |\mathbf{k}\cdot\mathbf{B}_0|^2 v^2/c^2 = k^2|\mathbf{E}|^2\cos^2\theta$, where (1.46) and (1.47) have been used, and hence (1.47) and (1.48) imply

$$\frac{|\mathbf{E}|^2}{8\pi} : \frac{|\delta\mathbf{B}|^2}{8\pi} : \tfrac{1}{2}\eta v^2 = \frac{v_A^2}{c^2} : 1 : 1. \tag{1.49}$$

The neglect of the displacement current implies the neglect of corrections of order v_A^2/c^2, and when the displacement current is taken into account there is equipartition between the magnetic energy and the sum of the electric and kinetic energies. More generally, for any MHD waves there is equipartition between the sum of the magnetic and potential energies and the sum of the electric and kinetic energies, but only if the displacement current is included (Denisse and Delcroix, 1963; Melrose, 1975a).

The foregoing examples have one common feature: in each of the models only one type of wave motion exists. Further generalization involves models in which the wave equation leads to solutions for two or more wave modes. The general theory of such wave motions in a collisionless plasma is developed in Chapter 2, and the theory for a collision-dominated compressible fluid is outlined in Section 12.3.

1.4 PHYSICAL, ASTROPHYSICAL AND PLASMA CONSTANTS

Before proceeding with the development of plasma theory it is appropriate to collect together relevant numerical values. Throughout the c.g.s.-gaussian system of units is used, except where accepted convention or convenience suggests otherwise. Values of relevant physical and astrophysical constants are given in Tables 1.1 and 1.2 respectively.

TABLE 1.1
Physical Constants.

Physical quantity	Symbol	Value	Unit
speed of light *in vacuo*	c	3.00×10^{10}	cm s^{-1}
elementary charge	e	4.80×10^{-10}	e.s.u.
electron mass	m_e	9.11×10^{-28}	g
proton mass	m_p	1.67×10^{-27}	g
(Planck's constant)/2π	$\hbar$	1.05×10^{-27}	erg s
gravitational constant	G	6.67×10^{-8}	dyn-cm^2 g^{-2}
classical radius of electron	$r_0 = e^2/m_e c^2$	2.82×10^{-13}	cm
Thomson cross-section	$(8\pi/3)r_0^2$	6.65×10^{-25}	cm^2

TABLE 1.2
Astrophysical Constants (Allen, 1973).

Quantity	Symbol	Value	Unit
parsec	pc	3.09×10^{18}	cm
light year	ly	9.46×10^{17}	cm
astronomical unit	AU	1.50×10^{13}	cm
radius of the Sun	$R_\odot$	6.96×10^{10}	cm
radius of the Earth	R_E	6.38×10^{8}	cm
radius of Jupiter	R_J	7.14×10^{9}	cm
mass of the Sun	$M_\odot$	1.99×10^{33}	g
luminosity of the Sun	$L_\odot$	3.82×10^{33}	erg s^{-1}

TABLE 1.3
Plasma Constants.

Quantity	Symbol	Value	Unit
plasma frequency	ω_p	$5.64\times10^4\ n_e^{1/2}$	s^{-1}
	f_p	$8.98\times10^3\ n_e^{1/2}$	Hz
electron gyrofrequency	Ω_e	$1.76\times10^7\ B$	s^{-1}
	f_B	$2.80\times10^6\ B$	Hz
thermal speed of electrons	V_e	$3.89\times10^5\ T_e^{1/2}$	cm s^{-1}
ion sound speed	v_s	$9.09\times10^3\ T_e^{1/2}$	cm s^{-1}
Alfvén speed	v_A	$2.18\times10^{11}\ Bn_e^{-1/2}$	cm s^{-1}
Debye length for electrons	λ_{De}	$6.90\ T_e^{1/2}n_e^{-1/2}$	cm
collision frequency	ν_0	$13.7(\ln \Lambda_c)n_eT_e^{-3/2}$	s^{-1}

Several of the relevant plasma parameters have already been defined. Some additional plasma parameters are defined below and various interrelations between them are pointed out. Numerical values of the plasma parameters are summarized in Table 1.3, and some illustrative values of the basic parameters for plasmas of astrophysical interest are given in Table 1.4.

Additional Plasma Parameters

It is convenient to consider plasma parameters in groups according to their dimensions.

The plasma parameters which are frequencies include the plasma frequencies $\omega_{p\alpha}$ for each species α. These are defined by (1.17). There is another characteristic frequency associated with each species. This is the gyrofrequency

$$\Omega_\alpha := \frac{|q_\alpha|\mathbf{B}}{m_\alpha c}, \tag{1.50}$$

which is the frequency at which the (nonrelativistic) particles of species α spiral about the magnetic field lines. In astrophysical plasmas, the gyrofrequency of electrons (the highest gyrofrequency) is typically less than the plasma frequency, with occasional exceptions. A third class of frequencies consists of collision frequencies. The basic collision frequency may be taken to be, cf. (7.13) below,

$$\nu_0 = \frac{\omega_p^2 e^2}{m_e V_e^3} \ln \Lambda_c \tag{1.51}$$

where the Coulomb logarithm $\ln \Lambda_c$ is given explicitly by (1.57a, b) below.

TABLE 1.4

Object	Description	Number density	Temperature	Magnetic field
Solar corona	The region $\gtrsim 10^4$ km above the photosphere joining on continuously to the interplanetary medium. The corona is homogeneous on a large scale, with overdense regions called streamers, and underdense regions called holes.	(i) $n_e \approx 10^9$ cm^{-3} at base of corona (ii) $n_e \approx 2\times10^8 (r/R_0)^{-6}$ cm^{-3} at greater heights (iii) Density up to ten times larger in streamers.	$(1.5\pm0.5)\times10^6$ K	(i) Mean field of 1 to 2 G. (ii) Closed magnetic regions, and isolated flux tubes sometimes extend into corona. (iii) The expansion of the corona draws out the magnetic field causing neutral planes to develop.
Solar wind	The region from where the expansion of solar corona becomes super-Alfvénic to an outer boundary with the interstellar medium (at ≈ 50 AU?) Mean expansion speed at 1 AU (variable by factor >2) 300 km s^{-1}.	Mean (variable by factor >10) $n_e \approx 10\, r^{-2}$ cm^{-3} with r in AU.	Mean (variable by factor >5) $T_p \approx 4\times10^4$ K, $T_{\parallel p} \approx 2T_{\perp p}$, $T_e \approx 7\times10^4$ to 1.5×10^5 K, $T_p > T_e$ in high speed streams (from holes).	Mean (variable by factor >2) $5\gamma = 5\times10^{-5}$ G
Magnetosphere	The region above the ionosphere and inside the magnetopause ($r \lesssim 10\, R_E$). The plasmasphere ($r \lesssim 3\, R_E$) is in diffusive equilibrium with ionosphere. The plasma trough ($r \gtrsim 3\, R_e$ and above polar regions) is populated from the solar wind.	$n_e \approx 10^3$ cm^{-3} in plasmasphere, $n_e \approx 1$ cm^{-3} in plasma trough.	In the plasmasphere $T_e \approx 10^3$ K. In the ring current $T_e \approx 10^5$ K.	Earth's field is roughly dipolar, with $B = 0.31\left(\frac{r}{R_E}\right)^{-3}$ G in equatorial plane out to near the magnetopause.
Interstellar medium	Three-component medium, (i) hot, diffuse, highly ionized background, (ii) cold dense clouds, (iii) warm edges to clouds.	(i) $n_e \approx n_p \approx 3\times10^{-3}$ cm^{-3} (ii) $n_e \approx 3\times10^{-2}$ cm^{-3}, $n_H \approx 10$ cm^{-3} (iii) $n_e \approx 3\times10^{-2}$ cm^{-3}, $n_H \approx 0.1$ cm^{-3}	(i) 5×10^5K, (ii) 10–100K, (iii) 10^4K.	Mean value 3×10^{-6}G
Synchrotron source	(1) Supernova remnants; cosmic-ray electrons in spiral arms. (2) Plasmoids ejected from galaxies; galactic nuclei.	unknown; upper limits implied by any observed Faraday depolarization or an inferred Alfvén speed.	unknown and of little relevance.	poorly known, 10^{-5} to 10^{-3} G.

The plasma parameters which are speeds include the thermal speeds V_α, as defined by (1.16), the Alfvén speed v_A, as defined by (1.46), and the sound speed. In a collision dominated plasma the sound speed is given by (1.36). However, in most cases of interest here the plasma is collisionless, and then the quantity which plays the role of the sound speed is the *ion sound speed*

$$v_s := \omega_{pi}\lambda_{De}, \tag{1.52}$$

where ω_{pi} is the ion plasma frequency. (The reason for the definition (1.52) is related to the sonic character of ion sound waves, cf. (2.113) below.) Thus the sound speed is ω_{pi}/ω_p times the thermal speed V_e of electrons. (In a plasma consisting of a mixture of ions, ω_{pi} in (1.52) is to be interpreted as the square root of the sum of the squares of the plasma frequencies for each species of ion.)

The plasma parameters which are lengths include the Debye lengths $\lambda_{D\alpha} := V_\alpha/\omega_{p\alpha}$ and the gyroradii of thermal particles V_α/Ω_α. The most important length is the Debye length λ_D defined by (1.18). One could also define mean free paths by dividing V_α by the collision frequency for species α.

Inter-relations Between Plasma Parameters

Plasmas tend to be charge neutral (because if they were not enormous electric forces would be present) and, with virtually no exceptions, plasmas of astrophysical interest are composed of negatively charged electrons and positively charged ions. Each ion may be assumed to have charge $q_i = Z_i e$ and mass $m_i = A_i m_p$, where e and m_p are the charge and mass of the proton. Charge neutrality requires

$$n_e = \sum_i Z_i n_i, \tag{1.53}$$

where the sum is over all species of ion. The relation

$$\frac{\omega_{pe}^2}{\Omega_e} = \sum_i \frac{\omega_{pi}^2}{\Omega_i} \tag{1.54}$$

then follows from (1.53).

The ratio of the Alfvén speed to the speed of light is related to the other plasma parameters. The mass density of the plasma is $\eta = \Sigma_\alpha m_\alpha n_\alpha$, and, consequently, (1.46) implies

$$\frac{c^2}{v_A^2} = \sum_\alpha \frac{\omega_{p\alpha}^2}{\Omega_\alpha^2} \approx \sum_i \frac{\omega_{pi}^2}{\Omega_i^2}, \tag{1.55}$$

where the final approximate expression corresponds to neglecting the contribution of the electrons to the mass density.

Another dimensionless quantity is the ratio of the thermal pressure to the magnetic pressure. This ratio is called the β of the plasma:

$$\beta := \frac{\Sigma_\alpha n_\alpha T_\alpha}{B^2/8\pi} = 2\sum_\alpha \frac{V_\alpha^2}{c^2}\frac{\omega_{p\alpha}^2}{\Omega_\alpha^2}. \tag{1.56}$$

The value of β is important in the following qualitative connection. The magnetic field can confine a plasma for $\beta<1$, and for $\beta>1$ the plasma can drag along the magnetic field. (The magnetic field and the plasma must move together in an infinitely conducting medium.) Conversely, if it is known that the magnetic field confines the plasma, then β must be less than unity. From yet another viewpoint, β is roughly the ratio v_s^2/v_A^2, and for $\beta<1$ pressure forces and related disturbances propagate at the Alfvén speed, while for $\beta>1$ they propagate at the sound speed. In astrophysical applications the pressure exerted by suprathermal particles may be significant, and should be added to the thermal pressure when discussing confinement.

The Coulomb Logarithm

The Coulomb logarithm appears in connection with Coulomb interactions or "collisions". There is no exact expression for the Coulomb logarithm, and the following approximation is due to Sivukhin (1966), cf. (7.27). The "classical" value

$$\ln \Lambda_c = 9.1 - \tfrac{1}{2}\ln n + \tfrac{3}{2}\ln T, \tag{1.57a}$$

where n is per cubic centimeter and T in kelvins, applies for all collisions at sufficiently high temperatures. For electron–ion collisions (1.57a) applies for $T(=T_e)<1.5\times10^5$ K, and for $T>1.5\times10^5$ K it is to be replaced by

$$\ln \Lambda_c = 15.1 - \tfrac{1}{2}\ln n + \ln T. \tag{1.57b}$$

TABLE 1.5
Conversion Factors.

Ratio	Value	Inverse
ergs/Kelvins	1.38×10^{-16}	7.24×10^{15}
electron volts/Kelvins	8.62×10^{-5}	1.16×10^{4}
electron volts/ergs	6.24×10^{11}	1.60×10^{-12}
electron volts/hertz	4.14×10^{-15}	2.42×10^{14}
seconds of arc/radians	2.06×10^{5}	4.85×10^{-6}
seconds of time/years	3.16×10^{7}	3.17×10^{-8}

Conversion Factors

Several conversion factors are given in Table 1.5. These include factors which allow one to convert energies to temperatures. In using the conversion factors, units are to be treated like algebraic quantities. For example, the temperature corresponding to the rest energy 0.511 MeV for an electron is given by

$$(0.511 \text{ MeV}) \times (10^6 \text{ eV/MeV}) \times (1{\cdot}16 \times 10^4 \text{ K/eV}) = 5.93 \times 10^9 \text{ K}.$$

PROBLEM SET 1

Answers to relevant questions are given at the end of the text. More difficult questions are marked with an asterisk.

1.1 A radio source appears to be eliptical with major axes 1″ arc and 2″ arc. Estimate $\Delta\Omega_s$ in steradians.

1.2 An intense solar radio burst at 100 MHz has a flux density of 10^{-18} Wm^{-2} Hz^{-1}. The source is circular with an angular diameter of 10′ arc. Estimate the brightness temperature.

1.3 An HII region (a region of ionized hydrogen surrounding a hot star, e.g., an O or B star) is 3 pc in diameter and at a temperature of 10^4 K. Estimate the flux density at 1 GHz ($=10^9$ Hz) if it is a distance of 3 kpc.

1.4 (a) Show that when a pressure term is included in the fluid Eq. (1.20) for electrons, and an adiabatic equation of state $Pn_e^{-\gamma}=\text{constant}$ is assumed, the dispersion relation reduces to

$$\omega^2 = \omega_p^2 + \gamma k^2 V_e^2. \tag{P1-1}$$

(Kinetic theory gives a dispersion relation of this form with $\gamma = 3$.)

(b) Show that with this dispersion relation (P1-1) the group velocity of Langmuir waves is

$$\mathbf{v}_g = \frac{\gamma k V_e^2}{\omega}\,\boldsymbol{\kappa}, \quad \boldsymbol{\kappa} := \mathbf{k}/k. \tag{P1-2}$$

1.5 Find the directions of the fluid velocity $\mathbf{v}$, the electric field $\mathbf{E}$ and the magnetic field $\delta\mathbf{B}$ in an Alfvén wave for given $\mathbf{k}$.

1.6 Show that the group velocity of Alfvén waves is directed along the background magnetic field.

1.7 Show that when the displacement current is included in (1.26b) the dispersion relation for Alfvén waves becomes

$$\omega^2 = \frac{k^2 v_A^2 \cos^2\theta}{1 + v_A^2/c^2}. \tag{P1-3}$$

1.8 Find conversion factors for (i) Ω_e/ω_p to v_A in kilometers per second, (ii) in minutes of arc to kilometers for the separation of two points on the Sun's surface, and (iii) speeds in parsecs per million years to kilometers per second.

CHAPTER 2

Waves in Plasmas

The theory of waves in plasmas is developed in this chapter with two aims in mind. The first is to outline the properties of the particular wave modes which are of most relevance in plasma astrophysics. The other is to develop a general description of wave properties, the purpose being to allow the equations describing particle–wave interactions to be written in such a way that they apply to any species of particle and to waves in any mode.

2.1 MATHEMATICAL PRELIMINARIES

Two mathematical tools which we need are Fourier transforms and tensor algebra.

Fourier Transforms

The Fourier transform $\tilde{F}(\mathbf{k}, \omega)$ of a function $F(\mathbf{r}, t)$ is defined by

$$\tilde{F}(\mathbf{k}, \omega) := \int \mathrm{d}t \int \mathrm{d}^3\mathbf{r} \exp\left[-i(\mathbf{k}\cdot\mathbf{r}-\omega t)\right] F(\mathbf{r}, t), \tag{2.1}$$

where the integral is over all time and space. (In Cartesian coordinates $\mathrm{d}^3\mathbf{r}$ means $\mathrm{d}x\,\mathrm{d}y\,\mathrm{d}z$.) The tilde on $\tilde{F}$ is to denote that it is a different mathematical function to F. Later we shall drop the tilde and denote the Fourier transform simply as $F(\mathbf{k}, \omega)$. The inverse transform is

$$F(\mathbf{r}, t) = \int \frac{\mathrm{d}\omega}{2\pi} \int \frac{\mathrm{d}^3\mathbf{k}}{(2\pi)^3} \exp\left[i(\mathbf{k}\cdot\mathbf{r}-\omega t)\right] \tilde{F}(\mathbf{k}, \omega), \tag{2.2}$$

where the integrals are over all $\omega(-\infty$ to $\infty)$ and all $\mathbf{k}$-space. (In Cartesian coordinates $\mathrm{d}^3\mathbf{k}$ means $\mathrm{d}k_x\,\mathrm{d}k_y\,\mathrm{d}k_z$.)

It is appropriate to comment on the existence of Fourier transforms. The theorem used in defining Fourier transforms is the Fourier integral theorem. A statement of it is that for any function $f(x)$ such that (i) $f(x)$ is sectionally continuous in $-\infty<x<\infty$, (ii) $f(x)$ is defined as

$$\lim_{\Delta\to 0} \tfrac{1}{2}\{f(x_0+\Delta)-f(x_0-\Delta)\}$$

at each point of discontinuity, and (iii)

$$\int_{-\infty}^{\infty} \mathrm{d}x\,|f(x)|$$

exists, the following identity holds,

$$f(x)=\frac{1}{2\pi}\int_{-\infty}^{\infty} \mathrm{d}y \int_{-\infty}^{\infty} \mathrm{d}z\, f(z) \exp\left[\pm iy(z-x)\right]. \tag{2.3}$$

In physical applications, one is often interested in the Fourier transform of functions which do not satisfy the conditions required for the Fourier integral theorem to apply. It is convenient to separate functions into two classes. The first class is those whose Fourier transforms exist. The second class consists of functions whose Fourier transforms do not exist as such, but whose Fourier transforms may be defined in a limiting sense. Functions which exist only in such a limiting sense are sometimes called generalized functions. The most familiar example of a generalized function is the Dirac δ-function. The 1-dimensional and 3-dimensional δ-functions may be defined by

$$\delta(\omega):=\frac{1}{2\pi}\int \mathrm{d}t \exp(\pm i\omega t), \quad \delta^3(\mathbf{k}):=\frac{1}{(2\pi)^3}\int \mathrm{d}^3\mathbf{r} \exp(\pm i\mathbf{k}\cdot\mathbf{r}). \tag{2.4a, b}$$

That is, the δ-function is the Fourier transform of unity. (Note that $\delta^3(\mathbf{k})$ means, in Cartesian coordinates for example, $\delta(k_x)\delta(k_y)\delta(k_z)$.) Although the Fourier transform of unity does not exist according to the Fourier integral theorem, it may be defined by taking a sequence of functions, each of whose Fourier transform does exist, and whose limit is unity. The δ-function is then defined as the limit of the sequence of Fourier transforms. A further example of a generalized function (the Fourier transform of the step function) is discussed in Section 3.1 below.

Another way in which the conditions for the Fourier integral theorem may fail to be satisfied is when functions remain finite at infinity. In practice it is useful to imagine idealized systems in which fluctuations remain finite at infinity, and the nonexistence of Fourier transforms would then prove restrictive. Indeed the very concept of a plane wave involves a supposedly infinite wavetrain. The Fourier transforms of functions which do not vanish at infinity may be defined in a limiting sense by truncating the functions. The idea is the same as that introduced above, namely to consider a sequence of functions each of whose Fourier transform exists, and such that the limit of the sequence is the function of interest. A truncation involves a sequence each element of which is equal to the given function over an arbitrarily long time (and/or large volume) and zero otherwise. The arbitrary long time (and/or volume) is then allowed to approach infinity. In practice one simply assumes that relevant truncations have been made. The only occasion when one needs to appeal explicitly to the truncation is when the square of a δ-function appears.

The Dirac δ-function

For present purposes an adequate definition of the Dirac δ-function is that it is the (generalized) function such that

$$\int_{x_1}^{x_2} \mathrm{d}x\, f(x)\delta(x-a) = f(a) \tag{2.5}$$

applies for all functions which are continuous at $x=a$ and for all ranges of integration $x_1 < a < x_2$. Sequences of functions which may be used to define the δ-function include

$$\delta(x) = \lim_{\varepsilon\to 0} \frac{\varepsilon}{\pi(x^2+\varepsilon^2)} = \lim_{\varepsilon\to 0} \frac{e^{-x^2/\varepsilon}}{\varepsilon\sqrt{\pi}} = \lim_{\tau\to\infty} \frac{\sin x\tau}{\pi x}. \tag{2.6}$$

Consider a function $f(x)$ which has simple zeros at $x=c_1, c_2, \ldots$. The integral

$$I := \int_{-\infty}^{\infty} \mathrm{d}x\, F(x)\, \delta(f(x)) \tag{2.7}$$

may be evaluated as follows. Firstly the integral is split into a sum of integrals

$$I = \sum_i I_i, \tag{2.8a}$$

$$I_i := \int_{c_i-\Delta x}^{c_i+\Delta x} \mathrm{d}x\, F(x)\, \delta(f(x)), \tag{2.8b}$$

where Δx is arbitrarily small and where the sum is over all zeros. In the neighborhood of $x=c_i$, $F(x)$ may be replaced by $F(c_i)$ and $f(x)$ by $f'(c_i)(x-c_i)$. Then by changing the variable of integration to $f'(c_i)x$ in (2.8b), and using

(2.5), one finds

$$I=\sum_i \frac{F(c_i)}{|f'(c_i)|}, \tag{2.9}$$

where f' denotes $\mathrm{d}f/\mathrm{d}x$.

The square of the δ-function is evaluated by using the truncated function

$$f(t)=\begin{cases}1 & \text{for} \quad |t|<\tau/2,\\ 0 & \text{for} \quad |t|>\tau/2,\end{cases} \tag{2.10}$$

i.e. using

$$2\pi\,\delta(\omega)=\lim_{\tau\to\infty}\int_{-\tau/2}^{\tau/2}\mathrm{d}t\,\mathrm{e}^{\pm i\omega t}, \tag{2.11}$$

which is equivalent to the final sequence in (2.6). Then one has

$$\begin{aligned}[2\pi\,\delta(\omega)]^2&=\lim_{\tau\to\infty}\int_{-\tau/2}^{\tau/2}\mathrm{d}t\int_{-\tau/2}^{\tau/2}\mathrm{d}t'\,\mathrm{e}^{\pm i\omega(t+t')}\\ &=\lim_{\tau\to\infty}\frac{1}{2}\int_{-\tau}^{\tau}\mathrm{d}(t-t')\int_{-\tau}^{\tau}\mathrm{d}(t+t')\,\mathrm{e}^{\pm i\omega(t+t')}\\ &=\lim_{\tau\to\infty}\tau\int_{-\tau}^{\tau}\mathrm{d}t\,\mathrm{e}^{\pm i\omega t}\\ &=\lim_{\tau\to\infty}[\tau 2\pi\,\delta(\omega)],\end{aligned}$$

i.e.

$$[\delta(\omega)]^2=\lim_{\tau\to\infty}\frac{\tau}{2\pi}\,\delta(\omega). \tag{2.12}$$

Similarly one has

$$[\delta^3(\mathbf{k})]^2=\lim_{V\to\infty}\frac{V}{(2\pi)^3}\,\delta^3(\mathbf{k}), \tag{2.13}$$

where V is the normalization volume.

Properties of Fourier Transforms

Elementary properties of Fourier transforms include the following.

The reality condition If $F(\mathbf{r}, t)$ is real then $\tilde{F}(\mathbf{k}, \omega)$ satisfies

$$\tilde{F}(\mathbf{k}, \omega)=\tilde{F}^*(-\mathbf{k}, -\omega), \tag{2.14}$$

where * denotes complex conjugation. The proof of (2.14) follows from (2.1) by complex conjugating and noting that $F^*(\mathbf{r}, t)$ is equal to $F(\mathbf{r}, t)$ by assumption.

The power theorem If $F(\mathbf{r}, t)$ and $G(\mathbf{r}, t)$ are real then one has

$$\int dt \int d^3\mathbf{r}\, F(\mathbf{r}, t)G(\mathbf{r}, t) = \int \frac{d\omega}{2\pi} \int \frac{d^3\mathbf{k}}{(2\pi)^3}\, \tilde{F}(\mathbf{k}, \omega)\tilde{G}^*(\mathbf{k}, \omega). \tag{2.15}$$

The proof of (2.15) is as follows. Write down $F(\mathbf{r}, t)$ and $G(\mathbf{r}, t)$ in terms of the Fourier transforms using (2.2), i.e.

$$F(\mathbf{r}, t)G(\mathbf{r}, t) = \int \frac{d\omega}{2\pi} \int \frac{d^3\mathbf{k}}{(2\pi)^3} \int \frac{d\omega'}{2\pi} \int \frac{d^3\mathbf{k}'}{(2\pi)^3}$$
$$\exp\left[i\{(\mathbf{k}+\mathbf{k}')\cdot\mathbf{r} - (\omega+\omega')t\}\right] \tilde{F}(\mathbf{k}, \omega)\tilde{G}(\mathbf{k}', \omega').$$

The integrals over time and space give δ-functions, cf. (2.4a, b), and hence

$$\int dt \int d^3\mathbf{r} F(\mathbf{r}, t)G(\mathbf{r}, t) = \int \frac{d\omega}{2\pi} \int \frac{d^3\mathbf{k}}{(2\pi)^3} \int \frac{d\omega'}{2\pi} \int \frac{d^3\mathbf{k}'}{(2\pi)^3}$$
$$\times (2\pi)^3 \delta^3(\mathbf{k}+\mathbf{k}')2\pi\delta(\omega+\omega')\tilde{F}(\mathbf{k}, \omega)\tilde{G}(\mathbf{k}', \omega').$$

The result (2.6) then follows by integrating over ω' and $\mathbf{k}'$ using the δ-functions, and, finally, by using the reality condition (2.14) to rewrite $\tilde{G}(-\mathbf{k}, -\omega)$ as $\tilde{G}^*(\mathbf{k}, \omega)$.

A corollary of the power theorem is the result

$$\int dt \int d^3\mathbf{r}\, |F(\mathbf{r}, t)|^2 = \int \frac{d\omega}{2\pi} \int \frac{d^3\mathbf{k}}{(2\pi)^3}\, |\tilde{F}(\mathbf{k}, \omega)|^2. \tag{2.16}$$

This is sometimes called Rayleigh's theorem, and it is also sometimes called Parseval's formula by analogy with the corresponding formula for Fourier series.

The convolution theorem The Fourier transform of the convolution of two functions, e.g. of

$$H(\mathbf{r}, t) := \int dt' \int d^3\mathbf{r}'\, F(\mathbf{r}-\mathbf{r}', t-t')G(\mathbf{r}', t'), \tag{2.17a}$$

is the product of the Fourier transforms, i.e.

$$\tilde{H}(\mathbf{k}, \omega) = \tilde{F}(\mathbf{k}, \omega)\tilde{G}(\mathbf{k}, \omega). \tag{2.17b}$$

Conversely, the Fourier transform of a product of two functions, e.g. of

$$J(\mathbf{r}, t) := F(\mathbf{r}, t)G(\mathbf{r}, t) \tag{2.18a}$$

is the convolution of their Fourier transforms, i.e.

$$\tilde{J}(\mathbf{k}, \omega) = \int \frac{d\omega'}{2\pi} \int \frac{d^3\mathbf{k}'}{(2\pi)^3} \tilde{F}(\mathbf{k}-\mathbf{k}', \omega-\omega')\tilde{G}(\mathbf{k}', \omega'). \tag{2.18b}$$

The proof of (2.17b), say, is as follows. Substitute for F and G in terms of their Fourier transforms,

$$F(\mathbf{r}-\mathbf{r}', t-t')G(\mathbf{r}', t') = \int \frac{d\omega}{2\pi} \int \frac{d^3\mathbf{k}}{(2\pi)^3} \int \frac{d\omega'}{2\pi} \int \frac{d^3\mathbf{k}'}{(2\pi)^3}$$
$$\times \exp\left[i(\mathbf{k}\cdot(\mathbf{r}-\mathbf{r}')-\omega(t-t'))\right] \exp\left[i(\mathbf{k}'\cdot\mathbf{r}'-\omega't')\right] \tilde{F}(\mathbf{k}, \omega)\tilde{G}(\mathbf{k}', \omega').$$

On integrating over t' and $\mathbf{r}'$ using (2.4a, b), the integral over ω' and $\mathbf{k}'$ may be performed to find

$$\int dt' \int d^3\mathbf{r}'\, F(\mathbf{r}-\mathbf{r}', t-t')G(\mathbf{r}', t') = \int \frac{d\omega}{2\pi} \int \frac{d^3\mathbf{k}}{(2\pi)^3} \exp\left[i(\mathbf{k}\cdot\mathbf{r}-\omega t)\right]$$
$$\times \tilde{F}(\mathbf{k}, \omega)\tilde{G}(\mathbf{k}, \omega),$$

from which (2.17b) follows.

The result (2.18a, b) may be generalized to apply to more than two functions. Specifically, the Fourier transform of

$$J(\mathbf{r}, t) := F_1(\mathbf{r}, t)F_2(\mathbf{r}, t) \cdots F_n(\mathbf{r}, t) \tag{2.19a}$$

is

$$\tilde{J}(\mathbf{k}, \omega) = \int d\lambda^{(n)} \tilde{F}_1(\mathbf{k}_1, \omega_1)\tilde{F}_2(\mathbf{k}_2, \omega_2) \cdots \tilde{F}_n(\mathbf{k}_n, \omega_n), \tag{2.19b}$$

where

$$d\lambda^{(n)} := \frac{d\omega_1 d^3\mathbf{k}_1}{(2\pi)^4} \frac{d\omega_2 d^3\mathbf{k}_2}{(2\pi)^4} \cdots \frac{d\omega_n d^3\mathbf{k}_n}{(2\pi)^4} (2\pi)^4 \delta(\omega-\omega_1-\omega_2 \cdots -\omega_n)$$
$$\times \delta^3(\mathbf{k}-\mathbf{k}_1-\mathbf{k}_2 \cdots -\mathbf{k}_n) \tag{2.20}$$

denotes the n-fold convolution integral.

Fourier Transform of Differential Operators

For our immediate purpose, the following Fourier transforms are required in order to Fourier transform Maxwell's equations,

$$\frac{\partial}{\partial t} F(\mathbf{r}, t) \underset{\text{FT}}{\longrightarrow} -i\omega\tilde{F}(\mathbf{k}, \omega), \tag{2.21a}$$

$$\text{grad } F(\mathbf{r}, t) \underset{\text{FT}}{\longrightarrow} i\mathbf{k}\tilde{F}(\mathbf{k}, \omega), \tag{2.21b}$$

$$\text{div } \mathbf{A}(\mathbf{r}, t) \underset{\text{FT}}{\longrightarrow} i\mathbf{k}\cdot\tilde{\mathbf{A}}(\mathbf{k}, \omega), \tag{2.21c}$$

$$\text{curl } \mathbf{A}(\mathbf{r}, t) \underset{\text{FT}}{\longrightarrow} i\mathbf{k}\times\tilde{\mathbf{A}}(\mathbf{k}, \omega), \tag{2.21d}$$

where $\underset{\text{FT}}{\longrightarrow}$ denotes taking the Fourier transform. These relations may be derived as follows. In deriving (2.21a) say, one starts from (2.2) and differentiates with respect to t to find

$$\frac{\partial}{\partial t} F(\mathbf{r}, t) = \int \frac{d\omega}{2\pi} \int \frac{d^3\mathbf{k}}{(2\pi)^3} \tilde{F}(\mathbf{k}, \omega) \frac{\partial}{\partial t} \exp\left[i(\mathbf{k}\cdot\mathbf{r} - \omega t)\right].$$

The derivations of (2.21b, c, d) are similar, and are facilitated by using the tensor notation developed below.

Tensor Notation

The (Cartesian) tensor notation used below may be summarized by the following rules and definitions:

1) Each element in a tensor equation is *either* a kernel symbol with none, one or more indices, *or* a product of such symbols.

2) The indices are latin subscripts which denote Cartesian components, usually understood to be labelled 1, 2, 3 or x, y, z. Indices may have affixes, e.g. primes as superscripts or numbers as subscripts. and two indices are the same only if they have the same affix.

3) No index may appear more than twice in a given element. Indices which appear once are called *free* indices, and those which appear twice are called *dummy* indices.

4) Summation (over all Cartesian components) over dummy indices is implied. This *summation convention* is due to Einstein.

5) A tensor equation relates elements with the same number and kind of free indices.

6) The *rank* of a tensor is equal to the number of its free indices. Scalars are tensors of rank zero and vectors are tensors of rank one.

7) A matrix representation is possible for tensor equations which involve tensors of rank no higher than two. In such representations vectors are represented by column or row matrices and second rank tensors are represented by square matrices with the first written index labelling rows and the second written index labelling columns.

8) A second rank tensor t_{ij} is *symmetric* if it satisfies $t_{ij}=t_{ji}$, *anti-symmetric* if it satisfies $t_{ij}=-t_{ji}$, *hermitian* if it satisfies $t_{ij}=t^*_{ji}$ and *anti-hermitian* if it satisfies $t_{ij}=-t^*_{ji}$.

9) The unit tensor or Kronecker delta is defined by

$$\delta_{ij}:=\begin{cases}1 & \text{for} \quad i=j,\\ 0 & \text{for} \quad i\neq j.\end{cases} \tag{2.22}$$

Note the trace of δ_{ij},

$$\delta_{ss}=\delta_{11}+\delta_{22}+\delta_{33}=3. \tag{2.23}$$

10) The permutation symbol, which acts like a third rank tensor in tensor equations, is defined by

$$\varepsilon_{ijk}:=\begin{cases}1 & \text{for } ijk \text{ an even permutation of 123},\\ -1 & \text{for } ijk \text{ an odd permutation of 123},\\ 0 & \text{otherwise},\end{cases} \tag{2.24}$$

where the Cartesian components are assumed to be labelled 1, 2 and 3. The permutation symbol is completely anti-symmetric in the sense that it reverses sign when any two neighboring indices are interchanged.

A basic identity is

$$\begin{aligned}\varepsilon_{abc}\varepsilon_{ijk}=&\delta_{ai}\delta_{bj}\delta_{ck}+\delta_{ak}\delta_{bi}\delta_{cj}+\delta_{aj}\delta_{bk}\delta_{ci}\\ &-\delta_{bi}\delta_{aj}\delta_{ck}-\delta_{bk}\delta_{ai}\delta_{cj}-\delta_{bj}\delta_{ak}\delta_{ci}.\end{aligned} \tag{2.25}$$

It then follows from (2.23) that one has

$$\varepsilon_{ibc}\varepsilon_{ijk}=\delta_{bj}\delta_{ck}-\delta_{bk}\delta_{cj}, \tag{2.26a}$$

and, using (2.23) again,

$$\varepsilon_{ijc}\varepsilon_{ijk}=2\delta_{ck}, \quad \varepsilon_{ijk}\varepsilon_{ijk}=6. \tag{2.26b, c}$$

Vector Algebra

The use of tensor notation in vector algebra is based on the following identifications

$$\mathbf{A}\cdot\mathbf{B}=A_sB_s,\quad (\mathbf{A}\times\mathbf{B})_i=\varepsilon_{ijk}A_jB_k, \tag{2.27a, b}$$

and

$$(\text{grad } F)_i=\frac{\partial F}{\partial x_i},\quad \text{div } \mathbf{A}=\frac{\partial A_s}{\partial x_s},\quad (\text{curl } \mathbf{A})_i=\varepsilon_{ijk}\frac{\partial A_k}{\partial x_j}. \tag{2.28a, b, c}$$

Note also that the operator ($\mathbf{A}\cdot$grad) becomes the operator $A_i\partial/\partial x_i$. One may then derive the usual formulae of vector algebra using the identities (2.23), (2.26a) and the trivial identity $\delta_{is}A_s=A_i$. For example, one has

$$\begin{aligned}(\text{curl } \mathbf{A}\times\mathbf{B})_i&=\varepsilon_{ijk}\frac{\partial}{\partial x_j}(\varepsilon_{krs}A_rB_s)\\ &=(\delta_{ir}\delta_{js}-\delta_{is}\delta_{jr})\left(A_r\frac{\partial B_s}{\partial x_j}+B_s\frac{\partial A_r}{\partial x_j}\right)\\ &=A_i\frac{\partial B_s}{\partial x_s}+B_s\frac{\partial A_i}{\partial x_s}-A_r\frac{\partial B_i}{\partial x_r}-B_i\frac{\partial A_r}{\partial x_r}\\ &=(\mathbf{A}\text{ div }\mathbf{B}+(\mathbf{B}\cdot\text{grad})\mathbf{A}-(\mathbf{A}\cdot\text{grad})\mathbf{B}-\mathbf{B}\text{ div }\mathbf{A})_i,\end{aligned}$$

which is a standard vector identity.

Similarly (2.21b, c, d) may be derived using (2.27a, b) and (2.28a, b, c). For example, (2.21d) follows from

$$\begin{aligned}[\text{curl } \mathbf{A}(\mathbf{r},t)]_i&=\varepsilon_{ijk}\frac{\partial}{\partial x_j}\int\frac{d\omega}{2\pi}\int\frac{d^3\mathbf{k}}{(2\pi)^3}\exp\left[i(\mathbf{k}\cdot\mathbf{r}-\omega t)\right]\tilde{A}_k(\mathbf{k},\omega)\\ &=\int\frac{d\omega}{2\pi}\int\frac{d^3\mathbf{k}}{(2\pi)^3}\exp\left[i(\mathbf{k}\cdot\mathbf{r}-\omega t)\right]\left[i\mathbf{k}\times\tilde{\mathbf{A}}(\mathbf{k},\omega)\right]_i.\end{aligned}$$

2.2 THE WAVE EQUATION

The wave equation for waves in a collisionless plasma is derived from Maxwell's equations with the current density identified with that due to the linear response of the plasma. The approach is similar to that used in the theory of waves in dielectrics, but with one important generalization. The generalization is that the response of the medium is allowed to be a function of both $\mathbf{k}$ and ω. Thus our starting point is the Fourier transformed versions of Maxwell's equations.

Maxwell's Equations

Maxwell's equations can always be written in terms of the electric field strength $\mathbf{E}(\mathbf{r}, t)$, the magnetic induction $\mathbf{B}(\mathbf{r}, t)$, with the charge density

$\rho(\mathbf{r}, t)$ and the current density $\mathbf{j}(\mathbf{r}, t)$ acting as sources for the fields:

$$\text{curl } \mathbf{E}(\mathbf{r}, t) = -\frac{1}{c}\frac{\partial \mathbf{B}(\mathbf{r}, t)}{\partial t}, \tag{2.29}$$

$$\text{div } \mathbf{B}(\mathbf{r}, t) = 0, \tag{2.30}$$

$$\text{curl } \mathbf{B}(\mathbf{r}, t) = \frac{4\pi}{c}\mathbf{j}(\mathbf{r}, t) + \frac{1}{c}\frac{\partial \mathbf{E}(\mathbf{r}, t)}{\partial t}, \tag{2.31}$$

$$\text{div } \mathbf{E}(\mathbf{r}, t) = 4\pi\rho(\mathbf{r}, t). \tag{2.32}$$

Two subsidiary equations are the equation for charge continuity

$$\frac{\partial \rho(\mathbf{r}, t)}{\partial t} + \text{div } \mathbf{j}(\mathbf{r}, t) = 0, \tag{2.33}$$

which follows from (2.31) and (2.32), and the equation for energy continuity

$$\frac{\partial}{\partial t}\left[\frac{|\mathbf{E}(\mathbf{r}, t)|^2}{8\pi} + \frac{|\mathbf{B}(\mathbf{r}, t)|^2}{8\pi}\right] + \text{div}\left[\frac{c}{4\pi}\mathbf{E}(\mathbf{r}, t) \times \mathbf{B}(\mathbf{r}, t)\right] = -\mathbf{j}(\mathbf{r}, t) \cdot \mathbf{E}(\mathbf{r}, t), \tag{2.34}$$

which follows from (2.29) and (2.31). On the basis of (2.34), and the fact that $-\mathbf{j} \cdot \mathbf{E}$ is the rate per unit volume at which work is done on the electromagnetic field, $E^2/8\pi$ and $B^2/8\pi$ are interpreted as the electric energy density and the magnetic energy density, respectively, and $(c/4\pi)\mathbf{E} \times \mathbf{B}$, called the Poynting vector, is interpreted as the electromagnetic energy flux density.

The basic equation (the wave equation) used in deriving the properties of waves in plasmas and in treating their emission and absorption may be derived from Maxwell's equations as follows. Firstly, Fourier transform (2.29) to (2.32) to find

$$\mathbf{k} \times \mathbf{E}(\mathbf{k}, \omega) = \frac{\omega}{c}\mathbf{B}(\mathbf{k}, \omega), \tag{2.35}$$

$$\mathbf{k} \cdot \mathbf{B}(\mathbf{k}, \omega) = 0, \tag{2.36}$$

$$\mathbf{k} \times \mathbf{B}(\mathbf{k}, \omega) = -\frac{4\pi i}{\omega}\mathbf{j}(\mathbf{k}, \omega) - \frac{\omega}{c}\mathbf{E}(\mathbf{k}, \omega), \tag{2.37}$$

$$\mathbf{k} \cdot \mathbf{E}(\mathbf{k}, \omega) = -4\pi i\rho(\mathbf{k}, \omega). \tag{2.38}$$

It is surprising at first sight to notice that one of these equations is redundant, specifically (2.36) is implied by (2.35). Of course we cannot have reduced four independent equations to three by a mathematical manipulation: the set (2.35) to (2.38) must contain less information than the set (2.29) to (2.32). Closer examination shows that the lost information concerns only the static fields and these must be specified independently. In plasma physics one is usually concerned either with an unmagnetized plasma (no static field) or a magnetized plasma (the static field is a magnetic field).

The second step is to reinterpret the three remaining independent equations as a basic equation and two subsidiary equations. One subsidiary equation is (2.35), which is henceforth interpreted as a definition of a subsidiary field vector $\mathbf{B}(\mathbf{k}, \omega)$ in terms of a basic field vector $\mathbf{E}(\mathbf{k}, \omega)$. The other subsidiary equation is

$$\omega\rho(\mathbf{k}, \omega) = \mathbf{k} \cdot \mathbf{j}(\mathbf{k}, \omega), \tag{2.39}$$

which is the Fourier transform of (2.33) and expresses charge continuity. Thus (2.39) is interpreted as defining a subsidiary quantity $\rho(\mathbf{k}, \omega)$ in terms of a basic quantity $\mathbf{j}(\mathbf{k}, \omega)$. The remaining basic equation is

$$\mathbf{k} \times \{\mathbf{k} \times \mathbf{E}(\mathbf{k}, \omega)\} + \frac{\omega^2}{c^2} \mathbf{E}(\mathbf{k}, \omega) = -\frac{4\pi i\omega}{c^2} \mathbf{j}(\mathbf{k}, \omega), \tag{2.40}$$

which is one form of the *wave equation.*

The Linear Response

The current $\mathbf{j}$ in (2.40) is separated into an induced (ind) part and an extraneous (ext) part. The induced part is assumed to describe the linear response of the medium to the electromagnetic disturbance. The disturbance is described by $\mathbf{E}(\mathbf{k}, \omega)$, which includes the subsidiary $\mathbf{B}(\mathbf{k}, \omega)$. Thus, by hypothesis, the response may be described by an equation of the form

$$j_i^{\text{ind}}(\mathbf{k}, \omega) = \sigma_{ij}(\mathbf{k}, \omega) E_j(\mathbf{k}, \omega). \tag{2.41}$$

The quantity $\sigma_{ij}(\mathbf{k}, \omega)$ is called the conductivity tensor. It is conventional to describe the response in terms of the dielectric tensor (sometimes called the equivalent dielectric tensor),

$$\varepsilon_{ij}(\mathbf{k}, \omega) := \delta_{ij} + \frac{4\pi i}{\omega} \sigma_{ij}(\mathbf{k}, \omega). \tag{2.42}$$

The wave equation (2.40), after rewriting in tensor notation, multiplying by c^2/ω^2 and including the induced part on the left hand side, becomes

$$\Lambda_{ij}(\mathbf{k}, \omega) E_j(\mathbf{k}, \omega) = -\frac{4\pi i}{\omega} j_i^{\text{ext}}(\mathbf{k}, \omega), \tag{2.43}$$

with

$$\Lambda_{ij}(\mathbf{k}, \omega) := \frac{k^2 c^2}{\omega^2} (\kappa_i \kappa_j - \delta_{ij}) + \varepsilon_{ij}(\mathbf{k}, \omega), \tag{2.44}$$

and with

$$\boldsymbol{\kappa}:=\frac{\mathbf{k}}{k}. \tag{2.45}$$

The wave equation in the form (2.43) is used extensively below.

The Dielectric Tensor: General Properties

There are three general properties of dielectric tensors. These reflect basic physical principles.

The first property relates to the separation into hermitian (h) and anti-hermitian (a) parts, i.e.

$$\varepsilon_{ij}(\mathbf{k},\omega)=\varepsilon^{\mathrm{h}}_{ij}(\mathbf{k},\omega)+\varepsilon^{\mathrm{a}}_{ij}(\mathbf{k},\omega),$$

with

$$\varepsilon^{\mathrm{h}}_{ij}(\mathbf{k},\omega)=\tfrac{1}{2}[\varepsilon_{ij}(\mathbf{k},\omega)+\varepsilon^{*}_{ji}(\mathbf{k},\omega)] \tag{2.46a}$$

and

$$\varepsilon^{\mathrm{a}}_{ij}(\mathbf{k},\omega)=\tfrac{1}{2}[\varepsilon_{ij}(\mathbf{k},\omega)-\varepsilon^{*}_{ji}(\mathbf{k},\omega)]. \tag{2.46b}$$

The hermitian part describes the reactive (or non dissipative) response and the anti-hermitian part describes the resistive (or dissipative) response. This may be seen by showing that the time-average of $\mathbf{j}^{\mathrm{ind}}\cdot\mathbf{E}$ is zero when only the hermitian part of ε_{ij} (and hence the anti-hermitian part of σ_{ij}) is retained in (2.41).

The second property is related to time reversal. Now, because the hermitian part describes the time-reversible part of the response, it must be an even function under time reversal. Similarly, because the anti-hermitian part describes the time-irreversible part of the response, it must be an odd function. For particles in a magnetic field time reversal corresponds to reversing not only the sense of t, but also the sense of the background magnetic field $\mathbf{B}$. Only then does the equation of motion

$$\frac{\mathrm{d}\mathbf{p}}{\mathrm{d}t}=q\frac{\mathbf{v}\times\mathbf{B}}{c}$$

transform into itself. Consequently, the time-reversal-invariance condition relates ε_{ij} to itself with the signs of ω and $\mathbf{B}$ reversed. Specifically time reversal invariance implies

$$\varepsilon^{\mathrm{h}}_{ij}(\mathbf{k},\omega)\big|_{\mathbf{B}}=\varepsilon^{\mathrm{h}}_{ij}(\mathbf{k},-\omega)\big|_{-\mathbf{B}}, \tag{2.47a}$$

$$\varepsilon^{\mathrm{a}}_{ij}(\mathbf{k},\omega)\big|_{\mathbf{B}}=-\varepsilon^{\mathrm{a}}_{ij}(\mathbf{k},-\omega)\big|_{-\mathbf{B}}. \tag{2.47b}$$

It is conventional to write (2.47a, b) as a single identity by using the reality condition, cf. (2.14),

$$\varepsilon_{ij}(\mathbf{k}, \omega)=\varepsilon_{ij}^{*}(-\mathbf{k}, -\omega). \tag{2.48}$$

The conventional form of (2.47a, b) is

$$\varepsilon_{ij}(\mathbf{k}, \omega)\big|_{\mathbf{B}}=\varepsilon_{ji}(-\mathbf{k}, \omega)\big|_{-\mathbf{B}}. \tag{2.49}$$

Relations like (2.49) which are implied by time-reversal invariance are called Onsager relations.

It is convenient to choose a particular coordinate system with axes related to the directions $\boldsymbol{\kappa}$ and

$$\mathbf{b}:=\mathbf{B}/B, \tag{2.50}$$

and then reversal of $\boldsymbol{\kappa}$ and $\mathbf{b}$ in (2.49) implies reversing the signs of the components along $\boldsymbol{\kappa}$ and $\mathbf{b}$. A specific choice of axes is

$$\mathbf{e}_1:=\frac{\mathbf{b}\times(\boldsymbol{\kappa}\times\mathbf{b})}{|\boldsymbol{\kappa}\times\mathbf{b}|}, \quad \mathbf{e}_2:=\frac{\mathbf{b}\times\boldsymbol{\kappa}}{|\mathbf{b}\times\boldsymbol{\kappa}|}, \quad \mathbf{e}_3:=\mathbf{b}. \tag{2.51}$$

Then, on reversing the signs of $\boldsymbol{\kappa}$ and $\mathbf{b}$, $\mathbf{e}_1$ and $\mathbf{e}_2$ reverse sign but $\mathbf{e}_2$ does not. Consequently, the components ε_{ij} with respect to the axes (2.51) change sign by $(-)^x$ with $x=0$ for $i=2=j$, with $x=1$ for $i=2$ and $j=1$ or 3, and for $j=2$ and $i=1$ or 3, and with $x=2$ for $i=1$ or 3 and $j=1$ or 3. In other words, (2.49) implies

$$\varepsilon_{12}(\mathbf{k}, \omega)=-\varepsilon_{21}(\mathbf{k}, \omega), \quad \varepsilon_{13}(\mathbf{k}, \omega)=\varepsilon_{31}(\mathbf{k}, \omega), \quad \varepsilon_{23}(\mathbf{k}, \omega)=-\varepsilon_{32}(\mathbf{k}, \omega). \tag{2.52}$$

An important implication is that for the hermitian part ε_{ij}^{h}, the components ε_{11}^{h}, ε_{22}^{h}, ε_{33}^{h} and $\varepsilon_{13}^{h}=\varepsilon_{31}^{h}$ are real, and $\varepsilon_{12}^{h}=\varepsilon_{21}^{h}$ and $\varepsilon_{23}^{h}=-\varepsilon_{32}^{h}$ are imaginary.

The third general property of the dielectric tensor is implied by the causal condition, and is expressed through the so-called Kramer.–Kronig relations. The causal condition is discussed in Section 3.1.

Classification of Media

All relevant information concerning the linear response of the plasma is assumed to be included in the dielectric tensor. As a consequence, the dielectric tensor plays a central role in the theory of waves (and of other fluctuations) in a plasma. In the discussion here, three aspects of the use of dielectric tensors are discussed. The first is the relation between mathematical properties of the dielectric tensor and the corresponding descriptive names applied to the medium. The second is the relation between the (equivalent) dielectric tensor as used in plasma physics and the description of the response of a medium in phenomenological electrodynamics. Finally, methods of calculating the dielectric tensor are outlined briefly.

In discussing descriptive names used to describe media it is important to recognize that these names arose in the context of phenomenological electrodynamics. The simplest description of the response of a dielectric, according to phenomenological electrodynamics, is in terms of a dielectric constant ε, which is a function of neither ω nor $\mathbf{k}$. The dielectric tensor is then $\varepsilon_{ij}=\varepsilon\delta_{ij}$. Media in which the dielectric tensor is a function of ω are said to be *dispersive*. Dispersion is related to the existence of natural frequencies in the medium. In fact all media are dispersive, but in some (such as various glasses) the response is only a weak function of frequency. One of the characteristic features of the response of a plasma is that it is a strong function of frequency, i.e. plasmas are strongly dispersive. If the response is also a function of $\mathbf{k}$ the medium is said to be *spatially dispersive* or *dispersive in space*. For most media of interest in phenomenological electrodynamics, and for plasmas in some limits, spatial dispersion is not important. The allowance for spatial dispersion is one of the important features of the use of the (equivalent) dielectric tensor in plasma physics. Spatial dispersion is related to the existence of natural lengths in the medium, and the most important relevant lengths in a plasma are the Debye lengths.

The following classifications apply to media which are not spatially dispersive (some, but not all of them, are relevant for spatially dispersive media). The medium is *isotropic* if the dielectric tensor can be written in the form $\varepsilon_{ij}(\omega)=\varepsilon(\omega)\delta_{ij}$. Media which are not isotropic are *anisotropic*. For media which are unaffected by any ambient magnetic field (or any other axial vector), the Onsager relations (2.49) imply $\varepsilon_{ij}(\omega)=\varepsilon_{ji}(\omega)$, i.e. the dielectric tensor is symmetric. For any second rank symmetric tensor there exists a choice of axes, called principal axes, in which the tensor is diagonal. An anisotropic medium is said to be *uniaxial* if any two of the resulting diagonal components are equal, and *biaxial* if all three are different. A medium which is affected by an ambient magnetic field (or any other axial vector such as the angular velocity of rotation for a rotating medium) is said to be *gyrotropic*. (As opposed to media with $\varepsilon_{ij}=\varepsilon_{ji}$, the natural modes of a gyrotropic medium are elliptically, rather than linearly, polarized.)

Now consider spatially dispersive media. If the medium is isotropic, then the dielectric tensor can be written in the form

$$\varepsilon_{ij}(\mathbf{k},\omega)=\varepsilon^{l}(k,\omega)\kappa_i\kappa_j+\varepsilon^{t}(k,\omega)(\delta_{ij}-\kappa_i\kappa_j). \tag{2.53}$$

The components ε^{l} and ε^{t} are called the longitudinal and transverse parts respectively. For a spatially dispersive anisotropic medium there is no general simplification of the dielectric tensor, apart from that implied by the Onsager relations (2.49).

The Phenomenological Description of the Response

In phenomenological electrodynamics it is conventional to separate the response into electric and magnetic parts. Rather than describe the response in terms of an induced current, one describes it in terms of a *polarization* **P** and a *magnetization* **M**. Two subsidiary field quantities are introduced,

$$\mathbf{D}:=\mathbf{E}+4\pi\mathbf{P}, \quad \mathbf{H}:=\mathbf{B}-4\pi\mathbf{M}, \tag{2.54}$$

called the electric induction and the magnetic field strength respectively. Phenomenological relations are introduced between **D** and **E** (for a dielectric), or between **H** and **B** (for a paramagnetic, diamagnetic or ferromagnetic medium), and even sometimes between **D** and **B** and **H** and **E** (for magneto-electric media). It would appear that this collection of relations is more general than the single relation used in plasma physics, but such is not the case. In fact, as conventionally used, the phenomenological description is less general.

The phenomenological description has developed from an expansion in multipole moments. The multipole expansion for the current density, i.e. of

$$\mathbf{j}(\mathbf{k},\omega)=\int dt \int d^3\mathbf{r}\,\mathbf{j}(\mathbf{r},t)\exp\left[-i(\mathbf{k}\cdot\mathbf{r}-\omega t)\right]$$

may be based on the expansion

$$\exp(-i\mathbf{k}\cdot\mathbf{r})=1-i\mathbf{k}\cdot\mathbf{r}+\cdots.$$

One finds

$$j_i(\mathbf{k},\omega)=-i\omega\,\mathrm{d}_i(\omega)+ic(\mathbf{k}\times\mathbf{m}(\omega))_i-\frac{\omega}{2}k_r q_{ri}(\omega)+\cdots, \tag{2.55}$$

where the terms are the electric dipole, magnetic dipole and electric quadrupole terms respectively. The polarization and magnetization are then defined as, respectively, the induced electric and magnetic dipole moments per unit volume. An electric quadrupole correction to the polarization needs to be taken into account for some media, and such media are said to be *optically active*.

The description based on an expansion in multipoles is inadequate in general. One reason is that there is no guarantee that the expansion converges. A more subtle reason is that the multipole expansion, as conventionally defined, is incomplete. Consider the electric quadrupole term (which is the lowest order term for which the following discussion is relevant). The quantity q_{ij} in (2.55) is defined by

$$q_{ij} := \int d^3\mathbf{r}\; x_i x_j \rho(\mathbf{r}). \tag{2.56a}$$

The trace of q_{ij} does not contribute to the electromagnetic field *in vacuo*, and, consequently, it is conventional to redefine the quadrupole moment to be trace free, specifically

$$d_{ij} := 3q_{ij} - q_{ss}\delta_{ij}. \tag{2.56b}$$

However, the trace does contribute to the particle–electromagnetic field interaction in a medium. For higher order electric and magnetic moments, there are further terms which are discarded, by convention, on the grounds that they do not contribute to the particle–field-interaction *in vacuo*, but which do contribute in a medium. Consequently, a conventional multipole expansion is incomplete when discussing electrodynamics in a medium.

The description used in phenomenological electrodynamics is therefore less general than the one used in plasma physics, despite the former's greater complexity. The phenomenological description can be made completely general by introducing four response tensors (between **D** and **H** on the one hand, and **E** and **B** on the other) and allowing each of these to be a function of ω and **k**. However, it can be shown that such a description is equivalent to the one used in plasma physics, e.g. O'Sullivan and Derfler (1973).

As an explicit illustration of the difference between the phenomenological description and that used in plasma physics, consider the following. An isotropic medium is described phenomenologically in terms of two constants ε and μ, specifically with $\mathbf{D} = \varepsilon\mathbf{E}$ and $\mathbf{B} = \mu\mathbf{H}$. The description of the same system in terms of the (equivalent) dielectric tensor used in plasma physics may be found by constructing **P** and **M** using (2.54a, b), and hence constructing the induced current using the counterpart (per unit volume) of (2.55). One may then express **B** in terms of **E** using (2.35), and define ε_{ij} using (2.41) and (2.42). One finds

$$\varepsilon_{ij}(\mathbf{k}, \omega) = \varepsilon\kappa_i\kappa_j + \left(\varepsilon + \frac{k^2c^2}{\omega^2}\frac{\mu - 1}{\mu}\right)(\delta_{ij} - \kappa_i\kappa_j),$$

where the result has been written in the form (2.53). In this case the phenomenological description is the simpler.

Calculations of Dielectric Tensors

The simplest useful theory for calculating the dielectric tensor for a plasma is cold plasma theory. Each species is assumed to be described by a number density n_α and a fluid velocity $\mathbf{u}_\alpha$, as well as their charge $q_\alpha = \epsilon_\alpha |q_\alpha|$, mass m_α, gyrofrequency Ω_α and plasma frequency $\omega_{p\alpha}$. The equation of fluid motion is

$$m_\alpha \frac{d\mathbf{u}_\alpha}{dt} = q_\alpha \mathbf{E} + q_\alpha \frac{\mathbf{u}_\alpha \times \mathbf{B}}{c}.$$

After Fourier transforming with respect to time, the equation of fluid motion becomes

$$(\omega\delta_{ij} - i\epsilon_\alpha \Omega_\alpha \varepsilon_{ijk} b_k)(\mathbf{u}_\alpha)_j = i\frac{q_\alpha}{m_\alpha} E_i, \tag{2.57}$$

where all arguments ω are omitted for convenience. On defining the tensor $\tau_{ij}^{(\alpha)}$ by

$$(\omega\delta_{ij} - i\epsilon_\alpha \Omega_\alpha \varepsilon_{ijk} b_k)\tau_{jl}^{(\alpha)}(\omega) := \omega\delta_{il}, \tag{2.58}$$

(2.57) may be solved for $\mathbf{u}_\alpha$. Hence one finds

$$j_i = \sum_\alpha q_\alpha n_\alpha (\mathbf{u}_\alpha)_i = \sum_\alpha \frac{i q_\alpha^2 n_\alpha}{m_\alpha \omega} \tau_{ij}^{(\alpha)} E_j, \tag{2.59}$$

where the sum is over all species of particle. The dielectric tensor follows from (2.42),

$$\varepsilon_{ij}(\omega) = \delta_{ij} - \sum_\alpha \frac{\omega_{p\alpha}^2}{\omega^2} \tau_{ij}^{(\alpha)}(\omega). \tag{2.60}$$

An explicit matrix representation for $\tau_{ij}^{(\alpha)}$ in the coordinate system by $\mathbf{b}$ along the z-axis is

$$\tau_{ij}^{(\alpha)}(\omega) = \begin{pmatrix} \dfrac{\omega^2}{\omega^2 - \Omega_\alpha^2} & \dfrac{i\epsilon_\alpha \omega \Omega_\alpha}{\omega^2 - \Omega_\alpha^2} & 0 \\ \dfrac{-i\epsilon_\alpha \omega \Omega_\alpha}{\omega^2 - \Omega_\alpha^2} & \dfrac{\omega^2}{\omega^2 - \Omega_\alpha^2} & 0 \\ 0 & 0 & 1 \end{pmatrix}. \tag{2.61}$$

A more general method of calculating the dielectric tensor is that based on kinetic theory. The relevant calculations are presented in Section 10.4 below. The resulting expression for the dielectric tensor for an unmagnetized plasma is

$$\varepsilon_{ij}(\mathbf{k}, \omega) = \delta_{ij} + \sum_\alpha \frac{4\pi q^2}{\omega^2} \int d^3\mathbf{p}\, \frac{v_i\{(\omega - \mathbf{k}\cdot\mathbf{v})\delta_{sj} + k_s v_j\}}{\omega - \mathbf{k}\cdot\mathbf{v} + iO} \frac{\partial f \mathbf{p})}{\partial p_s}, \tag{2.62}$$

where subscripts α are omitted and where the infinitesimal imaginary term iO is included to take account of the causal condition, see Section 3.1 below.

For completeness let us write down the counterpart of (2.62) for the case where an ambient magnetic field is present,

$$\varepsilon_{ij}(\mathbf{k}, \omega)=\delta_{ij}+\sum_{\alpha}\frac{4\pi q^2}{\omega^2}\int d^3\mathbf{p}\left[\frac{p_{\parallel}}{p_{\perp}}\left(v_{\perp}\frac{\partial}{\partial p_{\parallel}}-v_{\parallel}\frac{\partial}{\partial p_{\perp}}\right)f(\mathbf{p})b_i b_j\right.$$

$$\left.+\sum_{s=-\infty}^{\infty}\frac{V_i(s, \mathbf{p}, \mathbf{k})V_j^*(s, \mathbf{p}, \mathbf{k})}{\omega-s\Omega-k_{\parallel}v_{\parallel}+iO}\left\{\frac{\omega-k_{\parallel}v_{\parallel}}{v_{\perp}}\frac{\partial}{\partial p_{\perp}}+k_{\parallel}\frac{\partial}{\partial p_{\parallel}}\right\}f(\mathbf{p})\right], \quad (2.63)$$

where the quantity $\mathbf{V}(s, \mathbf{p}, \mathbf{k})$ is defined by (4.14) below.

The hermitian and anti-hermitian parts of (2.62) and (2.63) are obtained by interpreting the resonant denominator using the Plemelj formula, see (3.12) below. The finiteness of the anti-hermitian part, which arises from the semi-residue (i.e. from the δ-function term in (3.12)), implies the existence of collisionless damping of waves, cf. Section 6.1 below. Methods for calculating dielectric tensors are summarized in more detail in Appendix B.

2.3 FORMAL THEORY OF WEAKLY DAMPED WAVES

The properties of waves are found by solving the homogeneous wave equation, which is obtained from (2.43) by neglecting source terms. The right hand member of (2.43), i.e. the term involving $\mathbf{j}^{\text{ext}}$, is an explicit source term. There is also an implicit source term in the dissipative part of the response, i.e. in the term involving $\varepsilon_{ij}^{\text{a}}$ on the left hand side. In discussing the time-reversible wave properties this implicit source term is also neglected. It is included later when the damping of the waves is considered.

Dispersion Relations

Thus the homogeneous wave equation is

$$\Lambda_{ij}^{\text{h}}(\mathbf{k}, \omega)E_j(\mathbf{k}, \omega)=0. \quad (2.64)$$

The Onsager relations in the form (2.52) imply that Λ_{11}^{h}, Λ_{22}^{h}, Λ_{33}^{h} and $\Lambda_{13}^{\text{h}}=\Lambda_{31}^{\text{h}}$ are real and that $\Lambda_{12}^{\text{h}}=-\Lambda_{21}^{\text{h}}$ and $\Lambda_{23}^{\text{h}}=-\Lambda_{32}^{\text{h}}$ are imaginary. Hence (2.64) could be written as a set of three real simultaneous equations for E_1, $-iE_2$ and E_3. The solution of such a set of equations involves only elementary algebra.

The condition for a solution to exist is

$$\Lambda(\mathbf{k}, \omega)=0, \quad (2.65)$$

where

$$\Lambda(\mathbf{k}, \omega) := \det\left[\Lambda^{\mathrm{h}}_{ij}(\mathbf{k}, \omega)\right] \tag{2.66}$$

is the determinant (of the matrix representation) of $\Lambda^{\mathrm{h}}_{ij}$. Λ is necessarily real because the determinant of any hermitian matrix is real. Equation (2.65) is called the *dispersion equation.* Any specific solution of it is the *dispersion relation* for a specific wave mode. For formal purposes it is convenient to assume that the dispersion relation is an expression for ω in terms of $\mathbf{k}$ (i.e. that the components of $\mathbf{k}$ are the independent variables and ω is the dependent variable). The various wave modes are labelled σ, σ', σ'' etc. or σ_1, σ_2, σ_3, etc. Thus, for formal purposes, the dispersion relation for waves in the mode σ is written

$$\omega = \omega^{\sigma}(\mathbf{k})[= -\omega^{\sigma}(-\mathbf{k})], \tag{2.67}$$

where the identity in square brackets follows from the reality condition, which implies $\Lambda(\mathbf{k}, \omega) = \Lambda(-\mathbf{k}, -\omega)$. In practice, it is often more convenient to take ω and the angle θ between $\boldsymbol{\kappa}$ and $\mathbf{b}$ as the independent variables and the square of the refractive index, i.e. $n^2 := k^2c^2/\omega^2$, as the dependent variable. In this case the dispersion relation is written

$$n^2 = n_\sigma^2(\omega, \theta)[= n_\sigma^2(-\omega, \pi - \theta)]. \tag{2.68}$$

The dispersion relations in the form (2.67) and (2.68) are equivalent, in the sense that there is a one-to-one correspondence between them (provided that both ω and k are real).

Polarization Vectors

The dispersion equation is the condition for a solution of the wave equation (2.64) to exist. For each mode σ, there is a solution for the electric field $\mathbf{E}(\mathbf{k}, \omega^{\sigma}(\mathbf{k}))$. However, it is obvious that both the magnitude and phase of $\mathbf{E}(\mathbf{k}, \omega^{\sigma}(\mathbf{k}))$ remain arbitrary, i.e. they cannot be determined by (2.64). Consequently, it is convenient to solve for a unimodular vector $\mathbf{e}^{\sigma}(\mathbf{k})$ in the direction of $\mathbf{E}(\mathbf{k}, \omega^{\sigma}(\mathbf{k}))$. This vector is called the *polarization vector.* The polarization vector is one of the characteristic wave properties attributed to each mode. Given $\mathbf{e}^{\sigma}(\mathbf{k})$ one can find the direction of all other vectors associated with the waves, e.g. the magnetic vector is given by (2.35) and the induced current by (2.41).

The polarization vector is complex in general, and the requirement that it be unimodular is

$$\mathbf{e}^{\sigma *}(\mathbf{k}) \cdot \mathbf{e}^{\sigma}(\mathbf{k}) = 1. \tag{2.69}$$

Moreover, because the phase of $\mathbf{e}^\sigma(\mathbf{k})$ is arbitrary, we are free to choose a convenient phase. As has already been noted, (2.64) reduces to a set of three real equations (for a magnetized plasma with coordinate axes defined by (2.51)) for E_1, $-iE_2$ and E_3. Hence we may specify the phase such that e_1^σ and e_3^σ are real and e_2^σ is imaginary. It follows that specification of the polarization vector involves specification of two functions (e.g. e_1^σ and e_2^σ with e_3^σ determined by (2.69)) of the independent variables.

A convenient representation of the polarization vector is in terms of the two functions $K_\sigma(\mathbf{k})$ and $T_\sigma(\mathbf{k})$ in

$$\mathbf{e}^\sigma(\mathbf{k}) = \frac{K_\sigma(\mathbf{k})\boldsymbol{\kappa} + T_\sigma(\mathbf{k})\mathbf{t} + i\mathbf{a}}{\{K_\sigma^2(\mathbf{k}) + T_\sigma^2(\mathbf{k}) + 1\}^{1/2}}. \tag{2.70}$$

The unit vectors in (2.70) are defined by

$$\boldsymbol{\kappa} = (\sin\theta, 0, \cos\theta), \quad \mathbf{a} = (0, 1, 0), \quad \mathbf{t} = (\cos\theta, 0, -\sin\theta), \tag{2.71}$$

where again θ is the angle between $\boldsymbol{\kappa}$ and $\mathbf{b}$, and (A_1, A_2, A_3) denotes the components of $\mathbf{A}$ in the coordinate system defined by (2.51). In (2.70) K_σ describes the *longitudinal* part of the polarization and T_σ describes the *transverse* part. The waves are *longitudinal* for $K_\sigma = \infty$ and transverse for $K_\sigma = 0$, or for $T_\sigma = \infty$ and finite K_σ. Note that "longitudinal" and "transverse" refer to the direction of $\mathbf{E}$ relative to $\boldsymbol{\kappa}$. (One of the inconsistencies between the terminology of the magnetoionic theory and accepted plasma terminology is in the use of these two words: in the magnetoionic theory they refer to the relative direction of $\boldsymbol{\kappa}$ and $\mathbf{b}$. Modern terminology retains "parallel" and "perpendicular" for the relative directions of $\boldsymbol{\kappa}$ and $\mathbf{b}$.)

The transverse part of the polarization is elliptical in general. The quantity $|T_\sigma|$ is the axial ratio of this ellipse (the component along $\mathbf{t}$ divided by the component along $\mathbf{a}$). The sign T_σ determines the handedness, with right hand polarization corresponding to $T_\sigma > 0$ and left hand to $T_\sigma < 0$. It should be emphasized that the handedness is defined as *the screw sense of* $\mathbf{E}$ *relative to* $\mathbf{k}$. In particular, $T_\sigma = 0$ and $T_\sigma = \infty$ correspond to linear polarization along $\mathbf{a}$ and $\mathbf{t}$ respectively and $T_\sigma = 1$ and $T_\sigma = -1$ correspond to right hand circular and left hand circular polarization respectively.

Construction of the Polarization Vectors

A standard method for solving simultaneous linear algebraic equations starts with the construction of the matrix of the cofactors of the coefficients. The coefficients in (2.64) are $\Lambda_{ij}^{\mathrm{h}}$. Let λ_{ij} be the cofactor of $\Lambda_{ji}^{\mathrm{h}}$ (note the inversion of indices), i.e.

$$\lambda_{11} = \Lambda_{22}^{\mathrm{h}}\Lambda_{33}^{\mathrm{h}} - \Lambda_{23}^{\mathrm{h}}\Lambda_{32}^{\mathrm{h}}, \quad \lambda_{12} = -\Lambda_{12}^{\mathrm{h}}\Lambda_{33}^{\mathrm{h}} + \Lambda_{23}^{\mathrm{h}}\Lambda_{13}^{\mathrm{h}},$$

and so on. A prescription for constructing $\mathbf{e}^\sigma(\mathbf{k})$ is as follows: construct $\lambda_{ij}(\mathbf{k}, \omega)$ and insert the dispersion relation $\omega=\omega^\sigma(\mathbf{k})$ to find

$$\lambda^\sigma_{ij}(\mathbf{k}):=\lambda_{ij}(\mathbf{k},\omega^\sigma(\mathbf{k})). \tag{2.72}$$

Choose any constant vector, α say, with the desired phase (α_1 and α_3 real and α_2 imaginary here). The polarization vector is then given by

$$e^\sigma_i(\mathbf{k})=\frac{\lambda^\rho_{ij}(\mathbf{k})\alpha_j}{\{\lambda^\sigma_{rr}(\mathbf{k})\alpha_s{}^*\alpha_t\lambda^\sigma_{st}(\mathbf{k})\}^{1/2}}. \tag{2.73}$$

Alternatively, if α is chosen parallel to any of the axes, which are defined by (2.51), this prescription is equivalent to choosing any column of $\lambda^\sigma_{ij}(\mathbf{k})$ and normalizing it to unity to find $\mathbf{e}^\sigma(\mathbf{k})$.

The foregoing prescription is based on the fact that $\lambda^\sigma_{ij}(\mathbf{k})$ is of rank one and therefore may be written as the outer product of a vector with itself. The *order* of a square matrix is the number of its rows and columns. The rank may be defined by considering the determinants of all submatrices; a submatrix of order m is obtained by excluding any $n-m$ rows and any $n-m$ columns from the original matrix. A matrix is of *rank m* if the determinants of all submatrices of order greater than m vanish and if at least one sub-matrix of order m has a non-vanishing determinant. In general, if Λ is a matrix of order n and λ is the matrix of its cofactors, then when the determinant of Λ is zero, Λ is of rank $n-1$ and λ is of rank one, and when Λ is of rank $n-2$, λ vanishes identically. In the present case one has $n=3$, and $\lambda^\sigma_{ij}(\mathbf{k})$ is of rank one.

The statement that $\lambda^\sigma_{ij}(\mathbf{k})$ is of rank one is equivalent to the statement that one has

$$\lambda^\sigma_{ij}(\mathbf{k})\lambda^\sigma_{lm}(\mathbf{k})=\lambda^\sigma_{im}(\mathbf{k})\lambda^\sigma_{lj}(\mathbf{k}), \tag{2.74}$$

which follows by setting $\Lambda=0$ in the general result (Sitenko, 1967)

$$\lambda_{ij}\lambda_{lm}=\lambda_{im}\lambda_{lj}+\Lambda\varepsilon_{ilr}\varepsilon_{jms}\Lambda_{sr}, \tag{2.74}$$

for all i, j, l, m. It is straightforward to prove that for any matrix of rank one all the columns are proportional to each other, and all the rows are proportional to each other. Moreover, because $\lambda^\sigma_{ij}(\mathbf{k})$ is hermitian, the columns are proportional to the complex conjugates of the rows. The final step is to note that $\lambda_{ij}(\mathbf{k}, \omega)$ may be defined by

$$\Lambda^{\mathrm{h}}_{ij}(\mathbf{k}, \omega)\lambda_{ij}(\mathbf{k}, \omega)=\Lambda(\mathbf{k}, \omega)\delta_{ij}. \tag{2.75}$$

By comparison of (2.64) and (2.75) for $\Lambda(\mathbf{k}, \omega)=0$, one concludes that the columns of λ_{ij} must be proportional to the components of $\mathbf{E}$ and hence the rows of λ_{ij} to the complex conjugates of the components of $\mathbf{E}$. These conclusions may be summarized in the form

$$\lambda_{ij}^{\sigma}(\mathbf{k})=\lambda_{ss}^{\sigma}(\mathbf{k})e_i^{\sigma}(\mathbf{k})e_j^{\sigma\,*}(\mathbf{k}), \tag{2.76}$$

where the constant of proportionality $\lambda_{ss}^{\sigma}(\mathbf{k})$ has been determined by taking the trace and using (2.69). The result (2.73) then follows from (2.76) and (2.74).

The reality condition, together with the fact that $\lambda_{ij}(\mathbf{k}, \omega)$ is hermitian, implies

$$\lambda_{ij}(-\mathbf{k}, -\omega)=\lambda_{ji}(\mathbf{k}, \omega).$$

Hence (2.67) and (2.76) imply

$$\mathbf{e}^{\sigma}(-\mathbf{k})=\mathbf{e}^{\sigma\,*}(\mathbf{k}), \tag{2.77}$$

apart from an irrelevant phase factor.

It is possible for λ_{ij} to vanish identically for $\Lambda=0$, i.e. for λ_{ij} to be of rank zero. This is the case for transverse waves in an isotropic medium. There is then no characteristic polarization vector, but only a polarization tensor, p_{ij} say. When λ_{ij} is of rank zero, $\Lambda_{ij}^{\mathrm{h}}$ is itself of rank one, and p_{ij} is any hermitian tensor ($p_{ij}=p_{ij}^{*}$) which is orthogonal to the vector defined by $\Lambda_{ij}^{\mathrm{h}}$, i.e. defined by writing $\Lambda_{ij}^{\mathrm{h}}$ as the outer product of this vector with itself. Polarization tensors are discussed further in Sections 4.1 and 6.2.

Energetics in the Waves

The dispersion relation and the polarization vector do not provide all relevant information on the characteristic properties of a specific wave mode. Suppose we define the electric amplitude in the waves. It is then only a matter of calculation to find the electric energy in the waves (Section 3.1 below). In general, we require additional information to determine the total energy in the waves. Consequently, it is convenient to introduce a further characteristic wave property, denoted $R_E^{\sigma}(\mathbf{k})$, which is the ratio of electric to total energy in the waves. This quantity appears in formulae which describe particle–wave interactions.

The total energy consists of the electric energy, the magnetic energy and an energy attributed to induced particle motions. Let these be denoted $W_E^{\sigma}(\mathbf{k})$, $W_M^{\sigma}(\mathbf{k})$ and $W_P^{\sigma}(\mathbf{k})$ respectively. The total energy is then

$$W^{\sigma}(\mathbf{k})=W_E^{\sigma}(\mathbf{k})+W_M^{\sigma}(\mathbf{k})+W_P^{\sigma}(\mathbf{k}). \tag{2.78}$$

The electric energy may be defined in terms of the electric amplitude (this is done explicitly in Section 3.1 below) and the magnitude energy is related simply to the electric energy,

$$W_M^\sigma(\mathbf{k}) = \left\{\frac{kc}{\omega^\sigma(\mathbf{k})}\right\}^2 \{1 - |\boldsymbol{\kappa}\cdot\mathbf{e}^\sigma(\mathbf{k})|^2\} W_E^\sigma(\mathbf{k}). \tag{2.79}$$

However, before (2.78) has any meaning, we must also define either $W_P^\sigma(\mathbf{k})$, and regard (2.78) as a definition of $W^\sigma(\mathbf{k})$, or define $W^\sigma(\mathbf{k})$, and regard (2.78) as a definition of $W_P^\sigma(\mathbf{k})$.

In the case of the MHD waves (for an adiabatic equation of state), it is straightforward to define $W_P^\sigma(\mathbf{k})$ as the sum of the kinetic energy $W_K(\mathbf{k})$ plus the potential energy $W_\phi(\mathbf{k})$ (associated with pressure fluctuations). It is interesting that one then has an equipartition relation (Denisse and Delcroix, 1963; Melrose, 1975a) $W_M^\sigma(\mathbf{k}) + W_\phi^\sigma(\mathbf{k}) = W_E^\sigma(\mathbf{k}) + W_K^\sigma(\mathbf{k})$. However, it does not seem possible to define $W_P^\sigma(\mathbf{k})$ in a meaningful analogous way for waves in a collisionless plasma. When the medium is described in terms of a dielectric tensor, it seems that one must use (2.78) as a definition of $W_P^\sigma(\mathbf{k})$, with the total energy $W^\sigma(\mathbf{k})$ defined independently.

There are two steps in the identification of $W^\sigma(\mathbf{k})$. The first is the construction of an equation which may be interpreted as expressing continuity of energy. (It is possible to derive the relevant equation from (2.34), but this proves more difficult than constructing the equation as below.) The second is a specific calculation which shows how the wave energy is related to mechanical energy. In outline, one may proceed as follows. Suppose the existence of the anti-hermitian part of the dielectric tensor is taken into account in the wave equation. This may be done in two complementary ways. The first is to include $\varepsilon_{ij}^{\mathrm{a}}$ as a small correction in the dispersion equation. To first order in $\varepsilon_{ij}^{\mathrm{a}}(\mathbf{k};\omega)$ one has

$$\det\left[\Lambda_{ij}^{\mathrm{h}}(\mathbf{k},\omega) + \varepsilon_{ij}^{\mathrm{a}}(\mathbf{k},\omega)\right] = \Lambda(\mathbf{k},\omega) + \lambda_{ij}(\mathbf{k},\omega)\varepsilon_{ji}^{\mathrm{a}}(\mathbf{k},\omega). \tag{2.80}$$

Now damping of the waves is taken into account by allowing ω to have a small imaginary part $-i\gamma/2$, such that the wave energy (being proportional to the square of the amplitude) damps as $e^{-\gamma t}$. Then, to lowest order in the terms associated with wave damping, (2.80) gives

$$\frac{-i\gamma}{2}\frac{\partial\Lambda(\mathbf{k},\omega)}{\partial\omega} + \lambda_{ij}(\mathbf{k},\omega)\varepsilon_{ji}^{\mathrm{a}}(\mathbf{k},\omega) = 0, \tag{2.81}$$

which is to be evaluated at $\Lambda(\mathbf{k},\omega)=0$. We could have allowed $\mathbf{k}$ to have a small imaginary part, $-i\mu/2$ say, as well. Then, with the energy $W^\sigma(\mathbf{k})$ damping as $e^{-\gamma t}$ and the energy flux $\mathbf{F}^\sigma(\mathbf{k})$ damping as $\mathbf{e}^{-\mu\cdot\mathbf{r}}$ (2.81) would be of the form of an equation for continuity of energy, i.e. of the form

$$\frac{\partial W^\sigma(\mathbf{k})}{\partial t} + \operatorname{div}\mathbf{F}^\sigma(\mathbf{k}) = Q^\sigma(\mathbf{k}), \tag{2.82}$$

where $Q^\sigma(\mathbf{k})\,d^3\mathbf{k}/(2\pi)^3$ is the rate mechanical energy is transferred to waves in the mode σ in the range $d^3\mathbf{k}/(2\pi)^3$. An immediate consequence of constructing (2.82) in this way is that one has

$$\frac{\mathbf{F}^\sigma(\mathbf{k})}{W^\sigma(\mathbf{k})}=\left[-\frac{\partial\Lambda(\mathbf{k},\omega)}{\partial\mathbf{k}}\bigg/\frac{\partial\Lambda(\mathbf{k},\omega)}{\partial\omega}\right]_{\omega=\omega^\sigma(\mathbf{k})}=\frac{\partial\omega^\sigma(\mathbf{k})}{\partial\mathbf{k}}=:\mathbf{v}_g^\sigma(\mathbf{k}),\tag{2.83}$$

where the chain rule of partial differentiation has been used. The result (2.83) implies that the velocity of energy propagation is the group velocity. It should be emphasized that the "velocity of energy propagation" does not mean a signal velocity; about the only physical meaning which should be ascribed to the group velocity is in connection with geometric optics, see Section 6.1. Secondly, the rate of energy generation $Q^\sigma(\mathbf{k})$ follows by comparison of (2.81) and (2.82), and is given by

$$\frac{Q^\sigma(\mathbf{k})}{W^\sigma(\mathbf{k})}=\frac{2i\lambda_{ss}^\sigma(\mathbf{k})}{\left[\dfrac{\partial\Lambda(\mathbf{k},\omega)}{\partial\omega}\right]_{\omega=\omega^\sigma(\mathbf{k})}}\,e_i^{\sigma *}(\mathbf{k})e_j^\sigma(\mathbf{k})\varepsilon_{ij}^{a}(\mathbf{k},\omega^\sigma(\mathbf{k})),\tag{2.84}$$

where (2.76) has been used.

The ratio $Q^\sigma(k)/W^\sigma(\mathbf{k})$ is calculated explicitly in Section 3.1 and by comparing the result with (2.84) one identifies

$$R_E^\sigma(\mathbf{k}):=\frac{W_E^\sigma(\mathbf{k})}{W^\sigma(\mathbf{k})}=\frac{\lambda_{ss}^\sigma(\mathbf{k})}{\left[\omega\dfrac{\partial}{\partial\omega}\Lambda(\mathbf{k},\omega)\right]_{\omega=\omega^\sigma(\mathbf{k})}}.\tag{2.85}$$

Furthermore the ratio $-Q^\sigma(\mathbf{k})/W^\sigma(\mathbf{k})$ is equal to the absorption coefficient γ, which is now rewritten as $\gamma^\sigma(\mathbf{k})$ and identified explicitly as

$$\gamma^\sigma(\mathbf{k})=-2i\omega^\sigma(\mathbf{k})R_E^\sigma(\mathbf{k})e_i^{\sigma *}(\mathbf{k})e_j^\sigma(\mathbf{k})\varepsilon_{ij}^{a}(\mathbf{k},\omega^\sigma(\mathbf{k})).\tag{2.86}$$

Damping may be temporal or spatial, or some combination of the two, depending on the boundary conditions. If the damping is purely temporal, then the wave energy decays as $\exp[-\gamma^\sigma(\mathbf{k})t]$. For purely spatial damping, the energy flux decays as $\exp[-\gamma^\sigma(\mathbf{k})s/v_g^\sigma(\mathbf{k})]$, where s denotes distance along the ray path. More generally both temporal damping, at the rate γ, and spatial damping, at the rate μ, may occur with $\gamma^\sigma(\mathbf{k})=\gamma+\mu v_g^\sigma(\mathbf{k})$.

Alternative Forms for $R_E(\mathbf{k})$

The expression (2.85) is useful for formal purposes, but alternative expressions for $R_E^\sigma(\mathbf{k})$ are more useful for explicit evaluation. Two alternatives expressions may be obtained by considering

$$\frac{\partial}{\partial\omega}\{\omega^n\Lambda(\mathbf{k},\omega)\}=\frac{\partial}{\partial\omega}\left\{\frac{\omega^n}{3}\Lambda^{\text{h}}_{ij}(\mathbf{k},\omega)\lambda_{ji}(\mathbf{k},\omega)\right\},\tag{2.87}$$

with $n=1$ and 2. (The identity used in (2.87) follows from the trace of (2.75).) Now, because λ_{ij} is a homogeneous quadratic form in the components of Λ^{h}_{ij}, (2.87) gives

$$\frac{\partial}{\partial\omega}[\omega^n\Lambda(\mathbf{k},\omega)]=\lambda_{ji}(\mathbf{k},\omega)\frac{\partial}{\partial\omega}[\omega^n\Lambda^{\text{h}}_{ij}(\mathbf{k},\omega)]+n\omega^{n-1}\Lambda(\mathbf{k},\omega).\tag{2.88}$$

Hence, using (2.76) and (2.85) and $\omega^n\dfrac{\partial\Lambda}{\partial\omega}=\dfrac{\partial}{\partial\omega}(\omega^n\Lambda)$ for $\Lambda=0$, one finds

$$[R^\sigma_E(\mathbf{k})]^{-1}=e^{\sigma}_i{}^*(\mathbf{k})e^\sigma_j(\mathbf{k})\left[\frac{1}{\omega^{n-1}}\frac{\partial}{\partial\omega}\{\omega^n\Lambda^{\text{h}}_{ij}(\mathbf{k},\omega)\}\right]_{\omega=\omega^\sigma(\mathbf{k})}\tag{2.89}$$

For $n=1$, (2.89) with (2.44) gives

$$W^\sigma(\mathbf{k})=W^\sigma_M(\mathbf{k})+W^\sigma_E(\mathbf{k})\left[\frac{\partial}{\partial\omega}\{\omega\varepsilon^\sigma(\mathbf{k},\omega)\}\right]_{\omega=\omega^\sigma(\mathbf{k})},\tag{2.90}$$

with

$$\varepsilon^\sigma(\mathbf{k},\omega):=e^{\sigma}_i{}^*(\mathbf{k})e^\sigma_j(\mathbf{k})\varepsilon^{\text{h}}_{ij}(\mathbf{k},\omega),\tag{2.91}$$

and where (2.79) has been used. We may now identify $W^\sigma_P(\mathbf{k})$ by comparison of (2.78) and (2.90),

$$W^\sigma_P(\mathbf{k})=W^\sigma_E(\mathbf{k})\left[\frac{\partial}{\partial\omega}\{\omega\varepsilon^\sigma(\mathbf{k},\omega)\}-1\right]_{\omega=\omega^\sigma(\mathbf{k})}.\tag{2.92}$$

For $n=2$, (2.89) gives

$$R^\sigma_E(\mathbf{k})=\frac{1}{\left[\dfrac{1}{\omega}\dfrac{\partial}{\partial\omega}\{\omega^2\varepsilon^\sigma(\mathbf{k},\omega)\}\right]_{\omega=\omega^\sigma(\mathbf{k})}}.\tag{2.93}$$

This form is particularly useful for evaluating $R^\sigma_E(\mathbf{k})$ in practice. In particular, for waves in a non-spatially-dispersive medium, (2.93) implies

$$R^\sigma_E(\mathbf{k})=\frac{1}{1-|\boldsymbol{\kappa}\cdot\mathbf{e}^\sigma(\mathbf{k})|^2}\frac{1}{2n_\sigma\dfrac{\partial}{\partial\omega}(\omega n_\sigma)},\tag{2.94}$$

where the independent variables are now chosen to be ω and θ. (The result (2.94) follows from (2.93) and $n_\sigma^2(1-|\boldsymbol{\kappa}\cdot\mathbf{e}^\sigma|^2)=\varepsilon^\sigma$, which is implied directly by the wave equation.)

Some Useful Formulas

Explicit expression for Λ and λ_{ij} in terms of Λ_{ij} are (arguments $\mathbf{k}$, ω and the superscript h on $\Lambda^{\mathrm{h}}_{ij}$ and $\varepsilon^{\mathrm{h}}_{ij}$ are omitted here)

$$\Lambda=\tfrac{1}{6}\varepsilon_{ijk}\varepsilon_{rst}\Lambda_{ri}\Lambda_{sj}\Lambda_{tk}, \tag{2.95a}$$

$$\varepsilon_{ijk}\Lambda=\varepsilon_{rst}\Lambda_{ri}\Lambda_{sj}\Lambda_{tk}, \tag{2.95b}$$

$$\Lambda=\tfrac{1}{3}\Lambda_{rs}\Lambda_{st}\Lambda_{tr}-\tfrac{1}{2}\Lambda_{tt}\Lambda_{rs}\Lambda_{sr}+\tfrac{1}{6}(\Lambda_{tt})^3, \tag{2.95c}$$

$$\lambda_{ij}=\tfrac{1}{2}\varepsilon_{iab}\varepsilon_{jrs}\Lambda_{ra}\Lambda_{sb}, \tag{2.96a}$$

$$\lambda_{ij}=\tfrac{1}{2}\delta_{ij}\{(\Lambda_{rr})^2-\Lambda_{rs}\Lambda_{sr}\}+\Lambda_{ir}\Lambda_{rj}-\Lambda_{rr}\Lambda_{ij}. \tag{2.96b}$$

Explicit expressions in terms of $n^2=k^2c^2/\omega^2$ and ε_{ij} are

$$\Lambda=n^4\kappa_i\kappa_j\varepsilon_{ij}-n^2\{\kappa_i\kappa_j\varepsilon_{ij}\varepsilon_{ss}-\kappa_i\kappa_j\varepsilon_{is}\varepsilon_{sj}\}+\det[\varepsilon_{ij}], \tag{2.97}$$

$$\begin{aligned}\lambda_{ij}=n^4\kappa_i\kappa_j-n^2\{&\kappa_i\kappa_j\varepsilon_{ss}+\delta_{ij}\kappa_r\kappa_s\varepsilon_{rs}-\kappa_i\kappa_s\varepsilon_{sj}-\kappa_j\kappa_s\varepsilon_{is}\}\\ &+\tfrac{1}{2}\delta_{ij}\{(\varepsilon_{rr})^2-\varepsilon_{rs}\varepsilon_{sr}\}+\varepsilon_{is}\varepsilon_{sj}-\varepsilon_{ss}\varepsilon_{ij}.\end{aligned} \tag{2.98}$$

Representations in terms of the choice (2.51) of coordinate axes are

$$\Lambda_{ij}=\begin{bmatrix}\varepsilon_{11}-n^2\cos^2\theta & \varepsilon_{12} & \varepsilon_{13}+n^2\cos\theta\sin\theta\\ -\varepsilon_{12} & \varepsilon_{22}-n^2 & \varepsilon_{23}\\ \varepsilon_{13}+n^2\cos\theta\sin\theta & -\varepsilon_{23} & \varepsilon_{33}-n^2\sin^2\theta\end{bmatrix}, \tag{2.99}$$

$$\lambda_{ij}=\begin{bmatrix}n^4\sin^2\theta-n^2(\varepsilon_{22}\sin^2\theta+\varepsilon_{33})+\varepsilon_{22}\varepsilon_{33}+\varepsilon_{23}{}^2 & n^2\sin\theta(\varepsilon_{12}\sin\theta-\varepsilon_{23}\cos\theta)-\varepsilon_{12}\varepsilon_{33}-\varepsilon_{23}\varepsilon_{13} & n^4\cos\theta\sin\theta-n^2(\varepsilon_{22}\cos\theta\sin\theta-\varepsilon_{13})+\varepsilon_{12}\varepsilon_{23}-\varepsilon_{13}\varepsilon_{22}\\ -n^2\sin\theta(\varepsilon_{12}\sin\theta-\varepsilon_{23}\cos\theta)+\varepsilon_{12}\varepsilon_{33}+\varepsilon_{23}\varepsilon_{13} & -n^2A+\varepsilon_{11}\varepsilon_{33}-\varepsilon_{13}^2 & -n^2\cos\theta(\varepsilon_{12}\sin\theta-\varepsilon_{23}\cos\theta)-\varepsilon_{11}\varepsilon_{23}-\varepsilon_{12}\varepsilon_{13}\\ n^4\cos\theta\sin\theta-n^2(\varepsilon_{22}\cos\theta\sin\theta-\varepsilon_{13})+\varepsilon_{12}\varepsilon_{23}-\varepsilon_{13}\varepsilon_{22} & n^2\cos\theta(\varepsilon_{12}\sin\theta-\varepsilon_{23}\cos\theta)+\varepsilon_{11}\varepsilon_{23}+\varepsilon_{12}\varepsilon_{13} & n^4\cos^2\theta-n^2(\varepsilon_{11}+\varepsilon_{22}\cos^2\theta)+\varepsilon_{11}\varepsilon_{22}+\varepsilon_{12}^2\end{bmatrix}, \tag{2.100}$$

and

$$\Lambda=An^4-Bn^2+C \tag{2.101}$$

with

$$A=\varepsilon_{11}\sin^2\theta+\varepsilon_{33}\cos^2\theta+2\varepsilon_{13}\cos\theta\sin\theta, \tag{2.102a}$$

$$B = A\varepsilon_{22} + \varepsilon_{11}\varepsilon_{33} - \varepsilon_{13}^2 + (\varepsilon_{12}\sin\theta - \varepsilon_{23}\cos\theta)^2, \tag{2.102b}$$

$$C = \varepsilon_{11}\varepsilon_{22}\varepsilon_{33} + \varepsilon_{11}\varepsilon_{23}^2 + \varepsilon_{33}\varepsilon_{12}^2 - \varepsilon_{22}\varepsilon_{13}^2 + 2\varepsilon_{12}\varepsilon_{23}\varepsilon_{13}. \tag{2.102c}$$

Finally, explicit expressions for K_σ and T_σ may be obtained by inserting the dispersion relation $\omega = \omega^\sigma(\mathbf{k})$ in

$$K = \frac{i\{\varepsilon_{11}\varepsilon_{23}\cos\theta - \varepsilon_{12}\varepsilon_{33}\sin\theta + \varepsilon_{13}(\varepsilon_{12}\cos\theta - \varepsilon_{23}\sin\theta) + n^2(\varepsilon_{12}\sin\theta - \varepsilon_{23}\cos\theta)\}}{\varepsilon_{11}\varepsilon_{33} - \varepsilon_{13}^2 - n^2 A}, \tag{2.103a}$$

$$T = \frac{i\{(\varepsilon_{12}\varepsilon_{33} + \varepsilon_{23}\varepsilon_{13})\cos\theta + (\varepsilon_{23}\varepsilon_{11} + \varepsilon_{12}\varepsilon_{13})\sin\theta\}}{\varepsilon_{11}\varepsilon_{33} - \varepsilon_{13}^2 - n^2}. \tag{2.103b}$$

2.4 SPECIFIC WAVE MODES

For many purposes it suffices to have a knowledge of the properties of (a) the wave modes of an isotropic thermal plasma, and (b) the cold plasma modes. These modes, and also low-frequency modes in a collisionless plasma are discussed briefly here.

Isotropic Thermal Plasmas

The dielectric tensor for any isotropic medium may be expressed in the form (2.53), viz.

$$\varepsilon_{ij}(\mathbf{k}, \omega) = \varepsilon^l(k, \omega)\kappa_i\kappa_j + \varepsilon^t(k, \omega)(\delta_{ij} - \kappa_i\kappa_j).$$

For a plasma in which each species α is described by a nonrelativistic Maxwellian distribution, i.e. a velocity distribution proportional to $\exp(-v^2/2V_\alpha^2)$ explicit calculation using (2.62) gives

$$\varepsilon^l(k, \omega) = 1 + \sum_\alpha \frac{1}{k^2\lambda_{D\alpha}^2}[1 - \phi(y_\alpha) + i\sqrt{\pi}\, y_\alpha \exp(-y_\alpha^2)], \tag{2.104a}$$

$$\varepsilon^t(k, \omega) = 1 + \sum_\alpha \frac{\omega_p^2}{\omega^2}[-\phi(y_\alpha) + i\sqrt{\pi}\, y_\alpha \exp(-y_\alpha^2)]. \tag{2.104b}$$

The function $\phi(y) - i\sqrt{\pi}\, y \exp(-y^2)$ is proportional to the plasma dispersion function of Fried and Conte (1961), with

$$\phi(y) := 2y\exp(-y^2)\int_0^y dt\,\exp(t^2) = \begin{cases} 2y^2 - \frac{4}{3}y^4 + \cdots & \text{for } |y^2| \ll 1 \\ 1 + (1/2y^2) + (3/4y^4) + \cdots & \text{for } |y^2| \gg 1 \end{cases} \tag{2.105}$$

and with

$$y_\alpha := \omega/\sqrt{2}\, kV_\alpha. \tag{2.106}$$

Retaining only the hermitian part of ε_{ij} corresponds to neglecting the imaginary parts of ε^l and ε^t here. Then the dispersion equation reduces to

$$\{\mathrm{Re}\ \varepsilon^{\mathrm{l}}(k, \omega)\}\{n^2 - \mathrm{Re}\ \varepsilon^{\mathrm{t}}(k, \omega)\}^2 = 0, \tag{2.107}$$

where "Re" denotes the real part. The solutions of $\mathrm{Re}\ \varepsilon^{\mathrm{l}}(k, \omega) = 0$ describe *longitudinal waves*, i.e. waves with $\mathbf{e} = \boldsymbol{\kappa}$. (Longitudinal waves are sometimes called "electrostatic waves".) The solutions of $n^2 = \mathrm{Re}\ \varepsilon^{\mathrm{t}}(k, \omega)$ describe *transverse waves*, i.e. waves with $\mathbf{e} \cdot \boldsymbol{\kappa} = 0$. Transverse waves have two independent states of polarization for a single dispersion relation. As pointed out above, in general it is not possible to describe the polarization of transverse waves in terms of a single polarization vector, and a polarization tensor, or some equivalent formalism is required, see Sections 3.2, 4.1 and 6.2.

In any isotropic plasma (provided only that relativistic and quantum effects are unimportant) the properties of transverse waves may be taken to be

$$n^2 = 1 - \frac{\omega_p^2}{\omega^2}, \quad \mathbf{e} \cdot \boldsymbol{\kappa} = 0, \quad R_E^{\mathrm{t}} = \tfrac{1}{2}. \tag{2.108}$$

Damping of the waves can result from collisions, see Section 6.1, but there is no collisionless damping of transverse waves in an isotropic plasma. (The nonvanishing imaginary part of ε^{t} in (2.104b) would appear to imply the existence of collisionless damping, but for $\omega/k > c$ this imaginary part is nonphysical, being due to the finite probability of nonphysical particles with $v > c$ in a nonrelativistic Maxwellian distribution.)

The dispersion relation for *Langmuir waves* may be obtained from $\mathrm{Re}\ \varepsilon^{\mathrm{l}}(k, \omega) = 0$ by retaining only the contribution of the electrons. One finds that a solution exists only for $\phi(y_e) \gtrsim 1$, i.e. for $\omega/k > 1.8 V_e$. For $y_e \gg 1$ one finds

$$\omega = \omega^{\mathrm{l}}(k) \approx [\omega_p^2 + 3k^2 V_e^2]^{1/2}, \quad \mathbf{e} = \boldsymbol{\kappa}, \quad R_E^{\mathrm{l}}(k) = \frac{\omega^2}{2\omega_p^2}, \tag{2.109}$$

with $\omega = \omega^{\mathrm{l}}(k)$. The imaginary part of ε^{l} leads to Landau damping (Landau, 1946) with

$$\gamma^{\mathrm{l}}(k) = \left(\frac{\pi}{2}\right)^{1/2} \frac{\omega^4}{k^3 V_e^3} \exp\left[-\frac{\omega^2}{2k^2 V_e^2}\right], \tag{2.110}$$

again with $\omega = \omega^{\mathrm{l}}(k)$. For $\omega/k \gg V_e$ collisional damping (Section 6.1) becomes more important than Landau damping.

The only other waves which can exist in an isotropic thermal plasma are *ion sound waves*. For simplicity, suppose there is only one ionic species. Then for

$$\sqrt{2}\,kV_i \ll \omega \ll \sqrt{2}\,kV_e, \tag{2.111}$$

(2.104a) gives

$$\mathrm{Re}\,\varepsilon^{l}(k,\omega) \approx 1 - \frac{\omega_{pi}^2}{\omega^2} + \frac{1}{k^2\lambda_{De}^2}. \tag{2.112}$$

The parameters ω_{pi} and λ_{De} combine into $v_s = \omega_{pi}\lambda_{De}$, and the wave properties are

$$\omega = \omega^s(k) \approx \frac{kv_s}{(1+k^2\lambda_{De}^2)^{1/2}}, \quad \mathbf{e} = \boldsymbol{\kappa}, \quad R_E^s(k) = \frac{\omega^2}{2\omega_{pi}^2}, \tag{2.113}$$

with $\omega = \omega^s(k)$. Actually, this wave mode is called the ion sound mode only for $k^2\lambda_{De}^2 \ll 1$, when one has $\omega^s(k) \approx kv_s$. For $k^2\lambda_{De}^2 \gg 1$ one has $\omega^s(k) \approx \omega_{pi}$ and the mode is called the ion plasma mode.

Collisionless damping of ion sound waves is due to both electrons and ions:

$$\gamma^s(k) = \gamma_e^s(k) + \gamma_i^s(k), \tag{2.114a}$$

$$\gamma_e^s(k) = \left(\frac{\pi}{2}\right)^{1/2} \omega \frac{v_s}{V_e}, \tag{2.114b}$$

$$\gamma_i^s(k) = \left(\frac{\pi}{2}\right)^{1/2} \frac{\omega^4}{k^3V_i^3} \exp\left[-\frac{\omega^2}{2k^2V_i^2}\right], \tag{2.114c}$$

with $\omega = \omega^s(k)$. Ion sound waves are strongly damped (by ions) except for $v_s \gg V_i$, i.e. for

$$Z_iT_e \gg T_i. \tag{2.115}$$

Consequently, ion sound waves can be said to exist only when (2.115) is satisfied. In astrophysical plasmas the dominant ionic species is singly charged (being protons in a fully ionized gas, and singly ionized carbon and oxygen in partially ionized regions for example) and $Z_i = 1$ implies that ion sound waves can exist only for $T_e \gg T_i$.

Cold Plasmas

The dielectric tensor for a cold plasma follows from (2.60) and (2.61), which may be rewritten in the form (Stix 1962)

$$\varepsilon_{ij}(\omega) = \begin{pmatrix} S & -iD & 0 \\ iD & S & 0 \\ 0 & 0 & P \end{pmatrix}, \tag{2.116}$$

with

$$S:=\tfrac{1}{2}(R_+ + R_-), \quad D:=\tfrac{1}{2}(R_+ - R_-),$$

$$R_\pm := 1-\sum_\alpha \frac{\omega_{p\alpha}^2}{\omega^2}\frac{\omega}{\omega \pm \epsilon_\alpha \Omega_\alpha}, \quad P:=1-\sum_\alpha \frac{\omega_{p\alpha}^2}{\omega^2}, \tag{2.117}$$

where $\epsilon_\alpha := q_\alpha/|q_\alpha|$ is the sign of the charge of species α. (The other quantities in (2.117) are defined in Section 1.3.) Cold plasmas are not spatially dispersive, i.e. ε_{ij} does not depend on $\mathbf{k}$.

The matrix representations Λ_{ij} and λ_{ij} in this case are, cf. (2.99) and (2.100),

$$\Lambda_{ij} = \begin{pmatrix} S-n^2\cos^2\theta & -iD & n^2\sin\theta\cos\theta \\ iD & S-n^2 & 0 \\ n^2\sin\theta\cos\theta & 0 & P-n^2\sin^2\theta \end{pmatrix}, \tag{2.118}$$

and

$$\lambda_{ij} = \begin{pmatrix} (S-n^2)(P-n^2\sin^2\theta) & iD(P-n^2\sin^2\theta) & -(S-n^2)n^2\sin\theta\cos\theta \\ -iD(P-n^2\sin^2\theta) & -n^2A+PS & iDn^2\sin\theta\cos\theta \\ -(S-n^2)n^2\sin\theta\cos\theta & -iDn^2\sin\theta\cos\theta & n^4\cos^2\theta - n^2S(1+\cos^2\theta)+S^2-D^2 \end{pmatrix}. \tag{2.119}$$

The quantity Λ is given by (2.101), viz.

$$\Lambda = An^4 - Bn^2 + C,$$

with, cf. (2.102a, b, c),

$$\begin{aligned} A&=P\cos^2\theta + S\sin^2\theta, \\ B&=(S^2-D^2)\sin^2\theta + PS(1+\cos^2\theta), \\ C&=P(S^2-D^2). \end{aligned} \tag{2.120}$$

For the cold plasma waves, the dispersion equation $\Lambda=0$ is a quadratic equation for n^2, and the two solutions $n^2=n_\pm^2$ may be written down immediately. They are

$$n^2 = n_\pm^2 := \frac{B\pm F}{2A}, \tag{2.121}$$

with

$$F^2 := B^2-4AC = (PS-S^2+D^2)^2\sin^4\theta + 4P^2D^2\cos^2\theta. \tag{2.122}$$

By implication, the independent variables are chosen to be ω and θ when one solves for the (dependent) variable n^2, i.e. $n_\pm^2(\omega, \theta)$ are to be regarded as functions of ω and θ. It is relevant to note that F^2 is positive semi-definite, and hence the solutions $n_\pm^2$ are real. However, the solutions describe propagating waves only if the refractive index n is itself real, i.e. must have $n^2 > 0$. In regions with $n^2 < 0$ the waves are evanescent.

The polarization vectors for the cold plasma waves, when written in the form (2.70), viz.

$$\mathbf{e}^\pm = \frac{K_\pm \boldsymbol{\kappa} + T_\pm \mathbf{t} + i\mathbf{a}}{(K_\pm^2 + T_\pm^2 + 1)^{1/2}},$$

have

$$T_\pm = \frac{DP\cos\theta}{An_\pm^2 - PS}, \quad K_\pm = \frac{(P - n_\pm^2)D\sin\theta}{An_\pm^2 - PS}. \tag{2.123a, b}$$

In deriving (2.123a, b), the polarization vectors have been constructed using the middle column of (2.119). Finally, the ratio of electric to total energy is given by (2.94), viz.

$$R_E^\pm = \frac{1}{1 - |\boldsymbol{\kappa}\cdot\mathbf{e}^\pm|^2} \frac{1}{2n_\pm \partial(\omega n_\pm)/\partial\omega}. \tag{2.124}$$

The two modes are orthogonal in the sense that, for given ω and θ, their polarization ellipses are orthogonal,

$$T_+ T_- = -1. \tag{2.125}$$

(The minus sign implies that one is left hand and the other is right hand polarized.)

Note that (2.123a) implies that T^{-1} is a linear function of n^2. Now n^2 satisfies a quadratic equation, and hence T^{-1} must also satisfy a quadratic equation. This equation can also be written as a quadratic equation for T. It is often useful to solve the relevant equation, which is

$$T^2 - \frac{(PS - S^2 + D^2)\sin^2\theta}{PD\cos\theta} T - 1 = 0, \tag{2.126}$$

for $T = T_\pm$ and to find the solutions for n^2 and K in terms of T. The solutions of (2.126) are

$$T = T_\pm = \frac{(PS - S^2 + D^2)\sin^2\theta \pm \mathrm{F}}{2\mathrm{P}D\cos\theta} = \frac{-2PD\cos\theta}{(PS - S^2 + D^2)\sin^2\theta \mp F}. \tag{2.127}$$

One has

$$n^2 = \frac{P}{A}\left(S + \frac{D\cos\theta}{T}\right) = \frac{S^2 - D^2}{S - DT\cos\theta} \tag{2.128}$$

and

$$K=\frac{\sin\theta}{A}\{(P-S)T\cos\theta-D\}=\frac{\sin\theta}{P}\frac{(PS-S^2+D^2)T\cos\theta-PD}{S-DT\cos\theta}. \qquad (2.129)$$

The first-written equality in (2.128) follows by solving (2.123a) for n^2 in terms of T, and the first-written equality in (2.129) follows by substituting the resulting expression for n^2 in (2.123b). The second-written equalities following in similar manner after replacing An^2 by $B-C/n^2$, in accord with the quadratic equation for n^2.

Limiting Cases of the Cold Plasma Modes

Cold plasma theory was developed originally as an extension of the magnetoionic theory, which may be obtained by setting the masses of any ionic components equal to infinity. There are then only two plasma parameters, namely ω_p and Ω_e, and it is convenient to incorporate them into two "magnetoionic parameters"

$$X:=\frac{\omega_p^2}{\omega^2},\quad Y:=\frac{\Omega_e}{\omega}. \qquad (2.103\text{a, b})$$

The magnetoionic theory is then obtained by writing

$$R_\pm=1-\frac{X}{1+Y},\quad S=\frac{1-X-Y^2}{1-Y^2},$$

$$D=\frac{-XY}{1-Y^2},\quad P=1-X,\quad S^2-D^2=R_+R_-=\frac{(1-X)^2-Y^2}{1-Y^2} \qquad (2.131)$$

in the foregoing equations. The properties of the magnetoionic modes are discussed in detail in Section 12.1.

Two important limiting cases of the magnetoionic modes are the following. At high frequencies ($\omega\gg\omega_p$, Ω_e) the waves are nearly circularly polarized (quasi-circular limit) except for nearly perpendicular propagation (except for $|(\pi/2)-\theta|\lesssim\Omega_e/2\omega$), and one has

$$n_\sigma^2=1-\frac{\omega_p^2}{\omega^2}\left(1-\sigma\frac{\Omega_e}{\omega}|\cos\theta|\right),\quad T_\sigma=-\frac{\sigma\cos\theta}{|\cos\theta|}-\frac{\Omega_e\sin^2\theta}{2\omega\cos\theta},$$

$$K_\sigma=\frac{\omega_p^2\Omega_e\sin\theta}{\omega^3},\quad R_E^\sigma=\frac{1}{1-|\boldsymbol{\kappa}\cdot\mathbf{e}^\sigma|^2}\frac{1}{2n_\sigma\partial(\omega n_\sigma)/\partial\omega}, \qquad (2.132)$$

with $\sigma=$o for the o-mode (ordinary mode) and $\sigma=$x for the x-mode (extraordinary mode), corresponding to $\sigma=+1$ and $\sigma=-1$ respectively in (2.126). For $\omega\ll\Omega_e|\cos\theta|\ll\omega_p$ one of the magnetoionic waves becomes the *whistler mode* with

$$n_W^2 = \frac{\omega_p^2}{\omega\Omega_e|\cos\theta|}, \quad \mathbf{e}^W = \frac{(1, i|\cos\theta|, 0)}{(1+\cos^2\theta)^{1/2}}, \quad R_E^W = \frac{\omega\Omega_e}{2\omega_p^2}\frac{1+\cos^2\theta}{|\cos\theta|}. \tag{2.133}$$

The neglect of the ions ceases to be justified at frequencies lower than or comparable with the lower hybrid frequency, which may be approximated by

$$\omega_{LH} = \left\{\frac{1}{\omega_{pi}^2+\Omega_i^2} + \frac{1}{\Omega_e\Omega_i}\right\}^{-1/2}. \tag{2.134}$$

The hydromagnetic modes (for zero sound speed) are given by the low frequency limit of the cold plasma modes. The simplest approximation corresponds to $P = -\infty$, $D = 0$ and $S = 1 + c^2/v_A^2 \approx c^2/v_A^2$ (assuming $v_A^2 \ll c^2$), which applies only if the plasma is charge neutral. In this case the two cold plasma modes reduce to the Alfvén mode

$$n_A^2 = \frac{c^2}{v_A^2\cos^2\theta}, \quad \mathbf{e}^A = (1, 0, 0), \quad R_E^A = \frac{v_A^2}{2c^2}, \tag{2.135}$$

and the magnetoacoustic mode

$$n_M^2 = \frac{c^2}{v_A^2}, \quad \mathbf{e}^M = (0, i, 0), \quad R_E^M = \frac{v_A^2}{2c^2}. \tag{2.136}$$

Thermal corrections to the hydromagnetic waves in the collisionless regime are discussed immediately below. The hydromagnetic waves in the collision-dominated regime are treated in Section 12.3 using the MHD equations.

Waves in Warm Magnetized Plasmas

When both thermal motions and an ambient magnetic field are taken into account the variety of wave motions which the plasma can support is increased considerably. The wave modes of a warm magnetized plasma may be separated into three classes: thermally modified cold plasma modes, magnetically modified isotropic-plasma modes and intrinsically new modes. The Bernstein modes (electrostatic cyclotron modes) are perhaps the best known examples of the final class of modes. They are restricted to angles of propagation close to perpendicular to the field lines, except for ion modes near the lower hybrid frequency (2.134) and electron modes near the upper hybrid frequency $\omega_{UH} := (\omega_p^2 + \Omega_e^2)^{1/2}$. The upper hybrid modes may be regarded as a limiting case of the generalized Langmuir modes discussed in Section 12.1.

One class of warm plasma modes which can be treated relatively simply is the class of low-frequency modes. For a plasma with a single ionic species and for $\Omega_e \gg kV_e$, $\Omega_i \gg kV_i$, $v_A \gg V_i$ and

$$\sqrt{2}\,kV_i|\cos\theta| \ll \omega \ll \sqrt{2}\,kV_e|\cos\theta|,$$

the dielectric tensor for a thermal magnetized plasma may be approximated by (Appendix B, Akhiezer *et al.*, 1967; Sitenko, 1967)

$$\varepsilon_{ij} = \frac{c^2}{v_A^2}\begin{pmatrix} 1 & \dfrac{i\omega}{\Omega_i} & 0 \\ \dfrac{-i\omega}{\Omega_i} & 1 & \dfrac{-i\Omega_i}{\omega}\tan\theta \\ 0 & \dfrac{i\Omega_i}{\omega}\tan\theta & -\dfrac{\Omega_i^2}{\omega^2}\left(1-\dfrac{c^2}{n^2 v_s^2\cos^2\theta}\right) \end{pmatrix}$$
$$+i\sqrt{\frac{\pi}{2}}\frac{\omega_p^2}{\omega^2}\frac{c}{nV_e|\cos\theta|}\begin{pmatrix} 0 & 0 & 0 \\ 0 & 2\left(\dfrac{\omega n V_e\sin\theta}{\Omega_e c}\right)^2 & \dfrac{-i\omega}{\Omega_e}\tan\theta \\ 0 & \dfrac{i\omega}{\Omega_e}\tan\theta & \dfrac{c^2\cos^2\theta}{n^2 V_e^2} \end{pmatrix}. \qquad (2.137)$$

(The second term is the anti-hermitian part, due to electrons at $s=0$, cf. Section 4.1.) In this case one finds three modes with the following approximate properties for $\omega \ll \Omega_i$:

the Alfvén mode

$$n_A^2 = \frac{c^2}{v_A^2\cos^2\theta}\left(1+\frac{\omega^2}{\Omega_i^2}\operatorname{cosec}^2\theta\right), \quad \mathbf{e}^A = \left(1, -\frac{i\omega}{\Omega_i}\cot^2\theta, \frac{\omega^2}{\Omega_i^2}\frac{v_s^2}{v_A^2}\frac{1}{\sin\theta\cos\theta}\right),$$

$$R_E^A = \frac{v_A^2}{2c^2}, \quad \gamma^A = \omega\left(\frac{\pi}{2}Z_i\frac{m_e}{m_i}\right)^{1/2}\frac{v_s}{v_A}\frac{\omega^2}{\Omega_i^2}(\tan^2\theta+\cot^2\theta); \qquad (2.138)$$

the magnetoacoustic mode

$$n_M^2 = \frac{c^2}{v_A^2}\left(1-\frac{v_s^2}{v_A^2}\sin^2\theta - \frac{\omega^2}{\Omega_i^2}\cot^2\theta\right), \quad \mathbf{e}^M = \left(\frac{\omega}{\Omega_i}\operatorname{cosec}^2\theta, i, \frac{\omega}{\Omega_i}\frac{v_s^2}{v_A^2}\sin\theta\cos\theta\right),$$

$$R_E^M = \frac{v_A^2}{2c^2}, \quad \gamma^M = \omega\left(\frac{\pi}{2}Z_i\frac{m_e}{m_i}\right)^{1/2}\frac{v_s}{v_A}\frac{\sin^2\theta}{|\cos\theta|}; \qquad (2.139)$$

and the ion sound mode

$$n_s^2 = \frac{c^2}{v_s^2\cos^2\theta}\left(1+\frac{v_s^2}{v_A^2}\sin^2\theta + \frac{\omega^2}{\Omega_i^2}\tan^2\theta\right),$$

$$\mathbf{e}^s = \frac{\left(\sin\theta,\ -i\dfrac{\Omega_i}{\omega}\dfrac{v_s^2}{v_A^2}\sin\theta\cos^2\theta,\ \cos\theta\right)}{\left(1+\dfrac{\Omega_i^2}{\omega^2}\dfrac{v_s^4}{v_A^4}\sin^2\theta\cos^4\theta\right)^{1/2}},$$

$$R_E^s = \frac{v_A^2}{2c^2}\frac{\omega^2}{\Omega_i^2\cos^2\theta}\left(1+\frac{\Omega_i^2}{\omega^2}\frac{v_s^4}{v_A^4}\sin^2\theta\cos^4\theta\right),\quad \gamma^s=\omega\left(\frac{\pi}{2}Z_i\frac{m_e}{m_i}\right)^{1/2}. \tag{2.140}$$

The former two are warm plasma versions of the two cold plasma modes (2.135) and (2.136), and the final mode is a magnetically modified version of the ion sound mode (2.113). Although there is similarity between these modes, and the MHD modes (Section 12.3) it should be emphasized that the foregoing results apply only in the collisionless regime, i.e. all collision frequencies must be much less than ω and collisional mean free paths must be much longer than k^{-1}.

PROBLEM SET 2

Answers to relevant questions are given at the end of the text. More difficult questions are marked with an asterisk.

2.1 When a quadrupole correction, cf. (2.55), to the response of an isotropic medium, is included, the dielectric tensor is of the form

$$\varepsilon_{ij}(\mathbf{k},\omega)=\varepsilon(\omega)\delta_{ij}+k_r\eta_{rij}(\omega). \tag{P2-1}$$

Show that one must have

$$\eta_{rij}(\omega)=-\eta_{rji}(\omega). \tag{P2-2}$$

Hint: Use the Onsager relations in the form (2.49).

2.2* Find the equivalent dielectric tensor $\varepsilon_{ij}(\mathbf{k},\omega)$ when the response of the medium is assumed to be of the form

$$D_i=\varepsilon_{ij}E_j,\quad B_i=\mu_{ij}H_j, \tag{P2-3}$$

with ε_{ij} and μ_{ij} functions of ω but not of $\mathbf{k}$.

2.3 In a cold unmagnetized plasma each species has a mean fluid velocity $\mathbf{U}_\alpha$. Show that the dielectric tensor may be written in the form

$$\varepsilon_{ij}(\mathbf{k},\omega)=\delta_{ij}-\sum_\alpha\frac{\omega_{p\alpha}^2}{\omega^2}\left[\delta_{ij}+\frac{1}{\omega-\mathbf{k}\cdot\mathbf{U}_\alpha}(k_iU_{\alpha j}+k_jU_{\alpha i})+(k^2-\omega^2/c^2)U_{\alpha i}U_{\alpha j}/(\omega-\mathbf{k}\cdot\mathbf{U}_\alpha)^2\right] \tag{P2-4}$$

Remark: In this case, $\mathrm{d}/\mathrm{d}t$ is to be interpreted as $(\partial/\partial t)+\mathbf{U}_\alpha\cdot\mathrm{grad}$.

2.4 Include the effects of collisions in a cold magnetized electron plasma by adding a frictional force $-\nu m_e \mathbf{u}_e$ to the fluid equation for the electrons. Show that the resulting dielectric tensor may be written in the form (2.116) with, in place of (2.117),

$$R_\pm = 1 - \frac{\omega_p^2}{\omega(\omega+\Omega_e+i\nu)}, \quad P = 1 - \frac{\omega_p^2}{\omega(\omega+i\nu)}. \qquad \text{(P2-5)}$$

2.5* Generalize (2.61) by allowing each species of particle in a cold plasma to have an average fluid velocity U_α along the direction of the magnetic field. Hence show that the generalization of the cold plasma dielectric tensor (2.116) with (2.117) may be written

$$\varepsilon_{ij}(k,\omega) = \begin{pmatrix} S_1 & -iD_1 & S_2 \\ iD_1 & S_1 & iD_2 \\ S_2 & -iD_2 & P \end{pmatrix}, \qquad \text{(P2-6)}$$

with $a=1$ or 2 in

$$S_a := \tfrac{1}{2}(R_{a+}+R_{a-}), \quad D_a := \tfrac{1}{2}(R_{a+}-R_{a-}), \qquad \text{(P2-7)}$$

and with

$$R_{1\pm} = 1 - \sum_\alpha \frac{\omega_{p\alpha}^2}{\omega^2}\frac{\omega-k_\parallel U_\alpha}{\omega\pm\epsilon_\alpha\Omega_\alpha-k_\parallel U_\alpha},$$

$$R_{2\pm} = -\sum_\alpha \frac{\omega_{p\alpha}^2}{\omega^2}\frac{k_\perp U_\alpha}{\omega\pm\epsilon_\alpha\Omega_\alpha-k_\parallel U_\alpha},$$

$$P = 1 - \sum_\alpha \frac{\omega_{p\alpha}^2}{\omega^2}\left[\frac{\omega^2}{(\omega-k_\parallel U_\alpha)^2} + \frac{k_\perp^2 U_\alpha^2}{(\omega-k_\parallel U_\alpha)^2-\Omega_\alpha^2}\right] \qquad \text{(P2-8)}$$

2.6 A uniaxial crystal has dielectric tensor

$$\varepsilon_{ij}(\omega) = \begin{pmatrix} \varepsilon_\perp(\omega) & 0 & 0 \\ 0 & \varepsilon_\perp(\omega) & 0 \\ 0 & 0 & \varepsilon_\parallel(\omega) \end{pmatrix}, \qquad \text{(P2-9)}$$

where the principal axis is along the z-axis. Show that there are two wave modes, called the ordinary (o) and extraordinary (x) modes, with wave properties

$$n_o^2 = \varepsilon_\perp, \quad \mathbf{e}^o = (0, 1, 0), \quad R_E^o = \frac{1}{2n_o\partial(\omega n_o)/\partial\omega} \qquad \text{(P2-10a)}$$

and

$$n_x^2 = \frac{\varepsilon_\perp\varepsilon_\parallel}{\varepsilon_\perp\sin^2\theta+\varepsilon_\parallel\cos^2\theta}, \quad \mathbf{e}^x = \frac{(\varepsilon_\parallel\cos\theta, 0, -\varepsilon_\perp\sin\theta)}{(\varepsilon_\perp^2\sin^2\theta+\varepsilon_\parallel^2\cos^2\theta)^{1/2}},$$

$$R_E^x = \frac{\varepsilon_\perp^2\sin^2\theta+\varepsilon_\parallel^2\cos^2\theta}{(\varepsilon_\perp\sin^2\theta+\varepsilon_\parallel\cos^2\theta)^2}\frac{1}{2n_x\partial(\omega n_x)/\partial\omega}, \qquad \text{(P2-10b)}$$

where $\boldsymbol{\kappa}$ is assumed to be in the x–z plane.

2.7 (a) Show that the group velocity for waves in a magnetized plasma may be written in the form

$$\mathbf{v}_g^\sigma(\omega,\theta) = \frac{c}{\partial\{\omega n_\sigma(\omega,\theta)\}/\partial\omega}\left\{\boldsymbol{\kappa} - \frac{1}{n_\sigma(\omega,\theta)}\frac{\partial n_\sigma(\omega,\theta)}{\partial\theta}\mathbf{t}\right\}, \qquad \text{(P2-11)}$$

where $\mathbf{t}$ is defined by (2.71).

(b) Show that for a non-spatially-dispersive medium the group velocity is given by

$$\mathbf{v}_g^\sigma = 2cn_\sigma R_E^\sigma \, Re(\boldsymbol{\kappa} - \mathbf{e}^{\sigma *} \boldsymbol{\kappa} \cdot \mathbf{e}^\sigma). \tag{P2-12}$$

Hint: In this case the energy flux is given by the Poynting vector.

(c) Hence show that for any non-spatially-dispersive medium one has

$$R_E^\sigma = \frac{1}{1-|\boldsymbol{\kappa}\cdot\mathbf{e}^\sigma|^2} \frac{1}{2n_\sigma[\partial(\omega n_\sigma)/\partial\omega]}, \quad \frac{1}{n_\sigma}\frac{\partial n_\sigma}{\partial\theta} = \frac{K_\sigma T_\sigma}{1+T_\sigma^2}. \tag{P2-13a, b}$$

(d) Generalize (P2-11) for waves in a bi-axial crystal, i.e., to the case where n depends on the azimuthal angle ϕ.

2.8 Show that the properties of the cold plasma waves for (a) $\theta=0$, (b) $\theta=\pi/2$, and (c) at a cutoff are given by

$$\text{(a)}\ n_\pm^2 = R_\pm, \quad T_\pm = \pm 1, \quad K_\pm = 0, \quad R_E^\pm = \frac{1}{(1/\omega)\partial(\omega^2 R_\pm)/\partial\omega}. \tag{P2-14a}$$

$$\text{(b)}\ n_\text{o}^2 = P, \quad T_\text{o} = \infty, \quad K_\text{o} = \frac{(P-S)^2 - D^2}{PD}, \quad R_E^\text{o} = \tfrac{1}{2};$$

$$n_\text{x}^2 = \frac{S^2-D^2}{S}, \quad T_\text{x} = 0, \quad K_\text{x} = -\frac{D}{S}, \quad R_E^\text{x} = \frac{S^2+D^2}{\dfrac{S^2}{\omega}\dfrac{\partial}{\partial\omega}\left\{\dfrac{\omega^2(S^2-D^2)}{S}\right\}}, \tag{P2-14b}$$

$$\text{(c)}\ n^2 = 0, \quad \mathbf{e} = \frac{1}{\sqrt{2}}(1, \pm i, 0), \quad R_E = \left.\frac{1}{\omega\partial(\omega R_\pm)/\partial\omega}\right|_{R_\pm=0}$$

$$\text{or} \qquad n^2 = 0, \quad \mathbf{e} = (0, 0, 1), \quad R_E = \tfrac{1}{2}. \tag{P2-14c}$$

2.9* Use the Cayley–Hamilton theorem (a matrix $\mathbf{M}$ satisfies its own characteristic equation $\det(\mathbf{M}-\lambda 1)=0$) to establish the identities (2.95c) and (2.96b).

2.10 Establish the identities

$$\det[\lambda_{ij}] = \Lambda^2, \quad \lambda_{ss} = \tfrac{1}{2}(\Lambda_{ss})^2 - \tfrac{1}{2}\Lambda_{rs}\Lambda_{sr}, \quad \lambda_{ij}\lambda_{rs} = \lambda_{is}\lambda_{rj} + \Lambda\varepsilon_{irm}\varepsilon_{jsn}\Lambda_{nm}. \tag{P1-15}$$

2.11* The wave equation derived from the MHD equations for an adiabatic equation of state and no displacement current may be written in the form

$$\Gamma_{ij}(\mathbf{k}, \omega)\xi_j(\mathbf{k}, \omega) = 0,$$

where the fluid displacement $\boldsymbol{\xi}$ is such that its time-derivative is the fluid velocity, and with

$$\Gamma_{ij}(\mathbf{k}, \omega) = \omega^2\delta_{ij} - k^2 c_s^2 \kappa_i \kappa_j - k^2 v_A^2 \{\kappa_i\kappa_j - \boldsymbol{\kappa}\cdot\mathbf{b}(\kappa_i b_j + \kappa_j b_i) + (\boldsymbol{\kappa}\cdot\mathbf{b})^2 \delta_{ij}\}. \tag{P2-16}$$

Show that the solutions of the wave equation define three modes, called the Alfvén mode, with

$$\omega^2 = k^2 v_A^2 \cos^2\theta, \quad \boldsymbol{\xi}^A = (0, 1, 0), \tag{P2-17}$$

and fast (+) and slow(−) modes, with

$$\omega^2=k^2v_{\phi\pm}^2, \quad v_{\phi\pm}^2=\tfrac{1}{2}(v_A^2+c_s^2)\pm\tfrac{1}{2}[(v_A^2+c_s^2)^2-4v_A^2c_s^2\cos^2\theta]^{1/2},$$

$$\xi^\pm=(\sin\psi^\pm, 0, \cos\psi^\pm), \quad \tan\psi^\pm=\frac{c_s^2\sin\theta\cos\theta}{v_{\phi\pm}^2-v_A^2-c_s^2\sin^2\theta}=\frac{v_{\phi\pm}^2-c_s^2\cos^2\theta}{c_s^2\sin\theta\cos\theta}, \tag{P2-18}$$

where the coordinate axes are chosen as in (2.51).

Hint: Write Γ_{ij} as a matrix and construct the matrix of its cofactors.

2.12* Consider waves in a compressible isothermal atmosphere under the influence of gravity. The equation of hydrostatic support then implies $-\text{grad}\ p_0+\eta_0\mathbf{g}=0$, where g is the acceleration due to gravity.

(a) Show that in the absence of a magnetic field the wave equation may be written

$$\Gamma_{ij}(\mathbf{k},\omega)\xi_j(\mathbf{k},\omega)=0, \quad \Gamma_{ij}(\mathbf{k},\omega)=\omega^2\delta_{ij}+i(\gamma-1)g_ik_j+ik_jg_i-c_s^2k_ik_j. \tag{P-19}$$

(b) Show that the dispersion equation is

$$\omega^4-\omega^2(k^2c_s^2+\omega_1^2)+\frac{|\mathbf{k}\times\mathbf{g}|^2c_s^2}{g^2}\omega_2^2=0, \tag{P2-20}$$

$$\omega_1^2:=\frac{\gamma^2g^2}{4c_s^2}, \quad \omega_2^2=\frac{(\gamma-1)g}{c_s^2}. \tag{P2-21}$$

Hint: replace ξ by $\zeta:=\xi\exp[\mathbf{g}\cdot\mathbf{r}/2gH]$, $H:=c_s^2/\gamma g$.
(c) Show that in the presence of a magnetic field, the dispersion equation factorizes into the dispersion relation $\omega^2=(\mathbf{k}\cdot\mathbf{b})^2\mathbf{v}_A^2$ for Alfvén waves and

$$\omega^4-\omega^2\{k^2(v_A^2+c_s^2)-i\gamma\mathbf{k}\cdot\mathbf{g}\}+k^2c_s^2v_A^2(\mathbf{k}\cdot\mathbf{b})^2+(\gamma-1)(\mathbf{k}\times\mathbf{g})^2-i\gamma\mathbf{g}\cdot\mathbf{b}k^2v_A^2\mathbf{k}\cdot\mathbf{b}=0. \tag{P2-22}$$

Remark: The solutions of the dispersion equation in part (b) describe sound waves for $\omega>\omega_1$ and interval gravity waves for $\omega<\omega_2$ (the Väisälä–Brunt frequency). The solutions of the dispersion equation in part (c) describe magnetoacoustic-gravity waves.

2.13* Calculate the contribution to the dielectric tensor from a suprathermal distribution of electrons with a Lorentzian profile:

$$f(v)=\frac{\bar{V}}{\pi}\frac{1}{(v-v_s)^2+\bar{v}^2},$$

where $\bar{v}$ and v_s are constants and where v is a one-dimensional velocity.

Remark: Firstly, integrate over the perpendicular components of assuming $f(\mathbf{v})\propto f(v_\parallel)\delta(\mathbf{v}_\perp)$, then write v in place of $v_\parallel$

CHAPTER 3

Spontaneous Emission

In this chapter we shall be concerned with solving the inhomogeneous wave equation and with applying the solutions to treat various spontaneous emission processes. Spontaneous emission by individual particles into various plasma modes is of interest in itself, e.g. in incoherent emission processes. The theory of spontaneous emission is used as the basis for a more general theory of particle-wave interactions in Chapter 5.

3.1 SOLUTIONS OF THE INHOMOGENEOUS WAVE EQUATION

The inhomogeneous wave equation is simply (2.43) with the source terms retained, i.e.

$$\Lambda^{\mathrm{h}}_{ij}(\mathbf{k}, \omega)E_j(\mathbf{k}, \omega) = -\frac{4\pi i}{\omega} j_i^{\mathrm{ext}}(\mathbf{k}, \omega). \tag{3.1}$$

For present purposes the term associated with the anti-hermitian part of the dielectric tensor is regarded as a source term and is assumed to be included in the extraneous current. Specifically, $\mathbf{j}^{\mathrm{ext}}$ is assumed to include the term

$$j_i^{\mathrm{ext}}(\mathbf{k}, \omega) = \frac{-i\omega}{4\pi} \varepsilon^{\mathrm{a}}_{ij}(\mathbf{k}, \omega)E_j(\mathbf{k}, \omega) + \cdots. \tag{3.2}$$

In any specific problem the source term needs to be identified. Thus, (3.2) is the relevant source term when considering damping of the waves. When treating emission by particles, $\mathbf{j}^{\mathrm{ext}}$ is to be identified as the Fourier transform of the current density associated with the particle (Section 3.2).

The energy or power radiated is derived below by calculating the work done by the extraneous current against the electric field which it generates. This method might be contrasted with the conventional method used to treat emission *in vacuo*. The idea behind both methods is to calculate the rate of increase in the electromagnetic energy. The equation of continuity for electromagnetic energy is (2.34), viz.

$$\frac{\partial}{\partial t}\left[\frac{|\mathbf{E}|^2}{8\pi}+\frac{|\mathbf{B}|^2}{8\pi}\right]+\operatorname{div}\left[\frac{c}{4\pi}\mathbf{E}\times\mathbf{B}\right]=-\mathbf{j}\cdot\mathbf{E}.$$

In the former method the power radiated is found by integrating $-\mathbf{j}\cdot\mathbf{E}$ over all space. In the latter method the power escaping is found by integrating the radial component of the Poynting vector $(c/4\pi)\mathbf{E}\times\mathbf{B}$ over the surface of an infinite sphere. The two methods are equivalent in a general sense, but not in the particular sense that they lead directly to the same quantity. The former method gives the instantaneous power lost by the source (for radiation into unit solid angle in a given direction say). The latter method gives the power (per unit solid angle say) received by a fixed distant observer due to the emission by the source at the retarded time. This difference has led to some confusion and controversy concerning the relevant result to use in treating synchrotron emission (Epstein and Feldman 1967); it was Scheuer (1968) who presented the definitive clarification. The power radiated is the relevant quantity.

The former method, based on the use of the $-\mathbf{j}\cdot\mathbf{E}$, is rarely used to treat emission *in vacuo*, and the latter method is rarely used in plasma physics. The reasons are not hard to identify. The former method is relatively simple only when one uses Fourier transforms (or some equivalent formalism) and even then one encounters a singular (generalized) function. Such a method is naturally avoided in introductory courses in electromagnetic theory. The latter method is useful for emission *in vacuo* because it is relatively easy to calculate the fields due to the sources, e.g. using the Liénard–Weichert potentials, and hence to find the Poynting vector. However, for emission in a medium one has to resort to the use of Fourier transforms to calculate the fields due to the sources. Moreover, the Poynting vector does not give the total energy flux in the waves in general (specifically, not in a spatially dispersive medium) and one has an additional problem of relating the total energy flux to the electromagnetic energy flux. Consequently, the latter method is not convenient for treating emission in a medium.

The Electric Amplitude

An important preliminary step in deriving the energy or power radiated is to define the amplitude of a wave and to calculate the energy in the waves in terms of the amplitude. Now it does not matter just how the amplitude is defined provided one couples together (a) the definition of the amplitude, and (b) the relation between the amplitude and the energy. The amplitude itself is of no physical significance here. Let the (electric) amplitude for waves in the mode σ be $\mathbf{E}^\sigma(\mathbf{k})$ defined by writing the Fourier transform for the electric field in the form

$$[\mathbf{E}(\mathbf{k},\,\omega)]^\sigma =: \mathbf{E}^\sigma(\mathbf{k})2\pi\delta(\omega-\omega^\sigma(\mathbf{k})), \tag{3.3}$$

with $\mathbf{E}^\sigma(\mathbf{k})$ necessarily proportional to $\mathbf{e}^\sigma(\mathbf{k})$, and with $\omega^\sigma(-\mathbf{k})=-\omega^\sigma(\mathbf{k})$ and $\mathbf{E}^\sigma(-\mathbf{k})=\mathbf{E}^{\sigma*}(\mathbf{k})$. Note that negative frequencies are included in (3.3), but it is usually convenient to rewrite negative frequencies as positive frequencies (wherever possible) using the relations obtained by reversing the sign of $\mathbf{k}$.

Now the electric energy, averaged over an arbitrarily long time τ, is given by

$$\lim_{\tau\to\infty}\frac{1}{\tau}\int_{-\tau/2}^{\tau/2}\mathrm{d}t\int \mathrm{d}^3\mathbf{r}\,\frac{|\mathbf{E}(\mathbf{r},\,t)|^2}{8\pi}=\lim_{\tau\to\infty}\frac{1}{\tau}\int\frac{\mathrm{d}\omega}{2\pi}\int\frac{\mathrm{d}^3\mathbf{k}}{(2\pi)^3}\frac{|\mathbf{E}(\mathbf{k},\,\omega)|^2}{8\pi}. \tag{3.4}$$

On inserting (3.3), using (2.12) and noting that there are equal contributions from $\omega=\pm|\omega^\sigma(\mathbf{k})|$ one finds that the electric energy density in waves in the mode σ in an elemental range $\mathrm{d}^3\mathbf{k}/(2\pi)^3$ at $\mathbf{k}$ is

$$W_E^\sigma(\mathbf{k})=\frac{|\mathbf{E}^\sigma(\mathbf{k})|^2}{4\pi V}, \tag{3.5}$$

where V is the normalization volume. (In formulae used in applications V either cancels out or is absorbed in some other quantity such as the power radiated per unit volume.) The total energy density in the waves in the mode σ in the same range is

$$W^\sigma(\mathbf{k})=\frac{|\mathbf{E}^\sigma(\mathbf{k})|^2}{4\pi V R_E^\sigma(\mathbf{k})}. \tag{3.6}$$

Other relevant quantities may be related to the electric amplitude in similar fashion, e.g. the electromagnetic energy flux $\mathbf{F}_{EM}^\sigma(\mathbf{k})$ is

$$\mathbf{F}_{EM}^\sigma(\mathbf{k})=\mathrm{Re}\left[\frac{c}{4\pi V}\{\mathbf{E}^\sigma(\mathbf{k})\times\mathbf{B}^{\sigma*}(\mathbf{k})\}\right],\quad \mathbf{B}^\sigma(\mathbf{k}):=\frac{c\mathbf{k}\times\mathbf{E}^\sigma(\mathbf{k})}{\omega^\sigma(\mathbf{k})}. \tag{3.7}$$

(The total energy flux $\mathbf{F}^\sigma(\mathbf{k})=\mathbf{v}_g^\sigma(\mathbf{k})W^\sigma(\mathbf{k})$ is equal to $\mathbf{F}_{EM}(\mathbf{k})$ only if the medium is not spatially dispersive.)

The Absorption Coefficient

As the first example of the use of the inhomogeneous wave equation, let us solve for the rate work is done by the extraneous current (3.2) associated with the anti-hermitian part of the dielectric tensor. The work done is denoted $Q^\sigma(\mathbf{k})$ in (2.82) and is subsequently identified as $-\gamma^\sigma(\mathbf{k})W^\sigma(\mathbf{k})$ in writing down the absorption coefficient (2.86). Actually, in this section we have defined $W^\sigma(\mathbf{k})$ such that it is the energy density in the waves and consequently we should redefine $Q^\sigma(\mathbf{k})$ as the rate work is done per unit volume.

The time averaged power P (or energy $U = P\tau$) generated in the waves may be calculated using

$$P = -\lim_{\tau\to\infty} \frac{1}{\tau} \int_{-\tau/2}^{\tau/2} dt \int d^3\mathbf{r}\, \mathbf{j}^{\text{ext}}(\mathbf{r}, t)\cdot \mathbf{E}(\mathbf{r}, t)$$
$$= -\lim_{\tau\to\infty} \frac{1}{\tau} \int_{-\infty}^{\infty} \frac{d\omega}{2\pi} \int \frac{d^3\mathbf{k}}{(2\pi)^3} Re[\mathbf{j}^{\text{ext}}(\mathbf{k}, \omega)\cdot \mathbf{E}^*(\mathbf{k}, \omega)], \tag{3.8}$$

where the power theorem (2.15) has been used. In the present case $\mathbf{E}(\mathbf{k}, \omega)$ is to be identified as the electric field (3.3) associated with waves in the mode σ, and $\mathbf{j}^{\text{ext}}(\mathbf{k}, \omega)$ is given by (3.2). Then, after dividing by the volume of the system, with P/V identified as the integral of $Q^\sigma(\mathbf{k})$ over $\mathbf{k}$-space, (3.8) gives

$$\int \frac{d^3\mathbf{k}}{(2\pi)^3} Q^\sigma(\mathbf{k}) = \lim_{\tau\to\infty} \frac{1}{\tau V} \int_{-\infty}^{\infty} \frac{d\omega}{2\pi} \int \frac{d^3\mathbf{k}}{(2\pi)^3}$$
$$\times Re\left[-\frac{i\omega}{4\pi} e_i^{\sigma *}(\mathbf{k})\, e_j^\sigma(\mathbf{k}) \varepsilon_{ij}^{\text{a}}(\mathbf{k}, \omega) |\mathbf{E}^\sigma(\mathbf{k})|^2 \{2\pi\delta(\omega - \omega^\sigma(\mathbf{k}))\}^2\right]. \tag{3.9}$$

The square of the δ-function is rewritten using (2.12), the integral from $-\infty$ to ∞ over ω is converted into twice the integral from 0 to ∞ and is performed using the δ-function. It is then straightforward to identify $Q^\sigma(\mathbf{k})$. Finally, using (3.5) and (3.6), the expression (2.86) is obtained for the absorption coefficient $\gamma^\sigma(\mathbf{k}) := -Q^\sigma(\mathbf{k})/W^\sigma(\mathbf{k})$.

The Causal Condition

In treating emission processes below, a singular integral arises and is evaluated by appealing to the causal condition. Causality, namely the fact that cause precedes effect, implies that one is to evaluate the relevant singular integrals by giving the frequency an infinitesimal positive imaginary part. This is the so-called Landau prescription. Landau derived it by using a Laplace transform rather than a Fourier transform in time.

Rather than Laplace transform, the causal condition can be included explicitly in Fourier transforms by using the step function

$$\theta(t)=\begin{cases}1 & \text{for } t>0,\\ 0 & \text{for } t<0.\end{cases} \tag{3.10}$$

The Fourier transform of the step function is the generalized function

$$\tilde{\theta}(\omega)=\lim_{\eta\to 0}\frac{i}{\omega+i|\eta|}=:\frac{i}{\omega+i0}. \tag{3.11}$$

Singular integrals involving the generalized function (3.11) may be evaluated using the Plemelj formulas

$$\frac{1}{\omega-\omega_0\pm i0}=\mathscr{P}\frac{1}{\omega-\omega_0}\mp i\pi\delta(\omega-\omega_0), \tag{3.12}$$

where $\mathscr{P}$ denotes the Cauchy principal value. The contribution from the δ-function (when (3.12) is used to evaluate an integral) is called the semi-residue. Formally, the Plemelj formula may be derived as ($-i$ times) the Fourier transform of the identity

$$\theta(t)=\frac{1}{2}\left(\frac{t}{|t|}+1\right)$$

together with the fact that the Fourier transform of $t/|t|$ is $2i$ times the generalized function

$$\mathscr{P}\frac{1}{\omega}:=\lim_{\eta\to 0}\begin{cases}1/\omega & \text{for } |\omega|>|\eta|,\\ 0 & \text{for } |\omega|<|\eta|.\end{cases}$$

The causal condition may be applied to the response of any medium. The dielectric tensor must be the Fourier transform of a quantity (an integro-differential operator in general) which vanishes for negative times. Hence it must be equal to the convolution of itself with the step function, i.e.

$$\varepsilon_{ij}(\mathbf{k},\omega)=\frac{i}{2\pi}\int_{-\infty}^{\infty}\mathrm{d}\omega'\,\frac{\varepsilon_{ij}(\mathbf{k},\omega')}{\omega-\omega'+i0}. \tag{3.13}$$

This is the third general property of dielectric tensors mentioned in Section 2.2. It is conventional to express the causal condition on the dielectric tensor in terms of relations between the hermitian and anti-hermitian parts. By using the Plemelj formula (3.12) in (3.13), and then separating into hermitian and anti-hermitian parts, one finds

$$\varepsilon^{\mathrm{h}}_{ij}(\mathbf{k},\omega)=\frac{i}{\pi}\mathscr{P}\int_{-\infty}^{\infty}\mathrm{d}\omega'\,\frac{\varepsilon^{\mathrm{a}}_{ij}(\mathbf{k},\omega')}{\omega-\omega'}, \tag{3.14a}$$

$$\varepsilon^{\mathrm{a}}_{ij}(\mathbf{k}, \omega) = \frac{i}{\pi} \mathscr{P} \int_{-\infty}^{\infty} \mathrm{d}\omega' \, \frac{\varepsilon^{\mathrm{h}}_{ij}(\mathbf{k}, \omega')}{\omega - \omega'}, \tag{3.14b}$$

which are the Kramers–Kronig relations.

The Energy Radiated Spontaneously

Now we have all the tools required to evaluate the energy or power radiated spontaneously. Let $U^{\sigma}(\mathbf{k})$ be the energy radiated, and $P^{\sigma}(\mathbf{k})$ be the power radiated, each in the elemental range $\mathrm{d}^3\mathbf{k}/(2\pi)^3$ at $\mathbf{k}$. In some problems, e.g. radiation in a collision, the energy radiated is the relevant quantity, and in other problems where continuous or periodic emission is involved, the power radiated is relevant. Here, the two are related trivially, specifically by

$$P^{\sigma}(\mathbf{k}) = \lim_{\tau \to \infty} \frac{U^{\sigma}(\mathbf{k})}{\tau}, \tag{3.15}$$

and one chooses whichever is relevant. In the discussion here we assume $U^{\sigma}(\mathbf{k})$ is relevant.

The energy radiated is given by the right hand side of (3.8) with the factor $1/\tau$ omitted. The electric field is to be identified with that generated by the extraneous current itself. This involves solving (3.1). One may do so by contracting each side with say $\lambda_{si}(\mathbf{k}, \omega)$. Then using (2.75), viz. $\Lambda^{\mathrm{h}}_{ij}\lambda_{jl} = \Lambda\delta_{il}$, one finds

$$E_i(\mathbf{k}, \omega) = -\frac{4\pi i}{\omega} \frac{\lambda_{ij}(\mathbf{k}, \omega)}{\Lambda(\mathbf{k}, \omega)} j^{\mathrm{ext}}_j(\mathbf{k}, \omega). \tag{3.16}$$

Now when (3.16) is inserted in (3.8) the integral vanishes (because one is to take the real part of an apparently imaginary quantity). However, there are contributions from the zeros of $\Lambda(\mathbf{k}, \omega)$ when the Landau prescription is used to write

$$\Lambda(\mathbf{k}, \omega) \approx (\omega - \omega^{\sigma}(\mathbf{k}) + i0) \left[\frac{\partial \Lambda(\mathbf{k}, \omega)}{\partial \omega}\right]_{\omega = \omega^{\sigma}(\mathbf{k})} \tag{3.17}$$

in the neighborhood of the zero at $\omega = \omega^{\sigma}(\mathbf{k})$. Consequently, the integral separates into a number of discrete parts each associated with a particular zero, and hence with emission into a particular mode.

The explicit calculation of $U^{\sigma}(\mathbf{k})$ involves the following steps. First, in the RH member of (3.8) (without the factor $1/\tau$) rewrite the ω-integral as twice the integral from 0 to ∞. Second, insert (3.17) in (3.16), reexpress the singular function using the Plemelj formula and retain only the semi-residue (the other part does not contribute). Third, perform the ω-integral over the

δ-function. Finally, use (2.76) and (2.85) to find

$$U^\sigma(\mathbf{k}) = 4\pi R_E^\sigma(\mathbf{k})\left|\mathbf{e}^{\sigma *}(\mathbf{k}) \cdot \mathbf{j}^{\text{ext}}(\mathbf{k}, \omega^\sigma(\mathbf{k}))\right|^2. \tag{3.18}$$

The final formula (3.18) is used extensively below. It gives the energy radiated in an arbitrary mode σ by an arbitrary source described by the Fourier transform of the associated current density, called the extraneous current for short. For continous or periodic emission the extraneous current has a δ-function or sum of δ-functions which relate ω to the (Doppler shifted) natural frequencies of oscillation. Consequently, squares of δ-functions appear in (3.18), and when these are reexpressed using (2.12), the energy radiated becomes proportional to τ. The time-averaged power radiated, defined by (3.15), is then the meaningful quantity.

Emission of Transverse Waves

One important application for which (3.18) needs to be modified is for transverse waves in an isotropic medium or *in vacuo*. The problem in this case is that there is then no unique polarization vector. It is convenient to introduce the basis vectors $\mathbf{e}^1$ and $\mathbf{e}^2$ which span the plane orthogonal to $\boldsymbol{\kappa}$, i.e. such that $\mathbf{e}^1$, $\mathbf{e}^2$ and $\boldsymbol{\kappa}$ are mutually orthonormal. It can be shown that to retain all the information on the polarization of the emission one needs not only to calculate the emission with $\mathbf{e}^\sigma$ replaced by $\mathbf{e}^1$ and by $\mathbf{e}^2$ but also to calculate a cross-term, e.g. $(\mathbf{e}^1 \cdot \mathbf{j}^{\text{ext}})(\mathbf{e}^2 \cdot \mathbf{j}^{\text{ext}})^*$. To retain all this information it is appropriate to describe the energy radiated in terms of a polarization tensor

$$U^{\alpha\beta}(\mathbf{k}) = 4\pi R_E(\mathbf{k}) e_i^\alpha e_j^\beta j_i^{\text{ext}}(\mathbf{k}, \omega) j_j^{\text{ext}\,*}(\mathbf{k}, \omega), \tag{3.19}$$

where α and β run over 1 and 2. The techniques for using polarization tensors are developed in Section 3.2, Section 4.1 and Section 6.2. For the present, let us merely note that the total energy (per unit volume of $\mathbf{k}$-space) radiated is $U^{\alpha\alpha}(\mathbf{k})$, where the summation convention is used, i.e. $U^{\alpha\alpha}$ means $U^{11} + U^{22}$.

Multipole Emission

A multipole expansion of a current density $\mathbf{j}(\mathbf{r}, t)$ is related to its spatial Fourier transform

$$\begin{aligned}\mathbf{j}(\mathbf{k}, \omega) &= \int_{-\infty}^{\infty} dt \exp(i\omega t) \int d^3\mathbf{r}\, \mathbf{j}(\mathbf{r}, t) \exp(-i\mathbf{k} \cdot \mathbf{r}) \\ &= \int_{-\infty}^{\infty} dt \exp(i\omega t) \left[\frac{d}{dt}\mathbf{d}(t) + ic\mathbf{k} \times \mathbf{m}(t) + \cdots\right],\end{aligned} \tag{3.20}$$

where $\mathbf{d}(t)$ and $\mathbf{m}(t)$ are the electric and magnetic dipole moments. These are defined by

$$\mathbf{d}(t) := \int d^3\mathbf{r}\, \mathbf{r}\rho(\mathbf{r}, t), \quad \mathbf{m}(t) := \frac{1}{2c}\int d^3\mathbf{r}\, \mathbf{r}\times\mathbf{j}(\mathbf{r}, t) \qquad (3.21a, b)$$

The next term in the expansion (3.20) is the electric quadrupole term. The ith component of the quadrupole term may be written as $-\frac{1}{2}k_r\dot{q}_{ir}(t)$, where

$$q_{ij}(t) := \int d^3\mathbf{r}\, x_i x_j \rho(\mathbf{r}, t) \qquad (3.21c)$$

is a quadrupole moment. It is conventional to define "the quadrupole moment" such that its Cartesian tensor components are

$$d_{ij}(t) := 3q_{ij}(t) - \delta_{ij}q_{ss}(t). \qquad (3.22)$$

On inserting (3.20) in (3.18) one finds

$$U^\sigma(\mathbf{k}) = 4\pi R_E^\sigma(\mathbf{k})|\omega^\sigma(\mathbf{k})\mathbf{e}^{\sigma\,*}(\mathbf{k})\cdot\tilde{\mathbf{d}}(\omega^\sigma(\mathbf{k}))|^2, \qquad (3.23a)$$

$$U^\sigma(\mathbf{k}) = 4\pi R_E^\sigma(\mathbf{k})k^2c^2|\mathbf{e}^{\sigma\,*}(\mathbf{k})\times\tilde{\mathbf{m}}(\omega^\sigma(\mathbf{k}))|^2, \qquad (3.23b)$$

$$U^\sigma(\mathbf{k}) = 4\pi R_E^\sigma(\mathbf{k})\tfrac{1}{4}|\omega^\sigma(\mathbf{k})e_i^{\sigma\,*}(\mathbf{k})k_j\tilde{q}_{ij}(\omega^\sigma(\mathbf{k}))|^2 \qquad (3.23c)$$

for the electric dipole, magnetic dipole and electric quadrupole terms respectively. Note that in writing down (3.23a, b, c) it is assumed that there is no correlation between the different multipole moments. (The cross terms in $|\mathbf{j}(\mathbf{k}, \omega)|^2$, with $\mathbf{j}(\mathbf{k}, \omega)$ given by (3.20) are ignored.) It is an implicit assumption in a multipole approach that the currents associated with different multipole components are uncorrelated. Note also that although the trace of q_{ij} does not contribute to emission of transverse waves (the trace in (3.23c) gives a term $\frac{1}{3}q_{ss}\mathbf{e}^{\sigma\,*}(\mathbf{k})\cdot\mathbf{k}$, which is zero for transverse waves), the trace does contribute for $\mathbf{e}^{\sigma\,*}(\mathbf{k})\cdot\mathbf{k}\neq 0$. Because the traces are ignored in a conventional multipole expansion, the conventional expansion is inadequate for quadrupole and higher terms.

3.2 EMISSION BY PARTICLES: CERENKOV EMISSION

In this section the extraneous current is identified with that due to the motion of a charged particle. The particular case of emission due to a particle in constant rectilinear motion, called Cerenkov emission, is treated in detail.

The Current Associated with a Moving Charge

The general problem of emission by a charge may be treated by interpreting the extraneous current in (3.18) as the current associated with a particle in arbitrary motion. Let the particle have charge q and rest mass m. Its orbit may be described by a set of three equations for the three coordinates of the point charge as functions of time. It is convenient to write these equations as $x=x(t)$, $y=y(t)$ and $z=z(t)$, or equivalently as

$$\mathbf{r}=\mathbf{r}(t). \tag{3.24}$$

The (instantaneous) velocity of the particle is given by

$$\mathbf{v}(t)=\dot{\mathbf{r}}(t) \tag{3.25}$$

where the dot denotes differentiation with respect to time t. The momentum $\mathbf{p}$, Lorentz factor γ and energy ε of the particle, which may all be functions of time, are given by

$$\mathbf{p}=\gamma m\mathbf{v}, \quad \gamma=(1-v^2/c^2)^{-1/2}, \quad \varepsilon=\gamma mc^2. \tag{3.26}$$

The charge density for the point charge is $q\delta(x-x(t))\delta(y-y(t))\delta(z-z(t))$ which may be written

$$\rho(\mathbf{r}, t)=q\delta^3(\mathbf{r}-\mathbf{r}(t)) \tag{3.27}$$

The current density is

$$\mathbf{j}(\mathbf{r}, t)=q\mathbf{v}(t)\delta^3(\mathbf{r}-\mathbf{r}(t)) \tag{3.28}$$

The Fourier transform of (3.28) is

$$\begin{aligned}\mathbf{j}(\mathbf{k}, \omega)&=q\int_{-\infty}^{\infty} dt \int d^3\mathbf{r}\, \exp\left[-i(\mathbf{k}\cdot\mathbf{r}-\omega t)\right]\, \mathbf{v}(t)\delta^3(\mathbf{r}-\mathbf{r}(t))\\ &=q\int_{-\infty}^{\infty} dt\, \mathbf{v}(t)\, \exp\left[-i(\mathbf{k}\cdot\mathbf{r}(t)-\omega t)\right].\end{aligned} \tag{3.29}$$

On identifying the extraneous current in (3.18) with (3.29) one obtains a general expression for the energy radiated in an arbitrary wave mode due to a particle in an arbitrary orbit.

We shall use (3.29) in several ways. Firstly, several general results relating to spontaneous emission will be derived using it. Next, it will be applied to a particle in constant rectilinear motion. Then the resulting emission is Cerenkov emission. Next, in Section 3.3, the current is identified as that due to an electron passing near an ion. The resulting emission is bremsstrahlung. In Section 3.4 the current is identified with that associated with the perturbed motion of the particle in the field of a wave. The resulting emission is the scattered radiation due to Thomson scattering of the assumed initial waves. Further examples are gyromagnetic emission (Section 4.1), which is due to the spiralling motion of a particle in a magnetic field, nonlinear scattering (Section 10.4) and curvature and linear acceleration emission (Section 13.3).

Larmor Formula

First however, it is appropriate to derive the Larmor formula, which gives the power radiated by a non-relativistic accelerated charge *in vacuo*. For a non-relativistic charge, and for $\omega = kc$, the term $\mathbf{k}\cdot\mathbf{r}(t)$ in the exponent in (3.29) may be neglected. In this approximation the current $\mathbf{j}(\mathbf{k}, \omega)$ reduces to

$$\mathbf{j}(\mathbf{k}, \omega) = iq\mathbf{a}(\omega)/\omega, \tag{3.30}$$

where $\mathbf{a}(\omega)$ is the temporal Fourier transform of the acceleration $\ddot{\mathbf{r}}(t)$. The approximation (3.30) is equivalent to retaining only the electric dipole term in the expansion (3.20). Consequently, it is called the *dipole approximation.*

Now for emission *in vacuo* one has $R_E = \frac{1}{2}$ and $k = \omega/c$ in (3.19), and hence the total energy radiated is

$$U = \int \frac{d^3\mathbf{k}}{(2\pi)^3}\, U^{\alpha\alpha}(\mathbf{k}) = \frac{4\pi}{(2\pi c)^3}\int_0^\infty d\omega\, \omega^2 \langle U^{\alpha\alpha}(\boldsymbol{\kappa}\omega/c)\rangle, \tag{3.31}$$

where the angular brackets denote an average over all directions. On inserting (3.30) in (3.19), $\boldsymbol{\kappa}$ appears explicitly only in the form $|\boldsymbol{\kappa}\times\mathbf{a}(\omega)|^2$, cf. (3.49) below, whose average is $(2/3)|\mathbf{a}(\omega)|^2$. Hence (3.31) becomes

$$U = \frac{2q^2}{3\pi c^3}\int_0^\infty d\omega |\mathbf{a}(\omega)|^2. \tag{3.32}$$

The Larmor formula may be derived from (3.32) by writing U as the time-integral of the power radiated and rewriting the right hand side in a form which allows one to use the power theorem (2.16):

$$\int_{-\infty}^\infty dt\, P(t) = \frac{2q^2}{3c^3}\int_{-\infty}^\infty \frac{d\omega}{2\pi}|\mathbf{a}(\omega)|^2 = \frac{2q^2}{3c^3}\int_{-\infty}^\infty dt\,|\ddot{\mathbf{r}}(t)|^2. \tag{3.33}$$

The Larmor formula is

$$P(t) = \frac{2q^2}{3c^3}|\ddot{\mathbf{r}}(t)|^2. \tag{3.34}$$

The generalization of (3.31) for the energy radiated *in vacuo* by a charge in arbitrary motion is

$$U = \frac{q^2}{\pi c^3}\int_0^\infty d\omega\, \omega^2 \left\langle \left| \int_{-\infty}^\infty dt\, \boldsymbol{\kappa}\times\mathbf{v}(t)\exp\left[-i\omega(\boldsymbol{\kappa}\cdot\mathbf{r}(t)/c - t)\right]\right|^2 \right\rangle. \tag{3.35}$$

Emission by particles into waves in an arbitrary mode in a plasma may be treated using (3.18) with the current identified with (3.29).

Relativistic Accelerated Charge

The current associated with a relativistic charge due to its accelerated motion may be derived as follows. Let the force acting on it be $\mathbf{F}(\mathbf{r}, t; \mathbf{v})$ such that the equation of motion is

$$\frac{\mathrm{d}\mathbf{p}}{\mathrm{d}t} = \mathbf{F}(\mathbf{r}, t; \mathbf{v}). \tag{3.36}$$

The effect of the force $\mathbf{F}$ is to be treated as a perturbation. The unperturbed motion is assumed to be constant rectilinear motion. Writing

$$\mathbf{p}(t) = \mathbf{p}^{(0)} + \mathbf{p}^{(1)} + \cdots, \tag{3.37}$$

$$\mathbf{r}(t) = \mathbf{r}^{(0)}(t) + \mathbf{r}^{(1)}(t) + \cdots, \tag{3.38}$$

with

$$\mathbf{r}^{(0)}(t) = \mathbf{r}_0 + \mathbf{v}^{(0)}t, \tag{3.39}$$

the first order equation of motion becomes

$$\frac{\mathrm{d}\mathbf{p}^{(1)}(t)}{\mathrm{d}t} = \mathbf{F}(\mathbf{r}^{(0)}(t), t; \mathbf{v}^{(0)}) =: \mathbf{F}^{(1)}(t). \tag{3.40}$$

The first order perturbation in the trajectory is given by integrating the following equation twice with respect to t:

$$\ddot{r}_i^{(1)}(t) = \frac{1}{m\gamma^{(0)}} \left(\delta_{ij} - \frac{v_i^{(0)} v_j^{(0)}}{c^2} \right) F_j^{(1)}(t) \tag{3.41}$$

and the first order perturbation in the velocity is $\mathbf{v}^{(1)}(t) = \dot{\mathbf{r}}^{(1)}(t)$.

The first order current follows by expanding the general expression (3.29) for the current due to a moving charge and retaining only those terms of first order in the perturbing force,

$$\mathbf{j}^{(1)}(\mathbf{k}, \omega) = q \int_{-\infty}^{\infty} \mathrm{d}t \, \{\mathbf{v}^{(1)}(t) - i\mathbf{k} \cdot \mathbf{r}^{(1)}(t)\mathbf{v}^{(0)}\} \exp\left[-i(\mathbf{k} \cdot \mathbf{r}^{(0)}(t) - \omega t)\right]. \tag{3.42}$$

After inserting the explicit expression (3.39) for $\mathbf{r}^{(0)}(t)$ and partially integrating (once over the first term and twice over the second term in the integrand), one obtains

$$j_i^{(1)}(\mathbf{k}, \omega) = \frac{iq \exp\left[-i\mathbf{k} \cdot \mathbf{r}_0\right] g_{ji}(\mathbf{k}, \omega; \mathbf{v})}{(\omega - \mathbf{k} \cdot \mathbf{v})^2} \int_{-\infty}^{\infty} \mathrm{d}t \, \ddot{r}_j^{(1)}(t) \exp\left[i(\omega - \mathbf{k} \cdot \mathbf{v})t\right], \tag{3.43}$$

with

$$g_{ij}(\mathbf{k}, \omega; \mathbf{v}) := (\omega - \mathbf{k} \cdot \mathbf{v})\delta_{ij} + k_i v_j. \tag{3.44}$$

Emission by an accelerated relativistic charge may be treated by inserting (3.43) in (3.18).

Cerenkov Emission

Now let us turn to the details of specific emission mechanisms. The simplest of all is Cerenkov emission, which is due to a charge in constant rectilinear motion at greater than the phase speed of the emitted waves. The mechanism is somewhat analogous to the generation of sound waves in a gas by a supersonic particle, or to the generation of water waves by a boat. The emission by fast particles in a material medium was discovered by Cerenkov in 1934 (the history of the subject has been reviewed by Jelley (1958)).

An interesting facet of Cerenkov emission is that it is related to the phase speed of the waves and therefore provides a possible method for measuring the phase speed (provided it is less than the speed of light). The suggestion sometimes made that the phase velocity is only a mathematical construction and of no direct physical significance is simply untrue.

The current required to treat Cerenkov emission may be derived as follows. In constant rectilinear motion $\mathbf{v}$ does not depend on time and the orbit may be described by

$$\mathbf{r}=\mathbf{r}_0+\mathbf{v}t, \tag{3.45}$$

where $\mathbf{r}_0$ is an arbitrary initial point. On inserting (3.45) in (3.29) one obtains

$$\begin{aligned}\mathbf{j}(\mathbf{k},\omega)&=q\mathbf{v}\exp(-i\mathbf{k}\cdot\mathbf{r}_0)\int_{-\infty}^{\infty}dt\exp[i(\omega-\mathbf{k}\cdot\mathbf{v})t]\\&=2\pi q\mathbf{v}\exp(-i\mathbf{k}\cdot\mathbf{r}_0)\,\delta(\omega-\mathbf{k}\cdot\mathbf{v}),\end{aligned} \tag{3.46}$$

where (2.4a) has been used.

The condition $\omega=\mathbf{k}\cdot\mathbf{v}$ contained in the δ-function is an example of a wave-particle *resonance* condition. If one makes a Lorentz transformation to the frame in which the particle is at rest, this condition implies that the wave has zero frequency in that frame. Thus, this Cerenkov condition corresponds to a wave-particle resonance such that in its rest frame the particle sees a static spatially periodic electromagnetic field associated with the wave.

Cerenkov emission is a continuous process and hence the power radiated rather than the energy radiated is the relevant quantity. On inserting (3.46) in (3.15) with (3.18), and evaluating the square of the δ-function using (2.12), one finds for Cerenkov emission

$$P^{\sigma}(\mathbf{k})=8\pi^2q^2R_E^{\sigma}(\mathbf{k})|\mathbf{e}^{\sigma *}(\mathbf{k})\cdot\mathbf{v}|^2\delta(\omega^{\sigma}(\mathbf{k})-\mathbf{k}\cdot\mathbf{v}). \tag{3.47}$$

Two remarks concerning Cerenkov emission are appropriate. Cerenkov emission *in vacuo* is forbidden. The reason is that the δ-function in (3.47) can be satisfied only for $\omega/k < v$, and for electromagnetic waves *in vacuo* one has $\omega/k = c$. Similarly, Cerenkov emission of transverse waves in a plasma is forbidden because these waves have $\omega/k = c/n$, $n = (1 - \omega_p^2/\omega^2)^{1/2} < 1$. Cerenkov emission is possible only for waves with refractive index greater than unity, and then only for particles with $v > c/n$. The second remark concerns the conditions under which the motion of the particle may be regarded as constant rectilinear motion, and hence the emission treated using (3.47). Consider a particle spiralling in a magnetic field. The particle spirals at the gyrofrequency and its motion perpendicular to the field lines is a circle with a characteristic gyroradius. Constant rectilinear motion corresponds to zero gyrofrequency and infinite gyroradius. The emission by a spiralling charge may be treated as Cerenkov emission provided the gyrofrequency is much smaller than other relevant frequencies and the gyroradius is much larger than other relevant lengths. The only relevant frequencies are ω and $\mathbf{k}\cdot\mathbf{v}$ in (3.47), and the only relevant length is k^{-1}. In fact, one finds that the conditions $\Omega \ll \omega$ and $\Omega \ll kv$ suffice for emission by a spiralling charge to be regarded as Cerenkov emission. In such cases the particles are said to be "unmagnetized". Put another way, "unmagnetized" particles interact with waves as though their motion were constant rectilinear motion, despite the fact that they are actually spiralling in a magnetic field.

Transverse Waves in an Isotropic Dielectric

It is useful for illustrative purposes to consider Cerenkov emission in isotropic dielectrics such as glass, water and air. This process is of little direct interest in astrophysics, although optical Cerenkov radiation due to cosmic-ray showers in air can be detected under ideal conditions (cloudless, moonless nights).

The dielectric tensor for an isotropic dielectric is of the form $\varepsilon(\omega)\delta_{ij}$ where $\varepsilon(\omega)$ is the dielectric constant. The wave properties are

$$n = n(\omega) := \varepsilon^{1/2}(\omega), \quad \mathbf{e}\cdot\boldsymbol{\kappa} = 0, \quad R_E = \frac{1}{2n(\omega)\mathrm{d}(\omega n(\omega))/\mathrm{d}\omega}. \tag{3.48}$$

On inserting these properties in (3.47) one must decide how to interpret the polarization vector in (3.47). This problem always arises when considering emission of transverse waves (in a plasma or *in vacuo*). One approach is to note that the only contribution in (3.47) is from the component of $\mathbf{e}$ along $\mathbf{v}$, and because $\mathbf{e}$ must be orthogonal to $\boldsymbol{\kappa}$ it follows that Cerenkov emission must be (linearly) polarized along the projection of $\mathbf{v}$ on the plane orthogonal to $\boldsymbol{\kappa}$. In this way one finds $|\mathbf{e}\cdot\mathbf{v}|^2=v^2\sin^2\psi$ where ψ is the angle between $\mathbf{v}$ and $\boldsymbol{\kappa}$.

A more general approach is to use a polarization tensor to describe the emission. It follows from inspection of (3.19) and (3.47) that the polarization tensor $P^{\alpha\beta}$ for the power generated in Cerenkov emission is proportional to $\mathbf{e}^\alpha\cdot\mathbf{v}\mathbf{e}^\beta\cdot\mathbf{v}$. Now, because polarization tensors may be represented as hermitian 2×2 matrices, it follows that if the determinant of a polarization tensor vanishes, then it is of rank one. Any tensor of rank one may be written as the outer product of a vector (here in two dimensions) with itself. Thus, if the determinant of the polarization tensor vanishes, then there exists a polarization vector and the radiation may be said to be completely polarized. Conversely, if the determinant does not vanish then the radiation is said to be partially polarized.

Unpolarized radiation is described by a diagonal polarization tensor with equal diagonal components, i.e. one proportional to $\delta^{\alpha\beta}:=e_i^\alpha e_j^\beta\delta_{ij}$. It can be shown that an arbitrary polarization tensor may be separately uniquely into a part proportional to $\delta^{\alpha\beta}$ and a part whose determinant vanishes (Section 6.2). This procedure is interpreted as a separation of the emission into its unpolarized and polarized components. The polarized component may be described by a polarization vector, and one additional quantity, usually chosen to be the degree of polarization, is required to describe partially polarized radiation in general.

In the case of Cerenkov emission, the polarization tensor is proportional to $v^\alpha v^\beta$ with $v^\alpha:=\mathbf{e}^\alpha\cdot\mathbf{v}$. The determinant of this quantity obviously vanishes, and it is already written in the form of an outer product of a vector with itself. Hence one concludes that the polarization vector is parallel to the component of $\mathbf{v}$ in the plane orthogonal to $\boldsymbol{\kappa}$. For many purposes it is sufficient to consider the total power radiated, and this is obtained by summing over the two states of transverse polarization. The sum is found by calculating the power radiated with $\mathbf{e}=\mathbf{e}^1$ in (3.47) and then with $\mathbf{e}=\mathbf{e}^2$ and adding the two. Formally, the sum is achieved by using

$$\sum_{\textit{polarization}} |\mathbf{e}\cdot\mathbf{v}|^2=|\boldsymbol{\kappa}\times\mathbf{v}|^2, \tag{3.49}$$

which follows directly from the identity

$$e_i^1e_j^1+e_i^2e_j^2+\kappa_i\kappa_j=\delta_{ij}. \tag{3.50}$$

Using (3.49) one infers that $|\mathbf{e}\cdot\mathbf{v}|^2$ in (3.47) is to be replaced by $v^2\sin^2\psi$ where ψ is the angle between $\mathbf{v}$ and $\boldsymbol{\kappa}$, in agreement with the earlier method.

The evaluation of the power radiated in transverse waves is straightforward. The total power is given by

$$P=\int \frac{d^3\mathbf{k}}{(2\pi)^3}\, 8\pi^2 q^2 R_E |\boldsymbol{\kappa}\times\mathbf{v}|^2 \delta(\omega-\mathbf{k}\cdot\mathbf{v}), \tag{3.51}$$

with R_E given by (3.48) and with $k=n(\omega)\omega/c$. One can write (in spherical polar coordinates)

$$d^3\mathbf{k}=\frac{n^2(\omega)\omega^2}{c^3}\frac{d}{d\omega}(\omega n(\omega))\, d\omega\, d\cos\psi\, d\phi, \tag{3.52}$$

where ϕ is an azimuthal angle over which one can integrate immediately. Thus (3.51) reduces to

$$P=\frac{q^2v^2}{c^3}\int_0^\infty d\omega\,\omega^2 n(\omega)\int_{-1}^{+1} d\cos\psi\,\sin^2\psi\,\delta\!\left(\omega-n(\omega)\omega\frac{v}{c}\cos\psi\right). \tag{3.53}$$

The integral over the δ-function is of the form (2.7) and is performed using (2.9). The only zero of the argument of the δ-function in (3.53) occurs at $\cos\psi=c/vn(\omega)$ and only for $v>c/n(\omega)$ does this satisfy $\cos\psi\leqslant 1$. Hence (3.53) gives

$$P=\frac{q^2v}{c^2}\int_{v>c/n(\omega)} d\omega\,\omega\left(1-\frac{c^2}{v^2n^2(\omega)}\right) \tag{3.54}$$

which is the usual formula for the power in Cerenkov emission, first derived by Tamm and Frank (1937).

Langmuir Waves

As has already been pointed out, Cerenkov radiation of transverse waves in an isotropic plasma is forbidden because the refractive index is less than unity. However, Cerenkov emission of Langmuir waves (and of ion sound waves when they exist) does occur.

The power in Cerenkov emission in Langmuir waves may be found by inserting the wave properties (2.109) for Langmuir waves in the expression (3.47) for $P^\sigma(\mathbf{k})$ and integrating over $\mathbf{k}$-space. One finds

$$P^l(\mathbf{k})=\frac{4\pi^2q^2\{\omega^l(k)\}^4}{\omega_p^2k^2}\,\delta(\omega^l(k)-\mathbf{k}\cdot\mathbf{v}), \tag{3.55}$$

where $\boldsymbol{\kappa}\cdot\mathbf{v}=\mathbf{k}\cdot\mathbf{v}/k=\omega^l(k)/k$ has been used. For simplicity let us approximate the dispersion relation by $\omega^l(k)=\omega_p$ for $k\lambda_{De}<1$ and assume that no Langmuir waves exist for $k\lambda_{De}>1$. Then the power in Cerenkov emission is given by

$$P^l=4\pi^2q^2\omega_p^2\int\frac{d^3\mathbf{k}}{(2\pi)^3}\frac{\delta(\omega_p-\mathbf{k}\cdot\mathbf{v})}{k^2}. \tag{3.56}$$

The integral can be evaluated by writing

$$\mathrm{d}^3\mathbf{k}=k^2\,\mathrm{d}k\,\mathrm{d}\cos\psi\,\mathrm{d}\phi,$$

where ψ is the angle between $\mathbf{k}$ and $\mathbf{v}$. The integral over ϕ is trivial and that over $\cos\psi$ can be performed using the δ-function (with $\mathbf{k}\cdot\mathbf{v}=kv\cos\psi$), giving

$$P^l=\frac{q^2\omega_p^2}{v}\int\frac{\mathrm{d}k}{k}\,. \tag{3.57}$$

The upper limit of integration is fixed at $k=\lambda_{De}^{-1}=\omega_p/V_e$ by the assumption that no Langmuir waves exist for $k\lambda_{De}>1$. The lower limit of integration is determined by the condition $\cos\psi=\omega_p/kv<1$. Hence the final result is

$$P^l=\frac{q^2\omega_p^2}{v}\ln\frac{v}{V_e}\,. \tag{3.58}$$

The approximation made to the dispersion relation is unimportant for $\ln(v/V_e)\gg 1$.

Energy Losses by Fast Particles

An important energy loss mechanism for fast particles in an ionized gas is that due to Coulomb interactions, which are called collisions (Section 7.1). The rate of energy loss is closely related to that due to Cerenkov emission of Langmuir waves. Indeed, one can treat the energy losses as though they are due to the emission of longitudinal waves, with these being freely propagating Langmuir waves for $k\lambda_{De}\ll 1$ and strongly damped fluctuations associated with Coulomb interactions for $k\lambda_{De}\gg 1$. In this way an adequate expression for most purposes is obtained by assuming the rate of energy loss to be minus the expression on the right hand of (3.57) with the upper limit of integration extended to $k=p/\hbar$. (The limit $k=p/\hbar$ corresponds to cutting of the integral at the de Broglie wave-length; put another way, the cutoff wavenumber k implies a momentum transfer, via the electromagnetic field, of a momentum comparable with the initial momentum of the electron.) Thus one finds

$$\frac{\mathrm{d}\varepsilon}{\mathrm{d}t}=-\frac{q^2\omega_p^2}{v}\ln\frac{pv}{\hbar\omega_p}\,. \tag{3.59}$$

This formula is adequate for non-relativistic test particles and for ultra-relativistic test particles with Lorentz factors less than about m/m_e where m is the rest mass of the test particle. For Lorentz factors greater than m/m_e the term pv in the logarithmic factor should be replaced by $(\varepsilon m_e c^2)^{1/2}$ (Tsytovich, 1962).

The losses described by (3.59) are sometimes called "ionization losses" by comparison with analogous losses by a fast particle moving through an un-ionized gas and losing energy by ionizing the ambient gas along its path. Energy loss mechanisms for fast particles in astrophysics are discussed in Section 9.3, and have been discussed in detail by Ginzburg and Syrovatskii (1964) and Ginzburg (1967) for example.

3.3 BREMSSTRAHLUNG

"Bremsstrahlung" means braking radiation, but the most common usage of the name "bremsstrahlung" is to describe emission associated with Coulomb interactions between charged particles, or "collisions" as they are called in plasma physics. In astrophysical applications bremsstrahlung is sometimes referred to as free-free emission. The absorptive process which is the inverse of bremsstrahlung is called collisional damping or free-free absorption.

Bremsstrahlung is an important radiation mechanism over a wide range of frequencies in astrophysical applications. In particular, bremsstrahlung is the mechanism responsible for at least some, if not all, of the non-thermal x-ray emissions from the Sun, and it is the most likely mechanism for the emission from other x-ray sources (with some exceptions such as the x-ray emission from the Crab Nebula, which is synchrotron emission). At radio frequencies bremsstrahlung is important as a thermal but not as a non-thermal emission mechanism. Thermal radio emission from hot diffuse plasmas is dominated by bremsstrahlung in the following sense: the power radiated per unit volume in the absence of any absorption would be dominated by bremsstrahlung, and the absorption coefficient is dominated by the contribution from collisional damping.

The treatment of bremsstrahlung given in this section is a classical one. Quantum mechanical effects are unimportant in the emission and absorption of radio waves (radio emission is associated with distant encounters where classical physics is quite adequate). The emission and absorption of optical and x-ray radiation does require a quantum mechanical treatment. Nevertheless a combination of classical results and physical insight enables one to infer most of the quantum mechanical modifications without actually performing any quantum mechanical calculations.

In this section some results relating to the total power radiated in inter-particle encounters are derived first, and then the frequency distribution for single encounters and for emission from a thermal gas are derived from the classical case. Relevant quantum mechanical generalizations are quoted.

Results Derived from the Larmor Formula

The equation of motion for a non-relativistic electron (charge $-e$, mass m_e) passing by an ion (charge $Z_i e$, mass infinite) at the origin is

$$m_e \ddot{\mathbf{r}} = -\frac{Z_i e^2}{r^3}\mathbf{r}. \tag{3.60}$$

When this is inserted in the Larmor formula (3.34) the resulting expression describes the instantaneous power radiated during the encounter:

$$P = \frac{2}{3}\frac{e^2}{c^3}|\ddot{\mathbf{r}}|^2 = \frac{2Z_i^2 e^6}{3m_e^2 c^3 r^4}. \tag{3.61}$$

(Emission due to electron–electron encounters is forbidden in the dipole approximation, so that only electron–ion encounters need be considered for most purposes.) Two useful results may be derived from (3.61), e.g. Dawson (1968).

First suppose that electrons with number density n_e are present and that these electrons are distributed uniformly. The power radiated due to all encounters with a single ion may be obtained by multiplying (3.61) by $4\pi n_e r^2\,\mathrm{d}r$ and integrating over r. The resulting integral diverges due to contributions from arbitrarily small r. Let the integral be cut off at $r = r_{\min}$. The total power per unit volume in bremsstrahlung, $\mathrm{d}P/\mathrm{d}V$ say, may be found by multiplying the resulting value of the integral by the number density of ions (and summing over the different species if there is a mixture of ions present):

$$\frac{\mathrm{d}P}{\mathrm{d}V} = \frac{8\pi}{3}\sum_i \frac{Z_i^2 n_i n_e e^6}{m_e^2 c^3 r_{\min}}. \tag{3.62}$$

The sum in (3.62) is over the ionic species. The choice of the minimum distance $r_{\min}$ in (3.62) requires the introduction of additional physical ideas. From hindsight the following choice is appropriate. The classical treatment is invalid for small r due to quantum mechanical effects, and one characteristic distance associated with quantum mechanical effects is the de Broglie wavelength for thermal electrons. The choice

$$r_{\min} = \frac{\hbar}{m_e V_e}, \tag{3.63}$$

when inserted in (3.62), gives a result which differs from the results of much more detailed calculations by no more than a few per cent.

A second application of (3.61) is to the calculation of the energy radiated in a single encounter:

$$U := \int_{-\infty}^{\infty} dt\, P(t) = 2\int_{r_a}^{\infty} \frac{dr}{\dot{r}} \frac{2Z_i^2 e^6}{3m_e^2 c^3 r^4}, \tag{3.64}$$

where r_a is the distance of closest approach of the electron to the ion. To evaluate the integral in (3.64) it is necessary to find the orbit of the particle (which is an hyperbola). Omitting the details of the calculation, the result is

$$U = \frac{2Z_i^2 e^6}{3m_e^2 c^3 v b^3}\left[3\frac{b_0}{b} + \left(1 + 3\frac{b_0^2}{b^2}\right)\left\{\frac{\pi}{2} + \text{arc tan}\left(\frac{b_0}{b}\right)\right\}\right]. \tag{3.65}$$

In (3.65) b is the impact parameter (the minimum distance between the asymptotic orbit and the ion),

$$b_0 := \frac{Z_i e^2}{mv^2} \tag{3.66}$$

is the impact parameter which corresponds to a deflection through $\pi/2$, and v is the asymptotic speed of the electron. One could derive the power per unit volume from (3.65) as follows: multiply by $2\pi n_i v b\, db$, integrate over b (the result diverges and so the integral needs to be cut off) and sum over ionic species. This gives the average power radiated by an individual electron due to its encounters with the ions: the total power per unit volume is found by multiplying by n_e. With the b-integral cut off at r_{min} the result (3.62) is re-derived to within a factor of order unity.

The Straight-Line Approximation

It turns out that, except for high frequencies, the emission of bremsstrahlung from a plasma is dominated by the many distant encounters between electrons and ions rather than by the few close encounters. A distant encounter may be defined as one with impact parameter $b \gg b_0$, when the angle through which the electron is deflected is approximately $2b_0/b$ radians. For $b \gg b_0$ it is reasonable to use the *straight-line approximation* which corresponds to the assumption that the zeroth order motion is constant rectilinear motion and that the Coulomb force may be treated as a perturbing force.

Let the initial velocity of the electron be $\mathbf{v}$ and let $\mathbf{b}$ be a vector with magnitude equal to the impact parameter and with direction from the ion (which is regarded as immobile) to the nearest point on the unperturbed orbit (see Figure 3.1). The unperturbed orbit is

$$\mathbf{r}^{(0)}(t) = \mathbf{b} + \mathbf{v}t, \tag{3.67}$$

where the zero of time has been chosen for convenience. The perturbing

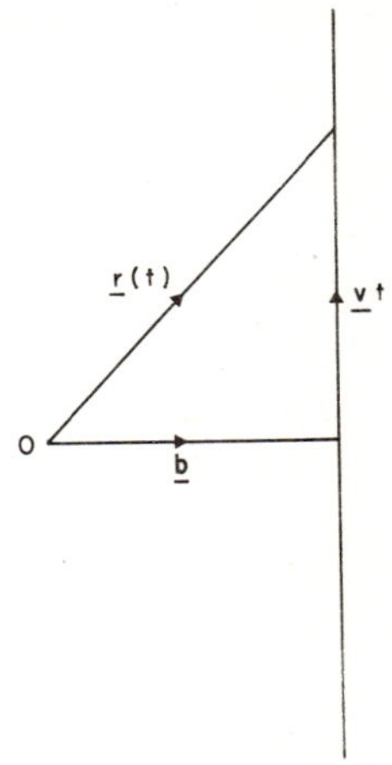

FIGURE 3.1 The orbit of an electron past an ion is drawn in the straight-line approximation. The velocity of the electron is $\mathbf{v}$ and $|\mathbf{b}|$ is its impact parameter (cf. Figure 7.3).

force follows from (3.40) with (3.60). Specifically, one inserts $\mathbf{r}^{(0)}(t)$ in place of $\mathbf{r}$ on the right hand side of (3.60) to find

$$\mathbf{F}^{(1)}(t) = -\frac{Z_i e^2 r^{(0)}(t)}{|r^{(0)}(t)|^3}. \tag{3.68}$$

In the following calculation the motion of the electron is assumed non-relativistic. The generalization to include relativistic effects is outlined at the end of this section.

For the non-relativistic case $\mathbf{F}^{(1)}(t)$ is m_e times the acceleration of the electron. This acceleration is to be inserted in the Larmor formula (3.34). The resulting expression for the power radiated, when integrated over time, leads to the following expression for the energy radiated in an encounter:

$$U = \frac{2e^2}{3c^3}\left(\frac{Z_i e^2}{m_e}\right)^2 \int_{-\infty}^{\infty} \frac{\mathrm{d}t}{(b^2+v^2t^2)^{3/2}} = \frac{\pi}{2}\left(\frac{2Z_i^2 e^6}{3m_e^2 c^3 v b^3}\right). \tag{3.69}$$

This result may be compared with (3.65) for $b \gg b_0$, i.e. with

$$U = \frac{\pi}{2}\left(\frac{2Z_i^2 e^6}{3m_e^2 c^3 v b^3}\right)\left(1 + \frac{8b_0}{\pi b} + \cdots\right). \tag{3.69'}$$

Evidently then, the straight-line approximation gives a reasonable approximation to the energy radiated for $b \gg (8/\pi) b_0$.

The frequency distribution of the radiation emitted in an isotropic plasma can be evaluated using

$$U=:\int_0^\infty d\omega U(\omega),\quad U(\omega)=\frac{2}{3\pi}\frac{e^2 n(\omega)}{c^3}|\mathbf{a}(\omega)|^2, \tag{3.70}$$

which generalizes (3.32) to transverse waves with properties (3.48), and where $q=-e$ has been inserted for electrons. The acceleration $\mathbf{a}(\omega)$ is given by

$$\mathbf{a}(\omega)=-\frac{Z_i e^2}{m_e}\int_{-\infty}^{\infty} dt\, e^{i\omega t}\frac{(\mathbf{b}+\mathbf{v}t)}{(b^2+v^2t^2)^{3/2}}$$
$$=-\frac{2Z_i e^2\omega}{m_e v^2}\left(\frac{\mathbf{b}}{b}K_1\left(\frac{b\omega}{v}\right)-i\frac{\mathbf{v}}{v}K_0\left(\frac{b\omega}{v}\right)\right), \tag{3.71}$$

where K_ν is a Macdonald function (Macdonald functions are modified Bessel functions, see Appendix A). The formula used in evaluating the integral in (3.71) is

$$K_\nu(xz)=\frac{\Gamma(\nu+\frac{1}{2})(2z)^\nu}{2x^\nu\Gamma(\frac{1}{2})}\int_{-\infty}^{\infty} dt\,\frac{e^{\pm ixt}}{(t^2+z^2)^{\nu+1/2}}. \tag{3.72}$$

Thus one finds

$$U(\omega)=\frac{8}{3\pi}\frac{n(\omega)Z_i^2e^6\omega^2}{c^3m_e^2v^4}\left[K_1^2\left(\frac{b\omega}{v}\right)+K_0^2\left(\frac{b\omega}{v}\right)\right]. \tag{3.73}$$

The Macdonald functions have the following approximations (Appendix A): for $x\ll 1$,

$$K_0(x)\approx\ln\left(\frac{2}{x}\right)-0.5772+\cdots,\quad K_1(x)\approx\frac{1}{x}, \tag{3.74a, b}$$

and, for $x\gg 1$,

$$K_0(x)\approx K_1(x)\approx\left(\frac{\pi}{2}\right)^{1/2}x^{-1/2}e^{-x}. \tag{3.74c}$$

Consequently (3.73) may be approximated by

$$U(\omega)\approx\frac{8}{3\pi}\frac{n(\omega)Z_i^2e^6\omega^2}{c^3m_e^2v^4}\begin{cases}\dfrac{v^2}{b^2\omega^2} & \text{for } \dfrac{b\omega}{v}\ll 1,\\[2ex] \pi\dfrac{v}{b\omega}e^{-2b\omega/v} & \text{for } \dfrac{b\omega}{v}\gg 1.\end{cases} \tag{3.75}$$

Thus the emission is roughly independent of frequency for frequencies much less than v/b. The frequency v/b is the inverse of the time $\Delta t=b/v$ characteristic of the interaction. For frequencies much greater than v/b there is negligible emission, as would be expected for frequencies $\omega \gg (\Delta t)^{-1}$ with Δt the characteristic time of an impulsive process.

The Quasi-continuous Emission from a Single Electron

The energy radiated in a single encounter is of little interest in practice because a given electron must be regarded as interacting with many ions simultaneously. The low frequency emission from any electron is more or less continous. The power per unit frequency, $P(\omega)$ say, radiated by an electron can be estimated as follows: multiply $U(\omega)$ by the number of encounters which the electron has with ions of species i and with impact parameters in the range $\mathrm{d}b$ at b, i.e. multiply $U(\omega)$ by $2\pi n_i v b\,\mathrm{d}b$ then integrate over b and sum over all ionic species:

$$P(\omega)=\sum_i 2\pi n_i v \int_0^\infty \mathrm{d}b\, bU(\omega). \tag{3.76}$$

When (3.73) is inserted in (3.76) the integral diverges due to contributions from $b \approx 0$. Let us cut off the integral in (3.76) at $b=b_1$ and leave the estimation b_1 as a separate problem. (In practice $b_1 = \max(b_0, r_{\min})$.) Evaluation of the integral then gives

$$P(\omega)=\sum_i \frac{16}{3}\frac{n(\omega)Z_i^2 n_i e^6}{c^3 m_e^2 v}\frac{b_1\omega}{v}K_0\left(\frac{b_1\omega}{v}\right)K_1\left(\frac{b_1\omega}{v}\right). \tag{3.77}$$

[The integral has been evaluated by using, c.f. Appendix A,

$$\frac{\mathrm{d}}{\mathrm{d}z}\{z^\nu K_\nu(z)\} = -z^\nu K_{\nu-1}(z), \quad K_{-\nu}(z)=K_\nu(z)$$

to find

$$\frac{\mathrm{d}}{\mathrm{d}z}\{zK_0(z)K_1(z)\} = -z\{K_0^2(z)+K_1^2(z)\},$$

and hence

$$\int_{z_1}^\infty \mathrm{d}z\, z\{K_0^2(z)+K_1^2(z)\} = z_1 K_0(z_1)K_1(z_1).]$$

Consider the low and high frequency limits of (3.77). For $\omega \ll v/b_1$ (3.77) reduces to

$$P(\omega) \approx \sum_i \frac{16}{3}\frac{n(\omega)Z_i^2 n_i e^6}{c^3 m_e^2 v}\ln\left(\frac{2}{\Gamma}\frac{v}{b_1\omega}\right), \tag{3.78a}$$

with $\ln \Gamma = 0.5772\ldots$, while for $\omega \gg v/b_1$ (3.77) reduces to

$$P(\omega) \approx \sum_i \frac{8\pi}{3} \frac{n(\omega) Z_i^2 n_i e^6}{c^3 m_e^2 v} \mathrm{e}^{-2b_1\omega/v} \tag{3.78b}$$

The high frequency approximation (3.78b) would not be expected to be very accurate, but the low frequency approximation (3.78a) would be expected to be a valid one. Nevertheless a surprising feature of (3.78a) is the divergence at zero frequency. This divergence is of no practical significance but its existence requires comment.

The divergence ($P(\omega)$ tends to infinity as $\ln(1/\omega)$ for $\omega \to 0$) appears to be incompatible with (3.75) which implies $U(\omega) \to$ constant for $\omega \to 0$ in any given encounter. The resolution of this apparent contradiction is that $\omega \to 0$ need not imply $b\omega/v \to 0$ as supposed in the interpretation of (3.75). Indeed emission from encounters with arbitrarily large b contributes at arbitrarily low frequencies $\omega \approx v/b$. The divergence is not important in practice because the refractive index $n(\omega) = (1 - \omega_p^2/\omega^2)^{1/2}$ implies that there is no emission of transverse waves at $\omega \leqslant \omega_p$. There is an additional reason (not that any is required) for rejecting (3.78a) for $\omega \lesssim \omega_p$. Encounters with $b \gtrsim \lambda_D$ are not relevant because the Coulomb field is shielded out at distances greater than the Debye length $\lambda_D = V_e/\omega_p$. Consequently, thermal electrons ($v \approx V_e$) which do not emit significantly at $\omega \ll V_e/b$ in any given encounter, could not emit significantly at $\omega \ll V_e/\lambda_D = \omega_p$.

Emission from a Thermal Plasma

The average power radiated per unit frequency per thermal electron may be found by averaging (3.77) over a thermal distribution of electrons. Let us denote the average over a Maxwellian distribution of electron by angular brackets $\langle\,\rangle$. One has

$$\left\langle \frac{1}{v} \right\rangle = \left(\frac{2}{\pi}\right)^{1/2} \frac{1}{V_e} \tag{3.79}$$

From (3.78a) it follows that at low frequencies $\langle P(\omega) \rangle$ is the average of $1/v$ times a logarithmic factor. This logarithmic factor is by tradition, written as $\pi/\sqrt{3}$ times the Gaunt factor $G(T_e, \omega)$ (Gaunt, 1930) where T_e is the electron temperature. In the classical case the Gaunt factor may be defined by writing

$$\left\langle \frac{1}{v} \frac{b_1\omega}{v} K_0\left(\frac{b_1\omega}{v}\right) K_1\left(\frac{b_1\omega}{v}\right) \right\rangle =: \left(\frac{2}{\pi}\right)^{1/2} \frac{1}{V_e} \frac{\pi}{\sqrt{3}} G(T_e, \omega). \tag{3.80}$$

However, having defined the Gaunt factor in this way we are free to regard its calculation as a separate problem, and to assume that where relevant quantum mechanical effects are taken into account.

Thus we are led to the following expression for the power radiated per thermal electron:

$$P(\omega)=\sum_i \frac{16}{3}\frac{n(\omega)Z_i^2 n_i e^6}{c^3 m_e^2}\left(\frac{2}{\pi}\right)^{1/2}\frac{1}{V_e}\frac{\pi}{\sqrt{3}}G(T_e,\omega). \tag{3.81}$$

The Gaunt factor has the following approximations. Firstly, for $\omega \ll T_e/\hbar$, it is related to the Coulomb logarithm $\ln \Lambda_c$ by

$$\frac{\pi}{\sqrt{3}}G(T_e,\omega)=\begin{cases}\ln \Lambda_c & \text{for} \quad \omega \leqslant \omega_p, \\ \ln \Lambda_c - \ln \omega/\omega_p & \text{for} \quad \omega > \omega_p.\end{cases} \tag{3.82}$$

Secondly, a quantum mechanical calculation in the Born approximation gives (for detailed discussions see Scheuer (1960) and the review by Osfer (1961) for example)

$$\frac{\pi}{\sqrt{3}}G(T_e,\omega)=K_0\left(\frac{\hbar\omega}{2T_e}\right)\exp\left(-\hbar\omega/2T_e\right)$$

$$\approx\begin{cases}\ln\left(\frac{4}{\Gamma}\frac{T_e}{\hbar\omega}\right) & \text{for} \quad \omega \ll \frac{T_e}{\hbar}, \\ \left(\frac{\pi T_e}{\hbar\omega}\right)^{1/2}\exp\left(-\hbar\omega/T_e\right) & \text{for} \quad \omega \gg \frac{T_e}{\hbar}.\end{cases} \tag{3.83}$$

Comparison of (3.82b) and (3.83a) indicates that the results are compatible provided that Λ_c is equal to $4m_e V_e^2/\Gamma\hbar\omega_p$, and, in view of the fact that only the logarithm of this quantity is of interest, this is a reasonable approximation, see Section 7.1.

Emission of Langmuir Waves

It should be pointed out that waves in any mode in a plasma can be emitted and absorbed as a result of collisions. In particular it is straightforward to proceed as above to derive the power radiated in Langmuir waves. Because Langmuir waves have frequencies close to the plasma frequency it is not appropriate to consider the power radiated per unit frequency range. A more relevant quantity is the power $P(k)$ per unit wavenumber, i.e. such that the total power radiated in Langmuir waves is

$$\int_0^\infty dk\, P(k)$$

Analogous to (3.77) the power radiated per electron is

$$P(k)=\sum_i \frac{8}{3}\frac{Z_i^2 n_i e^6}{m_e^2 v}\frac{k^2}{\omega_p^2}\ln\left(\frac{2}{\Gamma}\frac{v}{b_1\omega}\right) \tag{3.84}$$

where $\omega_p \ll v/b_1$ has been assumed. The result (3.84) is actually analogous to (3.78a). The average power per thermal electron is, analogous to (3.81),

$$P(k)=\sum_i \frac{8}{3}\frac{Z_i^2 n_i e^6}{m_e^2}\frac{k^2}{\omega_p^2}\left(\frac{2}{\pi}\right)^{1/2}\frac{1}{V_e}\frac{\pi}{\sqrt{3}}G(T_e, \omega_p). \tag{3.85}$$

Relativistic Effects

The current associated with the motion of an electron with arbitrary initial speed past an ion is

$$j_i(\mathbf{k}, \omega)= -\frac{2Z_i e^3 \exp(i\mathbf{k}\cdot\mathbf{b})\{(\omega-\mathbf{k}\cdot\mathbf{v})\delta_{ij}+v_i k_j\}}{m_e \gamma v^2(\omega-\mathbf{k}\cdot\mathbf{v})}$$
$$\times\left[i\frac{b_j}{b}K_1\left(\frac{b}{v}(\omega-\mathbf{k}\cdot\mathbf{v})\right)+\frac{1}{\gamma^2}\frac{v_j}{v}K_0\left(\frac{b}{v}(\omega-\mathbf{k}\cdot\mathbf{v})\right)\right] \tag{3.86}$$

This result follows from (3.41), (3.43) and (3.71) together with the fact that the acceleration along **b** is γ^{-1} times the non-relativistic value while that along **v** is γ^{-3} times the non-relativistic value. Qualitatively, an important relativistic effect is beaming of the emitted radiation into the forward direction (all but a fraction of the power goes into a forward cone of half-angle γ^{-1}). Due to this effect one may make the approximation

$$\omega-\mathbf{k}\cdot\mathbf{v}\approx\omega\left(1-\frac{v}{c}\right)\approx\frac{\omega}{2\gamma^2}, \quad \gamma:=(1-v^2/c^2)^{-1/2},$$

for emission *in vacuo*. It then follows from the arguments of the Macdonald functions in (3.86) that emission from a single encounter with impact parameter b extends to an upper frequency $\omega\approx 2\gamma^2 c/b$, which is much higher than the upper frequency $\omega\approx v/b$ in the non-relativistic case.

The current (3.86) was used by Melrose (1972a) to discuss low-frequency bremsstrahlung by relativistic electrons in a plasma. The plasma causes the emission at $\omega\lesssim\gamma\omega_p$ to be suppressed compared with emission *in vacuo*, which effect is analogous to Razin suppression for synchrotron emission (Section 4.3).

Bremsstrahlung by relativistic electrons does not seem to be important in astrophysics. However, an analogous process in which energetic ions emit bremsstrahlung through their interaction with thermal electrons may be important; this process has been called suprathermal proton bremsstrahlung (Bolt and Serlemitsos, 1969; Brown, 1970; Jones, 1971). A quantum method has been used to treat this case, specifically the Weizsäcker-Williams method (e.g. Heitler, 1954; Blumenthal and Gould, 1970). Quantum effects become important only when the emitted photon has an energy comparable with that of the electron (in the rest frame of the system), and in suprathermal proton bremsstrahlung one is interested in the emission of γ-rays.

Bremsstrahlung by electrons with energies up to about 50 keV is important in solar hard X-rays bursts. In this case the energy of the photon is comparable with the energy of the electron (the electron-ion encounters are close range and should not be treated classically). The relevant formula for the emission of photons with energy $\hbar\omega$ comparable with the kinetic energy ε_K of the electrons is the Bethe-Heitler formula: the cross-section per unit energy range of the emitted photon is (e.g. Heitler 1954)

$$\frac{d\sigma(\varepsilon_K, \omega)}{d\hbar\omega} = \frac{8\alpha r_0^2}{3} \frac{m_e c^2}{\hbar\omega\varepsilon_K} \ln\left[\frac{1+(1-\hbar\omega/\varepsilon_K)^{1/2}}{1-(1-\hbar\omega/\varepsilon_K)^{1/2}}\right], \tag{3.87}$$

where $\alpha \approx 1/137$ is the fine structure constant and r_0 is the classical radius of the electron. The cross-section (3.87) has been used extensively in discussions of solar hard X-ray bursts, e.g. Brown (1971, 1975).

3.4 THOMSON SCATTERING

Thomson scattering is the scattering of radiation by free electrons. Thomson scattering, Compton scattering and inverse Compton scattering are all essentially the same phenomenon. The three names tend to be used interchangeably in the astrophysical literature. From a historical viewpoint Thomson scattering is the classical theory of scattering by free electrons, Compton scattering is that of low-energy electrons by hard (i.e. high-energy) photons and inverse Compton scattering is that of soft (i.e. low-energy) photons by high energy electrons. Although it is conventional to use a quantum mechanical formalism and quantum mechanical concepts when describing Compton scattering and inverse Compton scattering, intrinsically quantum effects are rarely important in practice. Thus the classical derivation of Thomson scattering given here can be used to treat Compton scattering and inverse Compton scattering except where very hard photons and/or electrons are involved.

In most applications involving high-energy electrons one can justify neglecting the effects of any ambient plasma. An exception is for scattering involving Langmuir waves. However, for scattering by low-energy particles an ambient medium can affect the scattering itself, as well as affecting the properties of the waves. In fact, scattering of waves by thermal particles in a plasma reduces to Thomson scattering only in the limit of short wavelengths (large k-values). For waves with wavelengths greater than about the Debye length the charge inhomogeneities associated with the shielding clouds around individual particles act as scattering centers. This scattering is discussed in Section 10.4 where it is called "non-linear scattering". Briefly, the situation is this. For wavelengths much greater than the Debye length, i.e. for $k\lambda_D \ll 1$, the shielding cloud around a given particle with charge q leads to a scattering which is roughly equivalent to Thomson scattering by a quasi-particle with charge $|q|$ and with mass equal to that of the electron. The quasi-particle is to be regarded as co-moving with the given particle and scattering in phase with it. Consequently, scattering by an electron and its shielding cloud interfere destructively while for ions Thomson scattering is insignificant and only the scattering by the shielding cloud is significant. In summary: for $k\lambda_D \gg 1$ the scattering in a thermal plasma may be regarded as Thomson scattering by thermal electrons, and for $k\lambda_D \ll 1$ the scattering may be regarded as non-linear scattering by ions. (The condition $k\lambda_D = 1$ for transverse waves reduces to $\omega = \omega_p(c/V_e)$; scattering in a thermal plasma can be regarded as Thomson scattering only for $\omega \gg \omega_p(c/V_e)$.)

Because of the astrophysical interest in inverse Compton scattering it is appropriate to retain all relativistic effects when treating Thomson scattering. This is achieved here as follows. First calculate the perturbed motion of the particle and determine the first order current from (3.43). Then insert this current in (3.18) and identify the power radiated (3.15) as the power generated in scattered waves.

Scattering by a Particle at Rest in Vacuo

Firstly, let us consider the simpler problem of scattering by a particle at rest *in vacuo*. Assuming that the deviations in the orbit due to the effects of the waves may be treated as a perturbation, the first order equation of motion is

$$\ddot{\mathbf{r}}(t) = \frac{q}{m}\,\mathbf{E}'(\mathbf{r}_0, t) \tag{3.88}$$

where $\mathbf{r}_0$ is the initial position of the particle and $\mathbf{E}'(\mathbf{r}, t)$ is the electric field in the unscattered waves.

It should be pointed out that "Thomson scattering" is defined as scattering in the limit where the equation of motion is approximated by (3.88), which is the counterpart for $\mathbf{v} = 0$ of the first order equation of motion (3.41). If second order terms in the expansion in powers of E were retained, these would lead to "double Thomson scattering", see below. Double scattering is most easily interpreted in terms of quantum mechanical language: single scattering corresponds to one initial photon scattered into one final photon, double scattering corresponds either to one initial and two final photons or to two initial and one final photon, and so on for "multiple scattering" of higher orders.

Of course it may be that the perturbation expansion is unacceptable because the deviations in the orbit are substantial. For example if the particle were accelerated by the electric field in the wave from rest to a relativistic energy in less than half a wave period, then clearly the effect of the wave could not be treated as a perturbation. This case has been considered in an astrophysical context. At distances $\gg c/\Omega_R$ from a magnetized neutron star with angular speed Ω_R of rotation the magnetic field becomes a radiation field (corresponding to magnetic dipole radiation at the frequency Ω_R). Particles in such a field can be accelerated to relativistic energies in less than the time $\sim\Omega_R^{-1}$ for the case where the amplitude and frequency of the wave are as is estimated for pulsars. For a wave with frequency Ω_R (3.88) implies that acceleration to relativistic energies occurs for

$$\nu := \frac{\Omega_E}{\Omega_R} \gg 1 \tag{3.89}$$

where $\mathbf{B}'(\mathbf{r}, t)$ is the magnetic field in the wave, and where

$$\Omega_E := \left|\frac{q\mathbf{E}'}{mc}\right| = \left|\frac{q\mathbf{B}'}{mc}\right| \tag{3.89$'$}$$

is a frequency with the following interpretation. Once the particle becomes relativistic (with Lorentz factor γ) it executes a periodic, or quasi-periodic, motion in the fields of the wave; the characteristic frequency of this periodic motion is Ω_E/γ. (The shape of the orbit of a relativistic particle for $\Omega_E \gg \Omega_R$ depends on the polarization of the wave.) For $\Omega_E \gg \Omega_R$ the particle does not scatter the waves, in the sense implied by Thomson scattering, but it radiates at harmonics of the "effective gyrofrequency" Ω_E/γ in a way which is closely analogous to the emission of synchrotron radiation. This radiation mechanism was first suggested by Rees (1971) who explored the possibility that it might be an alternative to synchrotron radiation for the emission from the Crab Nebula. Rees' radiation mechanism is considered further at the end of Section 4.4.

Returning to (3.88), the power radiated in scattered waves is given by the Larmor formula (3.34):

$$P(t) = \frac{2}{3}\frac{q^4}{m^2c^3}|\mathbf{E}'(\mathbf{r}_0, t)|^2. \tag{3.90}$$

The average of (3.90) over time, together with the fact that the average of $|\mathbf{E}'(\mathbf{r}_0, t)|^2$ is 4π times the mean energy density W' in the waves at the point $\mathbf{r}_0$, gives

$$\bar{P} = \frac{8\pi}{3}\left(\frac{q^2}{mc^2}\right)^2 W'c. \tag{3.91}$$

The frequency distribution of the scattered waves is the same as that of the unscattered waves. (This follows from the fact the Fourier transform of the power radiated is proportional to the square of the Fourier transform of $\ddot{\mathbf{r}}(t)$ through (3.34), and so, in view of (3.88), to the frequency spectrum of the unscattered waves.)

For electrons the quantity

$$\sigma_T := \frac{8\pi}{3}\left(\frac{e^2}{m_e c^2}\right)^2 \tag{3.92}$$

is called the *Thomson cross-section.* Because of the square of the mass in (3.91) Thomson scattering by ions is negligible in comparison with that by electrons.

The Current Associated with Scattering by a Moving Particle

Now let us consider the more general case of a particle whose unperturbed motion is with a constant velocity $\mathbf{v}$. The relevant current is (3.42) and the force in (3.41) is to be identified as that due to the fields in the unscattered waves, in the mode σ' say:

$$\mathbf{F}(\mathbf{r}, t; \mathbf{v}) = q\left\{\mathbf{E}^{\sigma'}(\mathbf{r}, t) + \frac{\mathbf{v} \times \mathbf{B}^{\sigma'}(\mathbf{r}, t)}{c}\right\}. \tag{3.93}$$

When the fields are expressed in terms of their Fourier transforms (3.93) becomes

$$F_i(\mathbf{r}, t; \mathbf{v}) = q \int_{-\infty}^{\infty} \frac{d\omega'}{2\pi} \int \frac{d^3\mathbf{k}'}{(2\pi)^3} \frac{1}{\omega'} g_{ij}(\mathbf{k}', \omega'; \mathbf{v}) E_j^{\sigma'}(\mathbf{k}', \omega') \exp\left[i(\mathbf{k}' \cdot \mathbf{r} - \omega' t)\right] \tag{3.94}$$

with

$$g_{ij}(\mathbf{k}, \omega; \mathbf{v}) := (\omega - \mathbf{k} \cdot \mathbf{v})\delta_{ij} + k_i v_j. \tag{3.95}$$

With (3.94) the current (3.42) becomes

$$j_i^{(1)}(\mathbf{k}, \omega) = \frac{2\pi i q^2}{m} \int_{-\infty}^{\infty} \frac{d\omega'}{2\pi} \int \frac{d^3\mathbf{k}'}{(2\pi)^3} \frac{\exp\left[-i(\mathbf{k} - \mathbf{k}') \cdot \mathbf{r}_0\right]}{\omega'} a_{ij}(\mathbf{k}, \omega; \mathbf{k}', \omega'; \mathbf{v})$$
$$\times E_j^{\sigma'}(\mathbf{k}', \omega') \delta\{(\omega - \mathbf{k} \cdot \mathbf{v}) - (\omega' - \mathbf{k}' \cdot \mathbf{v})\}, \tag{3.96}$$

with

$$a_{ij}(\mathbf{k}, \omega; \mathbf{k}', \omega'; \mathbf{v}) := \frac{g_{ri}(\mathbf{k}, \omega; \mathbf{v})\left(\delta_{rs} - \dfrac{v_r v_s}{c^2}\right) g_{sj}(\mathbf{k}', \omega'; \mathbf{v})}{\gamma(\omega - \mathbf{k}\cdot\mathbf{v})^2} \tag{3.97}$$

The δ-function in (3.96) implies $\omega - \mathbf{k}\cdot\mathbf{v} = \omega' - \mathbf{k}'\cdot\mathbf{v}$. In view of this identity, the following symmetry relation obtains

$$a_{ij}(\mathbf{k}, \omega; \mathbf{k}', \omega'; \mathbf{v}) = a_{ji}(\mathbf{k}', \omega'; \mathbf{k}, \omega; \mathbf{v}). \tag{3.98}$$

A form of (3.97) which exhibits the symmetry property (3.98) is

$$a_{ij}(\mathbf{k}, \omega; \mathbf{k}', \omega'; \mathbf{v}) = \frac{1}{\gamma}\left[\delta_{ij} + \frac{k_i' v_j}{\omega' - \mathbf{k}'\cdot\mathbf{v}} + \frac{k_j v_i}{\omega - \mathbf{k}\cdot\mathbf{v}} + \frac{(\mathbf{k}\cdot\mathbf{k}' - (\omega\omega'/c^2))v_i v_j}{(\omega - \mathbf{k}\cdot\mathbf{v})(\omega' - \mathbf{k}'\cdot\mathbf{v})}\right]. \tag{3.99}$$

Power Generated in Scattered Waves

When (3.96) is inserted in (3.18) the following lengthy expression results for the power generated (3.15) in scattered waves in the mode σ:

$$\begin{aligned} P^\sigma(\mathbf{k}) = \lim_{\tau\to\infty} \frac{4\pi}{\tau} R_E^\sigma(\mathbf{k}) e_i^{\sigma*}(\mathbf{k}) e_j^\sigma(\mathbf{k}) \int_0^\infty d\omega\, \delta(\omega - \omega^\sigma(\mathbf{k})) \\ \times \left(\frac{2\pi q^2}{m}\right)^2 \int_{-\infty}^{\infty} \frac{d\omega'}{2\pi} \int \frac{d^3\mathbf{k}'}{(2\pi)^3} \int_{-\infty}^{\infty} d\omega'' \int \frac{d^3\mathbf{k}''}{(2\pi)^3} \frac{\exp\left[i(\mathbf{k}' - \mathbf{k}'')\cdot\mathbf{r}_0\right]}{\omega'\omega''} \\ \times a_{ir}(\mathbf{k}, \omega; \mathbf{k}', \omega'; \mathbf{v}) E_r^{\sigma'}(\mathbf{k}', \omega') a_{js}(\mathbf{k}, \omega; \mathbf{k}'', \omega''; \mathbf{v}) E^{\sigma'*}(\mathbf{k}'', \omega'') \\ \times \delta\{(\omega - \mathbf{k}\cdot\mathbf{v}) - (\omega' - \mathbf{k}'\cdot\mathbf{v})\}\delta\{(\omega - \mathbf{k}\cdot\mathbf{v}) - (\omega'' - \mathbf{k}''\cdot\mathbf{v})\}, \end{aligned} \tag{3.100}$$

where the factor

$$\int_0^\infty d\omega\, \delta(\omega - \omega^\sigma(\mathbf{k}))$$

implies that the remaining terms are to be evaluated for $\omega = \omega^\sigma(\mathbf{k}) > 0$, as in (3.18). The first step in the evaluation of (3.100) is to notice that the only contribution to the integral can be for $\mathbf{k}'' = \mathbf{k}'$. (The δ-functions imply $\omega'' = \omega'$ for $\mathbf{k}'' = \mathbf{k}'$.) If this were not so $P^\sigma(\mathbf{k})$ would depend on $\mathbf{r}_0$ and this would be unacceptable for the following reason. In evaluating the power radiated we have averaged over time at the outset. (Specifically the power radiated is the limit $\tau\to\infty$ of $U^\sigma(\mathbf{k})/\tau$ with $U^\sigma(\mathbf{k})$ given by (3.18).) Consequently, it would be inconsistent to derive a result which depends on $\mathbf{r}_0$ because $\mathbf{r}_0$ is the position of the particle at the arbitrarily chosen zero of time.

Now because the only contribution to the integral in (3.100) is for $\mathbf{k}''=\mathbf{k}'$, $\omega''=\omega'$ there is no loss in generality in supposing that the waves have random phases and in replacing the product of electric fields by the corresponding product averaged over the phases. Let this average be denoted by a bar. The average gives

$$\overline{E_r^{\sigma'}(\mathbf{k}',\omega')E_s^{\sigma'*}(\mathbf{k}'',\omega'')} = \lim_{V,\tau\to\infty} \frac{(2\pi)^4}{V\tau}\,\delta^3(\mathbf{k}'-\mathbf{k}'')\delta(\omega'-\omega'')$$
$$\times E_r^{\sigma'}(\mathbf{k}',\omega')E_s^{\sigma'*}(\mathbf{k}',\omega'). \qquad (3.101)$$

The electric fields are to be rewritten in terms of the amplitude using (3.3). The resulting square of the δ-function is rewritten using (2.12) and the modulus squared of the amplitude is rewritten using (3.5) and (3.6). There results

$$\overline{E_r^{\sigma'}(\mathbf{k}',\omega')E_s^{\sigma'*}(\mathbf{k}',\omega')} = \lim_{V,\tau\to\infty} V\tau e_r^{\sigma'}(\mathbf{k}')e_s^{\sigma'*}(\mathbf{k}')$$
$$\times R_E^{\sigma'}(\mathbf{k}')W^{\sigma'}(\mathbf{k}')2\pi\delta(\omega'-\omega^{\sigma'}(\mathbf{k}')). \qquad (3.102)$$

Finally, when (3.101) is inserted in (3.100) it is straightforward to reduce (3.100) to

$$P^{\sigma}(\mathbf{k}) = \frac{4(2\pi)^3 q^4}{m^2} R_E^{\sigma}(\mathbf{k}) \int \frac{d^3\mathbf{k}'}{(2\pi)^3} \frac{|a^{\sigma\sigma'}(\mathbf{k},\mathbf{k}';\mathbf{v})|^2}{\{\omega^{\sigma'}(\mathbf{k}')\}^2}$$
$$\times R_E^{\sigma'}(\mathbf{k}')W^{\sigma'}(\mathbf{k}')\delta\{(\omega^{\sigma}(\mathbf{k})-\mathbf{k}\cdot\mathbf{v})-(\omega^{\sigma'}(\mathbf{k}')-\mathbf{k}'\cdot\mathbf{v})\} \qquad (3.103)$$

with

$$a^{\sigma\sigma'}(\mathbf{k},\mathbf{k}',\mathbf{v}) := e_i^{\sigma*}(\mathbf{k})e_j^{\sigma'}(\mathbf{k}')a_{ij}(\mathbf{k},\omega^{\sigma}(\mathbf{k});\mathbf{k}',\omega^{\sigma'}(\mathbf{k}');\mathbf{v}). \qquad (3.104)$$

Several comments on (3.103) are appropriate. First, as is implied by (3.91), the power generated in scattered waves is proportional to the energy density rather than to the energy itself in the unscattered waves. Second, the condition $\omega-\mathbf{k}\cdot\mathbf{v}=\omega'-\mathbf{k}'\cdot\mathbf{v}$ implied by the δ-function is just the result obtained from the equality (initial frequency)=(final frequency) in the rest frame of the particle by making a Lorentz transformation to the frame in which the particle has velocity $\mathbf{v}$. Third, in the derivation of (3.103) the possibility that ω' might be negative has been ignored. It is conceivable that the equality

$$\omega^{\sigma}(\mathbf{k})-\mathbf{k}\cdot\mathbf{v}+\omega^{\sigma'}(\mathbf{k}')-\mathbf{k}'\cdot\mathbf{v}=0 \qquad (3.105)$$

might be satisfied. It is possible to construct examples in which (3.105) can be satisfied. (However, for the cases which are of possible practical interest the relevant scattering tends to be non-linear scattering rather than Thomson scattering.) The condition (3.105) does not correspond to a scattering process;

it corresponds to double emission or absorption. "Double emission", in quantum mechanical terminology, corresponds to the simultaneous emission of two photons. It is simple to show that double emission (and higher order multiple emissions) is forbidden when single photon ("Cerenkov") emission is forbidden; to be specific, double emission is forbidden when both waves have refractive index less than or equal to unity. In quantum mechanical language, double emission and Compton scattering are related by a "crossing symmetry". One says that double emission is a "crossed" process in relation to Compton scattering.

It is convenient to introduce a semi-classical formalism (see Section 5.1). In semi-classical language, a scattering process may be described in terms of the probability $w^{\sigma\sigma'}(\mathbf{p}, \mathbf{k}, \mathbf{k}')$ per unit time that a photon in the mode σ' in the range $d^3\mathbf{k}'/(2\pi)^3$ be scattered into a photon in the mode σ in the range $d^3\mathbf{k}/(2\pi)^3$. This probability may be derived by noting that the rate scattered photons are generated in the range $d^3\mathbf{k}/(2\pi)^3$ is $P^\sigma(\mathbf{k})/\hbar\omega^\sigma(\mathbf{k})$ and that the number density of unscattered photons in the range $d^3\mathbf{k}'/(2\pi)^3$ is $W^{\sigma'}(\mathbf{k}')/\hbar\omega^{\sigma'}(\mathbf{k}')$, where $\hbar$ is Planck's constant divided by 2π. Thus the probability per unit time for Thomson scattering by a particle with momentum $\mathbf{p}$ is

$$w^{\sigma\sigma'}(\mathbf{p}, \mathbf{k}, \mathbf{k}') = \frac{4(2\pi)^3 q^4}{m^2} \frac{R_E^\sigma(\mathbf{k}) R_E^{\sigma'}(\mathbf{k}') |a^{\sigma\sigma'}(\mathbf{k}, \mathbf{k}', \mathbf{v})|^2}{\omega^\sigma(\mathbf{k})\omega^{\sigma'}(\mathbf{k}')} \times \delta\{(\omega^\sigma(\mathbf{k}) - \mathbf{k}\cdot\mathbf{v}) - (\omega^{\sigma'}(\mathbf{k}') - \mathbf{k}'\cdot\mathbf{v})\}. \tag{3.106}$$

Thomson Scattering by a Spiralling Charge

The generalization of Thomson scattering to allow for the spiralling motion of the scattering charge in an ambient magnetic field involves an expansion in Bessel functions (Section 4.1). A detailed analysis leads to the following results (Melrose and Sy, 1972). The current (3.96) is replaced by

$$j_i^{(1)}(\mathbf{k}, \omega) = \sum_{s=-\infty}^{\infty} \frac{2\pi i q^2}{m} \int_{-\infty}^{\infty} \frac{d\omega'}{2\pi} \int \frac{d^3\mathbf{k}'}{(2\pi)^3} \exp[-i(\mathbf{k}-\mathbf{k}')\cdot\mathbf{r}_0] \times \frac{1}{\omega'} a_{ij}(s; \mathbf{k}, \omega; \mathbf{k}', \omega') E_j^{\sigma'}(\mathbf{k}', \omega') \times \delta\{\omega - \omega' - s\Omega - (k_\parallel - k'_\parallel)v_\parallel\}, \tag{3.107}$$

with the dependence on $v_\perp$ and $v_\parallel$ (see Section 4.1) implied, the quantity which generalizes (3.97) is

$$a_{ij}(s; \mathbf{k}, \omega; \mathbf{k}', \omega') := \sum_{t=-\infty}^{\infty} a_{ij}(s-t, \mathbf{k}, \omega; t, \mathbf{k}', \omega'), \tag{3.108}$$

with

$$a_{ij}(s, \mathbf{k}, \omega; s', \mathbf{k}', \omega') := \frac{\exp[-i\epsilon(s\psi - s'\psi')]}{\gamma\omega_s\omega'_{s'}}$$
$$\times\{\omega_s\omega'_{s'}J_s(z)J_{s'}(z')\tau_{ij}(\omega_s) + \omega_s J_s(z)\tau_{im}(\omega_s)k_m V_j^*(s', \mathbf{k}')$$
$$+\omega'_{s'}J_{s'}(z')V_i(s, \mathbf{k})k'_m\tau_{mj}(\omega_s)$$
$$+\left(k_l k'_m \tau_{lm}(\omega_s) - \frac{\omega\omega'}{c^2}\right)V_i(s, \mathbf{k})V_j^*(s', \mathbf{k}')\Big\}. \quad (3.109)$$

The quantity $\mathbf{V}(s, \mathbf{k}) := V(s, \mathbf{p}, \mathbf{k})$ is given by (4.13) below. Also in (3.109) one has

$$\tau_{ij}(\omega) := \frac{1}{\omega^2 - \Omega^2}\{\omega^2\delta_{ij} - \Omega^2 b_i b_j + i\epsilon\omega\Omega\epsilon_{ijk}b_k\}, \quad (3.110)$$

with $\mathbf{b} := \mathbf{B}/B$, $\epsilon := q/|q|$. (τ_{ij} is the solution of (2.58) and has the matrix representation (2.61).) Finally one has

$$\omega_s := \omega - s\Omega - k_\parallel v_\parallel, \quad \omega'_{s'} := \omega' - s'\Omega - k'_\parallel v_\parallel,$$
$$z := \frac{k_\perp v_\perp}{\Omega}, \qquad z' := \frac{k'_\perp v_\perp}{\Omega}. \quad (3.111)$$

The δ-function in (3.107), together with (3.108), implies $\omega_s = \omega'_{s'}$ in (3.109), and this has been used to exhibit the symmetry relation, c.f. (3.98),

$$a_{ij}(s, \mathbf{k}, \omega; s', \mathbf{k}', \omega') = a_{ji}^*(s', \mathbf{k}', \omega'; s, \mathbf{k}, \omega). \quad (3.112)$$

The counterpart of the probability (3.106) is

$$w^{\sigma\sigma'}(\mathbf{p}, \mathbf{k}, \mathbf{k}') = \sum_{s=-\infty}^{\infty} \frac{4(2\pi)^3 q^4}{m^2} \frac{|a^{\sigma\sigma'}(s, \mathbf{k}, \mathbf{k}')|^2}{\omega^\sigma(\mathbf{k})\omega^{\sigma'}(\mathbf{k}')}$$
$$\times R_E^\sigma(\mathbf{k})R_E^{\sigma'}(\mathbf{k}')\delta\{\omega^\sigma(\mathbf{k}) - \omega^{\sigma'}(\mathbf{k}') - s\Omega - (k_\parallel - k'_\parallel)v_\parallel\}, \quad (3.113)$$

with

$$a^{\sigma\sigma'}(s, \mathbf{k}, \mathbf{k}') := e_i^{\sigma*}(\mathbf{k})e_j^{\sigma'}(\mathbf{k}')a_{ij}(s; \mathbf{k}, \omega^\sigma(\mathbf{k}); \mathbf{k}', \omega^{\sigma'}(\mathbf{k}')). \quad (3.114)$$

Double Thomson Scattering

There are several extensions of Thomson scattering which have been mentioned already, specifically "nonlinear" scattering (Section 10.4), scattering by an intense wave (Section 4.4), scattering by a spiralling charge and double Thomson scattering. Let us discuss double Thomson scattering briefly.

In quantum mechanical language double Compton scattering is the scattering of one photon into two photons. It is sometimes stated that this process is intrinsically quantum mechanical, i.e. that it has no classical counterpart. This is only partially correct. Double Thomson scattering exists and is a classical counterpart to double Compton scattering, but it is only the crossed process of scattering of two photons into one (rather than of one into two) which has a classical counterpart. Double Thomson scattering may be treated by identifying the relevant current with the second order term in the perturbing force, cf. (3.42). This was done by Melrose (1972b).

PROBLEM SET 3

Answers to relevant questions are given at the end of the text. More difficult questions are marked with an asterisk.

3.1 Find an expression for the power radiated due to Cerenkov emission into ion sound waves by a particle in an isotropic plasma.

3.2 Estimate the minimum energy for an electron to emit Cerenkov radiation in air (refractive index $n=1+3\times 10^{-4}$).

3.3 Find the power radiated by an uncharged particle with (a) an electric dipole moment, or (b) a magnetic dipole moment in constant rectilinear motion at velocity $\mathbf{v}$ through an isotropic dielectric.

3.4 Find the power radiated by a particle in constant rectilinear motion along the principal axis of a uniaxial crystal, cf. problem 2.6.

3.5 Use the Larmor Formula to derive the expression (3.65) for the energy radiated in a single encounter between an electron and an ion, viz.

$$U=\frac{2Z_i^2e^6}{3m_e^2c^3vb^3}\left[3\frac{b_0}{b}+\left(1+\frac{3b_0^2}{b^2}\right)\left\{\frac{\pi}{2}+\arctan\left(\frac{b_0}{b}\right)\right\}\right] \tag{P3-1}$$

Hint: The orbit ($r=r(t)$ and $\theta=\theta(t)$) of the electron may be found by appealing to conservation of energy and angular momentum.

3.6 (a) Derive the current (3.86) for the motion of an electron with arbitrary speed past an ion in the straight-line approximation.
(b) Assume $K_1(x)=1/x\gg K_0(x)$ for $x<1$ and $K_1(x)=0=K_0(x)$ for $x>1$, and show that the power radiated into transverse waves, $n(\omega)=\{1-\omega_p^2/\omega^2\}^{1/2}$, is given by

$$P(\omega)=\frac{16}{3}\sum_i\frac{n_iZ_i^2e^6}{m_e^2c^3v}\frac{n(\omega)(1-v^2/c^2)}{1-n^2(\omega)v^2/c^2}\ln\left(\frac{b_2}{b_1}\right), \tag{P3-2}$$

with $b_2\approx\gamma^2v/\omega$ and $b_1\approx\hbar/m_ev$.

3.7 Calculate the power radiated per unit frequency in bremsstrahlung into the two magnetoionic modes for $\omega\gg\omega_p\gg\Omega_e$ due to a thermal distribution of electrons.

Remark: The sum of the power in the two modes must be equal to $P(\omega)$ as given by (3.81) to lowest order in Ω_e/ω_p and Ω_e/ω.

3.8* A particle with charge q and mass m is subject to a force

$$\mathbf{F}(t)=\mathbf{F}_0 \cos \omega_0 t \tag{P3-3}$$

directed along its velocity.

(a) Show that the current associated with the first order (in $\mathbf{F}_0$) perturbed motion of the particle is

$$j^{(1)}(\mathbf{k}, \omega)=\frac{i\pi q\omega \mathbf{F}_0}{m\gamma^3\omega_0^2}\{\delta(\omega+\omega_0-\mathbf{k}\cdot\mathbf{v})+\delta(\omega-\omega_0-\mathbf{k}\cdot\mathbf{v})\}. \tag{P3-4}$$

(b) Show that the power radiated *in vacuo* is

$$P=\frac{q^2|\mathbf{F}_0|^2}{3m^2c^3} \tag{P3-5}$$

(c) Show that the power radiated into Langmuir waves for $\omega_p \gg \omega_0$ is

$$P=\frac{q^2|\mathbf{F}_0|^2}{m^2v^3\gamma^6}\left(\frac{\omega_p}{\omega_0}\right)^4 \ln\left(\frac{v}{V_e}\right) \tag{P3-6}$$

3.9 (a) Show directly that the current associated with Thomson scattering by a particle at rest at the origin ($\mathbf{r}_0=0$) in a magnetized plasma is given by

$$j_i(\mathbf{k}, \omega)=\frac{iq^2}{m\omega^2}\tau_{ij}(\omega)E_j(\mathbf{k}, \omega). \tag{P3-7}$$

That is, derive (P3-7) from the equation of motion

$$m\frac{d\mathbf{v}}{dt}=q\mathbf{E}+q\frac{\mathbf{v}\times\mathbf{B}_0}{c}.$$

(b) Hence derive the probability for scattering by such a particle and show that it is equal to the relevant limit of (3.113).

3.10* A charge q moves in a Keplerian orbit with eccentricity e, semi-major axis R and period $2\pi/\omega_0$. Assume that the orbit is in the x-y plane, that the origin is at the center of force and that at $t=0$ the particles is on the x-axis. The orbit may then be described by, e.g., Watson (1944, p. 554),

$$\begin{aligned}\frac{x}{R}&=-\tfrac{3}{2}e+\sum_{n=1}^{\infty}\frac{2}{n}J_n'(ne)\cos n\omega_0 t,\\ \frac{y}{R}&=\frac{(1-e^2)^{1/2}}{e}\sum_{n=1}^{\infty}\frac{2}{n}J_n(ne)\sin n\omega_0 t,\\ z&=0.\end{aligned} \tag{P3-8}$$

(a) Using the Larmor formula (3.34) and (P3-8) show that the time-averaged power radiated is

$$P=\frac{q^2}{3c^3}R^2\omega_0^4\frac{(2+e^2)}{(1-e^2)^{5/2}}, \tag{P3-9}$$

where the Kapteyn series which appear are summed in (4.59a, b).

(b) Using the Larmor formula and the equation of motion, rederive (P3-9) directly by evaluating $\int_0^{2\pi/\omega_0} dt\, r^{-4}$ explicitly.

Hint: With $r^2 := x^2 + y^2$ and $\theta := \arctan(y/x)$, the total energy ε($\varepsilon < 0$ necessarily) and the constant k in the potential energy $U(r) = -k/r$ are related to the other parameters by

$$e = \left(1 + \frac{2\varepsilon l^2}{mk^2}\right)^{1/2}, \quad l = mr^2\dot{\theta},$$

$$\varepsilon = \tfrac{1}{2}m(\dot{r}^2 + r^2\dot{\theta}^2) - \frac{k}{r}, \quad R = -\frac{k}{2\varepsilon}, \quad \omega_0 = (k/mR^3)^{1/2}, \tag{P3-10}$$

and the semi-minor axis is equal to $R(1-e^2)^{1/2}$.

(c) Making the dipole approximation, calculate the extraneous current associated with the motion of the charge.

(d) Using the current found in (c), find the power radiated in the transverse waves in an isotropic plasma with refractive index $n(\omega) = (1 - \omega_p^2/\omega^2)^{1/2}$.

(e) Assuming the orbit to be nearly circular ($e \ll 1$), estimate the ratio of the power radiated at the second harmonic ($\omega = 2\omega_0$) to that at the fundamental ($\omega = \omega_0$) *in vacuo*.

CHAPTER 4

Gyromagnetic, Synchrotron and Inverse Compton Emissions

Synchrotron emission, which is gyromagnetic emission from relativistic electrons, is the accepted emission mechanism for supernova remnants, radio galaxies and quasars. Some radio emission from the Sun (and from flare stars), and from Jupiter is also due to synchrotron emission. The theory of synchrotron emission is developed in detail in this chapter. Inverse Compton emission, which is also of relevance to synchrotron sources, is discussed in Section 4.4, along with Rees' (1971) alternative mechanism ("nonlinear inverse Compton emission") for synchrotron sources.

4.1 GYROMAGNETIC EMISSION

Gyromagnetic emission is emission by particles due to their spiralling motion in a magnetic field. Gyromagnetic emission from non-relativistic particles is called *cyclotron* emission while that from ultrarelativistic particles is called *synchrotron* emission. For mildly relativistic particles the names cyclotron, synchrotron and even gyro-synchrotron are used. Gyromagnetic emission and the corresponding gyromagnetic absorption (Section 6.3) result from a resonant interaction between a wave and a spiralling charge. This resonance is called a gyromagnetic or cyclotron resonance. The gyro-resonance condition is the Doppler condition

$$\omega - s\Omega - k_{\parallel} v_{\parallel} = 0, \tag{4.1}$$

with $s=0, \pm 1, \pm 2, \ldots$, and where the particle is described by its gyrofrequency $\Omega := |q|B/mc\gamma$ and its component $v_{\parallel}$ of velocity along the magnetic field lines. Resonances at $s>0$ are said to be via the *normal Doppler effect*, those at $s<0$ via the *anomalous Doppler effect* and that at $s=0$ by the *Cerenkov effect*. For $s=0$ (4.1) is not equivalent to the condition $\omega=\mathbf{k}\cdot\mathbf{v}$ of Section 3.2 because $k_{\perp}$ and $v_{\perp}$ remain arbitrary in (4.1), and this leads to confusion over the meaning of "Cerenkov" effect.

The gyro-resonance interaction between waves and particles is of major importance in several quite different contexts. One important application is to synchrotron radiation, which is discussed in detail in Section 4.2 and Section 4.3. For waves with frequencies close to the gyrofrequency of a species of thermal particle the coupling between the waves and the particles is very strong and permits an efficient transfer of energy to or from the waves. This cyclotron resonance is the basis of one of the methods of heating laboratory plasmas. A third application of the gyroresonance is to the scattering of fast particles by low frequency waves. This so-called *resonant scattering* is discussed in Section 7.2. It is known to be important for the scattering of cosmic rays in the interplanetary medium and for energetic particles in the Earth's radiation belts, and it is probably important in a number of other contexts.

The discussion in this section involves four related aspects of gyroemission. Firstly, a general formula for gyroemission is derived in a form which is an obvious generalization of the formula (3.47) for Cerenkov emission. Secondly, the emissivity is defined and an explicit expression for it is given for waves in a non-spatially-dispersive medium. Thirdly, the polarization of gyromagnetic emission *in vacuo* and in an isotropic plasma is discussed and the emissivity is written as a polarization tensor. Finally, an explicit expression is derived for the power radiated *in vacuo* by a spiralling charge.

Spiralling Motion

The equation of motion for a charged particle spiralling in a magnetic field is

$$\frac{\mathrm{d}\mathbf{p}}{\mathrm{d}t}=q\frac{\mathbf{v}\times\mathbf{B}}{c}. \tag{4.2}$$

There are two independent constants of the motion. These may be chosen as the energy ε and the components $p_{\parallel}$ of $\mathbf{p}$ along $\mathbf{B}$,

$$\frac{\mathrm{d}\varepsilon}{\mathrm{d}t}=\mathbf{v}\cdot\frac{\mathrm{d}\mathbf{p}}{\mathrm{d}t}=0,\quad \frac{\mathrm{d}p_{\parallel}}{\mathrm{d}t}:=\frac{\mathrm{d}}{\mathrm{d}t}\left(\frac{\mathbf{p}\cdot\mathbf{B}}{B}\right)=\frac{\mathbf{B}}{B}\cdot\frac{\mathrm{d}\mathbf{p}}{\mathrm{d}t}=0.$$

Other related constants of the motion are

$$\gamma := \frac{\varepsilon}{mc^2}, \quad \Omega := \frac{|q|B}{\gamma mc}, \quad p_\perp := (p^2 - p_\parallel^2)^{1/2},$$

$$v_\parallel := \frac{p_\parallel}{\gamma m}, \quad v_\perp := \frac{p_\perp}{\gamma m},$$

$$\beta := \frac{v}{c}, \quad \alpha := \arc \tan \left(\frac{p_\perp}{p_\parallel}\right); \tag{4.3}$$

α is the *pitch angle*.

Let **B** lie along the 3-axis and write

$$\mathbf{p} =: (p_\perp \cos \phi, p_\perp \sin \phi, p_\parallel). \tag{4.4}$$

Then (4.2) reduces to

$$\frac{d\phi}{dt} = -\epsilon\Omega, \tag{4.5}$$

where $\epsilon := q/|q|$ is the sign of the charge. Positive (negative) charges spiral in a left hand (right hand) sense, i.e. the particle gyrates in a LH (RH) screw sense relative to **B**.

It is straightforward to solve for the orbit of the charge:

$$\mathbf{r}(t) = \mathbf{r}_0 + (R \sin (\phi_0 + \Omega t), \epsilon R \cos (\phi_0 + \Omega t), v_\parallel t), \tag{4.6}$$

where

$$R := \frac{v_\perp}{\Omega} = \frac{p_\perp c}{|q|B} \tag{4.7}$$

is the radius of gyration or the *gyroradius*, and where $\mathbf{r}_0$ and ϕ_0 are constants. ($\mathbf{r}_0$ is related to the position of the center of gyration at the time $t=0$.) The instantaneous velocity is

$$\mathbf{v}(t) := \dot{\mathbf{r}}(t) = (v_\perp \cos (\phi_0 + \Omega t), -\epsilon v_\perp \sin (\phi_0 + \Omega t), v_\parallel). \tag{4.8}$$

The Current due to a Spiralling Charge

The current (3.29) can be evaluated by first expanding in Bessel functions. Let us write

$$\begin{aligned} \mathbf{k} &=: (k_\perp \cos \psi, k_\perp \sin \psi, k_\parallel) \\ &=: k(\sin \theta \cos \psi, \sin \theta \sin \psi, \cos \theta). \end{aligned} \tag{4.9}$$

The exponential factor in (3.29), with (4.6) and (4.9), becomes

$$\exp \left[-i\{\mathbf{k} \cdot \mathbf{r}(t) - \omega t\}\right] =$$

$$\exp \left[-i\mathbf{k} \cdot \mathbf{r}_0 - i(\omega - k_\parallel v_\parallel)t - ik_\perp R \sin (\phi_0 + \epsilon\psi + \Omega t)\right]. \tag{4.10}$$

This can be expanded using the generating function for Bessel functions (set $t=\exp[i(\phi_0+\epsilon\psi+\Omega t)]$ and $z=k_\perp R$ in equation (A.17) of Appendix A):

$$\exp[-ik_\perp R\sin(\phi_0+\epsilon\psi+\Omega t)]=\sum_{s=-\infty}^{\infty}\exp[is(\phi_0+\ \psi+\Omega t)]\,J_s(k_\perp R). \tag{4.11}$$

Using (4.8) and

$$e^{\pm i\phi}\sum_{s=-\infty}^{\infty}e^{-is\phi}J_s(z)=\sum_{s=-\infty}^{\infty}e^{-is\phi}J_{s\pm1}(z),$$

one finds

$$\dot{\mathbf{r}}(t)\exp[-ik_\perp R\sin(\phi_0+\epsilon\psi+\Omega t)]=\sum_{s=-\infty}^{\infty}\exp[-is(\phi_0+\epsilon\psi+\Omega t)]\,\mathbf{V}(s,\mathbf{p},\mathbf{k}), \tag{4.12}$$

with

$$\mathbf{V}(s,\mathbf{p},\mathbf{k}):=\Big(\tfrac{1}{2}v_\perp\{e^{i\ \psi}J_{s-1}(z)+e^{-i\epsilon\psi}J_{s+1}(z)\},\\ -\frac{i\epsilon v_\perp}{2}\{e^{i\ \psi}J_{s-1}(z)-e^{-i\epsilon\psi}J_{s+1}(z)\},\ v_\parallel J_s(z)\Big) \tag{4.13}$$

and

$$z:=k_\perp R=\frac{k_\perp v_\perp}{\Omega}=\frac{k_\perp p_\perp c}{|q|B}. \tag{4.14}$$

It is then trivial to evaluate the current (3.29):

$$\mathbf{j}(\mathbf{k},\omega)=\\ 2\pi q\exp(-i\mathbf{k}\cdot\mathbf{r}_0)\sum_{s=-\infty}^{\infty}\exp[-is(\phi_0+\epsilon\psi)]\,\mathbf{V}(s,\mathbf{p},\mathbf{k})\,\delta(\omega-s\Omega-k_\parallel v_\parallel). \tag{4.15}$$

When treating emission by a single particle one is free to choose $\mathbf{r}_0=0$, $\phi_0=0$ and $\psi=0$. In particular, for $\psi=0$, which corresponds to the choice (2.51) of axes, (4.13) reduces to

$$\mathbf{V}(s,\mathbf{p},\mathbf{k})=\left(v_\perp\frac{s}{z}J_s(z),\ -i\epsilon v_\perp J_s'(z),\ v_\parallel J_s(z)\right), \tag{4.16}$$

where use has been made of the recursion relations (equations (A.17) of Appendix A) for the Bessel functions. In this case the outer product of **V** with its complex conjugate, i.e.

$$V_i(s,\mathbf{p},\mathbf{k})V_j^*(s,\mathbf{p},\mathbf{k})$$

is an hermitian tensor which satisfies the Onsager relations in the form (2.52). An interesting property of (4.13) or (4.16) is that the sum over s of this outer product is equal to the time average of the outer product $v_i(t)v_j(t)$ with $v(t)=\dot{r}(t)$:

$$\sum_{s=-\infty}^{\infty} V_i(s, \mathbf{p}, \mathbf{k})V_j^*(s, \mathbf{p}, \mathbf{k})=\overline{v_i(t)v_j(t)}, \tag{4.17}$$

where the bar denotes a time-average. The matrix representation of the right hand side of (4.17) is

$$\overline{v_i(t)v_j(t)}=\begin{pmatrix} \frac{1}{2}v_\perp^2 & 0 & 0 \\ 0 & \frac{1}{2}v_\perp^2 & 0 \\ 0 & 0 & v_\parallel^2 \end{pmatrix}. \tag{4.17'}$$

The result (4.17) is somewhat analogous to Parseval's formula for Fourier series.

The Power Radiated and the Emissivity

A general formula for the power radiated due to gyromagnetic emission may be obtained by substituting (4.15) in (3.18) and using (3.15):

$$P^\sigma(\mathbf{k})=\sum_{s=-\infty}^{\infty} 8\pi^2q^2R_E^\sigma(\mathbf{k})|\mathbf{e}^{\sigma*}(\mathbf{k})\cdot\mathbf{V}(s, \mathbf{p}, \mathbf{k})|^2\,\delta(\omega^\sigma(\mathbf{k})-s\Omega-k_\parallel v_\parallel). \tag{4.18}$$

In deriving this result a double sum of the form

$$\sum_{s=-\infty}^{\infty}\sum_{s'=-\infty}^{\infty} e^{i(s-s')\phi} f(s)g^*(s')\delta(\omega-s\omega_0)\delta(\omega-s'\omega_0)$$

must be evaluated in (3.18). The δ-functions give zero except for $s'=s$ when (2.12) implies

$$\sum_{s=-\infty}^{\infty} f(s)g^*(s)[\delta(\omega-s\omega_0)]^2=\lim_{\tau\to\infty}\frac{\tau}{2\pi}\sum_{s=-\infty}^{\infty} f(s)g^*(s)\delta(\omega-s\omega_0),$$

and the τ cancels with the τ^{-1} in (3.15).

The formula (4.18) may be used to derive the *emissivity*. The emissivity is defined as the power radiated per unit frequency range and per unit solid angle. (Angle here refers to wave angle and not to ray angle, i.e. the unit solid angle is about the wave normal direction $\boldsymbol{\kappa}$.) By implication the independent variables are the frequency ω and the wave angle θ. The dependent variable is usually chosen to be the refractive index $n_\sigma(\omega, \theta)$. Let $\eta^\sigma(s, \omega, \theta)$ be the emissivity at the sth harmonic for waves in the mode σ. From the definition of the emissivity one has

$$\int \frac{d^3\mathbf{k}}{(2\pi)^3} P^\sigma(\mathbf{k}) = \sum_{s=-\infty}^{\infty} 2\pi \int_{-1}^{+1} d\cos\theta \int_0^\infty d\omega\, \eta^\sigma(s, \omega, \theta). \tag{4.19}$$

Comparison of (4.18) and (4.19) allows one to identify the emissivity:

$$\eta^\sigma(s, \omega, \theta) = \frac{q^2 n_\sigma^2 \omega^2}{\pi c^3} \frac{\partial}{\partial\omega}(\omega n_\sigma) R_E^\sigma |\mathbf{e}^{\sigma *} \cdot \mathbf{V}(s, \mathbf{p}, \mathbf{k})|^2$$
$$\times \delta\{\omega(1 - n_\sigma \beta \cos\alpha \cos\theta) - s\Omega\}, \tag{4.20}$$

where a mixed notation is used with the arguments ω, θ or $\mathbf{k}$ omitted. For waves in a non-spatially-dispersive medium, which is the only case where the emissivity is useful, one has, cf. (2.94),

$$R_E^\sigma = \frac{1}{2(1 - |\boldsymbol{\kappa} \cdot \mathbf{e}^\sigma|^2) n_\sigma \partial(\omega n_\sigma)/\partial\omega}. \tag{4.21}$$

Then if $\mathbf{e}^\sigma$ is written in the form (2.70), the expression (4.20) reduces to

$$\eta^\sigma(s, \omega, \theta) = \frac{q^2 n_\sigma \omega^2 \beta^2 \sin^2\alpha}{2\pi c(1 + T_\sigma^2)}$$
$$\times \left| \frac{K_\sigma \sin\theta + T_\sigma(\cos\theta - n_\sigma\beta\cos\alpha)}{n_\sigma \beta \sin\alpha \sin\theta} J_s - J_s' \right|^2$$
$$\times \delta\{\omega(1 - n_\sigma\beta\cos\alpha\cos\theta) - s\Omega\}, \tag{4.22}$$

where the arguments ω, θ of n_σ, T_σ and K_σ are omitted, and where the argument of the Bessel functions is

$$\frac{\omega}{\Omega} n_\sigma \beta \sin\alpha \sin\theta = \begin{cases} s \dfrac{n_\sigma\beta \sin\alpha \sin\theta}{1 - n_\sigma\beta\cos\alpha\cos\theta} & \text{for} \quad s \neq 0, \\[2ex] \dfrac{\omega}{\Omega} \tan\alpha \tan\theta & \text{for} \quad s = 0 \end{cases} \tag{4.23}$$

An alternative form of the emissivity for $s \neq 0$ and for electrons ($\epsilon = -1$) is

$$\eta^\sigma(s, \omega, \theta) = \frac{e^2 s^2 \Omega_e^2 (1 - \beta^2)^{1/2} y_\sigma}{2\pi c n_\sigma (1 + T_\sigma^2) \sin^2\theta}$$
$$\times \left| \left(\frac{K_\sigma y_\sigma \sin\theta}{(1 - \beta^2)^{1/2}} + z_\sigma T_\sigma \right) J_s(s x_\sigma) + x_\sigma J_s'(s x_\sigma) \right|^2 \delta(\omega - s\Omega_e y_\sigma) \tag{4.24}$$

with

$$x_\sigma := \frac{n_\sigma \beta \sin\alpha \sin\theta}{1 - n_\sigma \beta \cos\alpha \cos\theta}, \tag{4.25a}$$

$$y_\sigma := \frac{(1-\beta^2)^{1/2}}{1-n_\sigma\beta\cos\alpha\cos\theta}, \tag{4.25b}$$

$$z_\sigma := \frac{\cos\theta - n_\sigma\beta\cos\alpha}{1-n_\sigma\beta\cos\alpha\cos\theta}. \tag{4.25c}$$

The Polarization of Transverse Waves at given s, ω and θ

The formulae (4.18), (4.22) and (4.24) describe the power radiated into waves with a specified polarization vector. For transverse waves in an isotropic medium there are two independent states of polarization. In general a polarization tensor is required to describe the polarization completely, and it is necessary to write the emissivity in transverse waves in the form of a polarization tensor. It is straightforward to generalize (4.18) to make it a polarization tensor:

$$P^{\alpha\beta}(\mathbf{k}) = \sum_{s=-\infty}^{\infty} 8\pi^2 q^2 R_E e_i^{*\alpha} e_j^{\beta} V_i(s, \mathbf{p}, \mathbf{k}) V_j^*(s, \mathbf{p}, \mathbf{k}) \delta(\omega - s\Omega - k_\parallel v_\parallel). \tag{4.26}$$

The notation in (4.26) is that introduced in (3.19).

However, because the radiation emitted at a given harmonic (and given ω and θ) is completely polarized, the polarization can be determined explicitly and there would appear to be no necessity to use polarization tensors. Let us therefore introduce the total emissivity (i.e. that summed over the two states of polarization) together with parameters which specify the state of polarization at given s, ω and θ.

The total emissivity at the sth harmonic may be obtained from (4.22) or (4.24) as follows: set $K=0$ (for transverse waves) and add the values of the emissivity for $T=0$ and $T=\infty$. The total emissivity is

$$\eta(s, \omega, \theta) = \frac{e^2 s^2 \Omega_e^2 (1-\beta^2)^{1/2} y}{2\pi c n(\omega)\sin^2\theta} [\{zJ_s(sx)\}^2 + \{xJ_s'(sx)\}^2] \delta(\omega - s\Omega_e y), \tag{4.27}$$

where x, y and z are defined by (4.25a, b, c) with n_σ replaced by the refractive index $n(\omega)$ of the transverse waves. The polarization vector may be found by noting that according to (4.26), the polarization is along the projection of $\mathbf{V}$ on the plane orthogonal to $\boldsymbol{\kappa}$. That is, the polarization vector is given by normalizing

$$\mathbf{V}(s, \mathbf{p}, \mathbf{k}) - \boldsymbol{\kappa\kappa}\cdot\mathbf{V}(s, \mathbf{p}, \mathbf{k})$$

to unity. In this case, the polarization may be specified by the single parameter T. Recall that $|T|$ is the axial ratio of the polarization ellipse, with $|T|>1$ corresponding to the major axis along the projection of the magnetic field lines on the plane orthogonal to $\boldsymbol{\kappa}$, and that the sign of T specifies the handedness of the polarization, with $T>0$ corresponding to RH polarization where the handedness is the screw sense of the electric vector relative to $\boldsymbol{\kappa}$. For emission at given s, ω and θ by a particle with speed βc, pitch angle α and whose charge has sign ϵ, T is given by

$$T=\frac{zJ_s(sx)}{-\epsilon x J_s'(sx)} \tag{4.28}$$

In particular (4.28) implies that for a non-relativistic ($\beta \ll 1$) electron ($\epsilon=-1$ the emitted radiation is RH polarized for $0\leqslant\theta\lesssim\pi/2$ and LH polarized for $\pi/2\lesssim\theta\leqslant\pi$. For an ultrarelativistic electron ($\beta\approx1$) the radiation is RH polarized for $0\leqslant\theta\lesssim\alpha$ and LH for $\alpha\lesssim\theta\leqslant\pi$. However, because nearly all the power is emitted at angles θ close to α, this result offers little insight into the handedness of the emitted synchrotron radiation.

It is interesting to show explicitly that the sum over the two components in any pair of orthogonal polarizations reduces to (4.27). This may be done by selecting

$$\mathbf{e}^1=\frac{(T_1\cos\theta,\, i,\, -T_1\sin\theta)}{(1+T_1^2)^{1/2}},\quad \mathbf{e}^2=\frac{(T_2\cos\theta,\, i,\, -T_2\sin\theta)}{(1+T_2^2)^{1/2}}, \tag{4.29a, b}$$

with $T_1T_2=-1$. The choice $T_1=-T_2=1$ represents a separation into two circularly polarized components, $T_1=0$ and $T_2=\infty$ represents two linearly polarized components and any other choice of real $T_1=-T_2^{-1}$ represents orthogonal elliptical polarizations. The choice (4.29a, b) with arbitrary $T_1=-T_2^{-1}$ is still not the most general choice because in (4.29a, b) the axes of the polarization ellipse are chosen along the directions $\boldsymbol{\kappa}\times\mathbf{b}$ and $\boldsymbol{\kappa}\times(\boldsymbol{\kappa}\times\mathbf{b})$. A general choice would be

$$\mathbf{e}^1=\frac{(\cos\theta(T_1\cos\phi-i\sin\phi),\, T_1\sin\phi+i\cos\phi,\, -\sin\theta(T_1\cos\phi+i\sin\phi))}{(1+T_1^2)^{1/2}} \tag{4.30a}$$

$$\mathbf{e}^2=\frac{(\cos\theta(T_2\cos\phi-i\sin\phi),\, T_2\sin\phi+i\cos\phi,\, -\sin\theta(T_1\cos\phi+i\sin\phi))}{(1+T_2^2)^{1/2}} \tag{4.30b}$$

again with $T_1T_2=-1$ and arbitrary ϕ. Detailed evaluations show that $\eta^{\alpha\alpha}:=\eta^{11}+\eta^{22}$ is independent of the choices of $T_1=-T_2^{-1}$ and of ϕ. More generally, when using a polarization-tensor notation, the information contained in a polarization tensor is unaffected by the choice of basis vectors (although the way that information is extracted is affected) and it is appropriate to make the simplest possible choice of $\mathbf{e}^1$ and $\mathbf{e}^2$.

The Emissivity as a Polarization Tensor

Although the emission at given s, ω and θ is completely polarized, the emission at given ω and θ involves a sum over s and it is not completely polarized. Consequently it is convenient to write the emissivity as a polarization tensor in order to retain all the relevant information on the polarization. Let the basis polarization vectors be chosen as (set $T_2=0$, $T_1=\infty$ in (4.29a, b))

$$\mathbf{e}^1=(\cos\theta, 0, -\sin\theta), \quad \mathbf{e}^2=(0, 1, 0), \tag{4.31}$$

where the fact that the overall phase is arbitrary is used to choose $\mathbf{e}^2$ real. The coordinate axes are specified by (2.51),

$$\mathbf{B}=B(0, 0, 1), \quad \boldsymbol{\kappa}=(\sin\theta, 0, \cos\theta). \tag{4.32}$$

Thus $\mathbf{e}^1$ describe the "projection along $\mathbf{B}$" and $\mathbf{e}^2$ the "projection orthogonal to $\mathbf{B}$" of the electric vector on the plane orthogonal to $\boldsymbol{\kappa}$. The polarization tensor for the emissivity may be written in the form

$$\eta^{\alpha\beta}(s, \omega, \theta)=\frac{n(\omega)q^2\beta_\perp^2\omega^2}{2\pi c} F^{\alpha\beta}(s, \theta)\delta\{\omega(1-n(\omega)\beta_\parallel \cos\theta)-s\Omega\}, \tag{4.33}$$

with

$$F^{\alpha\beta}(s, \theta):=\frac{1}{\beta_\perp^2c^2} e_i^\alpha e_j^\beta V_i(s, \mathbf{p}, \mathbf{k})V_j^*(s, \mathbf{p}, \mathbf{k}). \tag{4.34}$$

Using (4.16), and with $\mathbf{e}^1$ and $\mathbf{e}^2$ given by (4.31), the quantity $F^{\alpha\beta}$ reduces to

$$F^{\alpha\beta}=\begin{pmatrix} A^2 & -iAC \\ iAC & C^2 \end{pmatrix} \tag{4.35}$$

with

$$A:=(z/x)J_s(sx), \tag{4.36a}$$

$$C:=-\epsilon J_s'(sx), \tag{4.36b}$$

where the notation is essentially that used in (4.24) and (4.27):

$$x:=\frac{n(\omega)\beta_\perp \sin\theta}{1-n(\omega)\beta_\parallel \cos\theta}, \tag{4.37a}$$

$$z:=\frac{\cos\theta-n(\omega)\beta_\parallel}{1-n(\omega)\beta_\parallel \cos\theta}. \tag{4.37b}$$

The determinant of the polarization tensor vanishes, which reflects the fact that the emission for given s, ω, θ is completely polarized. (The argument is as follows: the polarization tensor is of second order and if its determinant vanishes it is of rank one, which implies that it may be written as the outer product of a vector with itself; this vector is the polarization vector.) The actual emission involves a sum over the emission at each harmonic s. Hence even for fixed ω and θ the actual emission need not be completely polarized.

For emission *in vacuo*, and also for that in a plasma with $n(\omega)<1$, emission at $s\leqslant 0$ is not possible. This can be seen from the fact that the δ-function in (4.33) implies

$$\omega=\frac{s\Omega}{1-n(\omega)\beta_{\parallel}\cos\theta}, \tag{4.38}$$

which permits $s<0$ only for $n(\omega)\beta_{\parallel}\cos\theta>1$, and *a forteriori* for $n(\omega)>1$. Emission occurs at $s=0$ only for $n(\omega)\beta_{\parallel}\cos\theta=1$, which also requires $n(\omega)>1$. The effect of an ambient plasma on gyromagnetic emission is to suppress the emission at low frequencies. This suppression is called the Razin effect, which is discussed in Section 4.3. In the remainder of this section the emission is assumed to occur *in vacuo*.

The Total Power Radiated in Vacuo

Consider the power radiated per particle *in vacuo* expressed as a polarization tensor:

$$P^{\alpha\beta}=\sum_{s=1}^{\infty}2\pi\int_{-1}^{+1}\mathrm{d}\cos\theta\int_{0}^{\infty}\mathrm{d}\omega\,\eta^{\alpha\beta}(s,\omega,\theta). \tag{4.39}$$

The $\alpha=\beta=1$ and $\alpha=\beta=2$ components may be calculated explicitly in two different ways. One way is to integrate over frequency, then over angle and, finally, to perform the sum over s. The other is to integrate over frequency, then perform the sum over s before integrating over angle. Thus the two ways involve opposite orders for performing the sum and the integral in

$$P^{\alpha\beta}=\sum_{s=1}^{\infty}\int_{-1}^{+1}\mathrm{d}\cos\theta\,\frac{q^2\beta_{\perp}^2 s^2\Omega^2}{c(1-\beta_{\parallel}\cos\theta)^3}F^{\alpha\beta}(s,\theta). \tag{4.40}$$

The Integral over Angle of Emission

Firstly let us perform the integral over $\cos\theta$ in (4.40) to find the power emitted at the sth harmonic. To facilitate this the following variables are introduced:

$$\cos\theta'=\frac{\cos\theta-\beta_{\parallel}}{1-\beta_{\parallel}\cos\theta}, \tag{4.41}$$

$$\beta':=\frac{\beta_{\perp}}{(1-\beta_{\parallel}^2)^{1/2}}, \tag{4.42}$$

with

$$\beta' \sin\theta' = \frac{\beta_\perp \sin\theta}{1-\beta_\parallel \cos\theta}. \tag{4.43}$$

The old limits of integration $\cos\theta = \pm 1$ transform into the new limits $\cos\theta' = \pm 1$. The quantities (4.36a, b) become

$$A = \frac{\cos\theta'}{\beta' \sin\theta'} J_s(s\beta' \sin\theta'), \tag{4.44a}$$

$$C = -\epsilon J_s'(s\beta' \sin\theta'). \tag{4.44b}$$

It might be noticed that the transformation from $\cos\theta$ to $\cos\theta'$ and from $\beta_\perp$, $\beta_\parallel$ to β', 0 is equivalent to a Lorentz transformation in the following sense. If $\beta_\perp$, $\beta_\parallel$ and θ are given in the laboratory frame (the "unprimed" frame) then β' and θ' are the values of $\beta_\perp$ and θ according to an observer in the frame $\alpha' = \pi/2$ (the "primed" frame), i.e. to an observer in whose frame the particle's orbit is a circular one. The Lorentz factor γ transforms into $\gamma' = \gamma(1-\beta_\parallel^2)^{1/2} = (1-\beta'^2)^{-1/2}$.

With (4.41) and (4.42), (4.40) reduces to

$$P^{\alpha\beta} = \sum_{s=1}^{\infty} \int_{-1}^{+1} d\cos\theta' \, \frac{q^2 s^2 \beta'^2 (1-\beta'^2)\Omega_0^2}{c} (1+\beta_\parallel \cos\theta') F^{\alpha\beta} \tag{4.45}$$

with $\Omega_0 := |q|B/mc$ and with

$$F^{11} = A^2 = \frac{\cos^2\theta'}{\beta'^2 \sin^2\theta'} J_s^2(s\beta' \sin\theta'), \tag{4.46a}$$

$$F^{12} = -F^{21} = -iAC = \frac{i\epsilon \cos\theta'}{\beta' \sin\theta'} J_s'(s\beta' \sin\theta') J_s(s\beta' \sin\theta')$$

$$= \frac{i\epsilon}{s\beta'} \frac{\cos\theta'}{\sin^2\theta'} \frac{\partial}{\partial\beta'} \{J_s^2(s\beta' \sin\theta')\}, \tag{4.46b}$$

$$F^{22} = C^2 = J_s'^2(s\beta' \sin\theta')$$

$$= -\frac{1}{\beta'^2 \sin^2\theta'} J_s^2(s\beta' \sin\theta') + \tfrac{1}{2}[J_{s-1}^2(s\beta' \sin\theta') + J_{s+1}^2(s\beta' \sin\theta')]. \tag{4.46c}$$

Both F^{11} and F^{22} are even functions of $\cos\theta'$ and the factor $(1+\beta_\parallel \cos\theta')$ in (4.45) may be replaced by unity for these terms. It follows that F^{11} and F^{22} depend on $\beta_\perp$ and $\beta_\parallel$ only through the quantity β' defined by (4.41). On the other hand $F^{12} = -F^{21}$ is an odd function of $\cos\theta'$ and the factor $(1+\beta_\parallel \cos\theta')$ in (4.45) may be replaced by $\beta_\parallel \cos\theta'$ for this term.

The integrals required in (4.45) are

$$\int_{-1}^{+1} d\cos\theta'\, J_t^2(s\beta'\sin\theta') = \frac{2}{\beta'}\int_0^{\beta'} dy\, J_{2t}(2sy), \tag{4.47a}$$

$$\int_{-1}^{+1} \frac{d\cos\theta'}{\sin^2\theta'}\, J_t^2(s\beta'\sin\theta') = 2\int_0^{\beta'} \frac{dy}{y}\, J_{2t}(2sy). \tag{4.47b}$$

With these integrals (4.45) reduces to

$$P^{\alpha\beta} = \sum_{s=1}^{\infty} \frac{2q^2\Omega_0^2}{c}\,\beta'^2(1-\beta'^2)s^2 G^{\alpha\beta}(s), \tag{4.48}$$

with

$$G^{11}(s) = \frac{1}{\beta'^2}\int_0^{\beta'}\frac{dy}{y}\, J_{2s}(2sy) - \frac{1}{\beta'^3}\int_0^{\beta'} dy\, J_{2s}(2sy), \tag{4.49a}$$

$$G^{12}(s) = -G^{21}(s) = \frac{i\epsilon\beta_{\parallel}}{2s\beta'^3}\int_0^{\beta'} dy\, J_{2s}(2sy), \tag{4.49b}$$

$$G^{22}(s) = \frac{1}{s\beta'}\, J'_{2s}(2s\beta') - \frac{1}{\beta'^2}\int_0^{\beta'}\frac{dy}{y}\, J_{2s}(2sy) + \frac{1}{\beta'}\int_0^{\beta'} dy\, J_{2s}(2sy). \tag{4.49c}$$

It is worth emphasizing that because P^{11} and P^{22} depend on $\beta_{\perp}$ and $\beta_{\parallel}$ only in the combination β', and because of the relation

$$p_{\perp} = \frac{mc\beta'}{(1-\beta'^2)^{1/2}}, \tag{4.50}$$

it follows that *P^{11} and P^{22} are independent of the value of $p_{\parallel}$*. On the other hand $P^{12} = -P^{21}$ does depend on $p_{\parallel}$.

The Power Radiated

The sum over s can be evaluated explicitly for the $\alpha=\beta=1$ and $\alpha=\beta=2$ components. The sums are those of Kapteyn series which were first evaluated explicitly by Schott (1912):

$$\sum_{s=1}^{\infty} s^2 J_{2s}(2sy) = \frac{y^2(1+y^2)}{2(1-y^2)^4} \qquad (0\leqslant y<1), \tag{4.51a}$$

$$\sum_{s=1}^{\infty} sJ'_{2s}(2sy) = \frac{y}{2(1-y^2)^2} \qquad (0\leqslant y<1). \tag{4.51b}$$

The remaining integrals over y are elementary and there results

$$P^{11} = \tfrac{1}{4}P - T, \quad P^{22} = \tfrac{3}{4}P - T, \tag{4.52a, b}$$

with P the total power radiated,

$$P:=P^{11}+P^{22}=\frac{2}{3}\frac{q^4B^2}{m^2c^3}\frac{\beta'^2}{1-\beta'^2}=\frac{2}{3}\frac{q^4B^2}{m^2c^3}\gamma^2\beta_\perp^2, \tag{4.53}$$

and with

$$T:=\frac{\beta'^2}{8}P=\frac{\beta_\perp^2}{8(1-\beta_\parallel^2)}P. \tag{4.54}$$

It is not possible to perform the sum over s in closed form for the remaining term $P^{12}=-P^{21}$ in (4.48).

Let us define a degree of linear polarization

$$r_l:=\frac{|P^{11}-P^{22}|}{P^{11}+P^{22}}=\frac{1}{2}+\frac{2T}{P}=\frac{2m^2c^2+3p_\perp^2}{4(m^2c^2+p_\perp^2)}. \tag{4.55}$$

Although r_l does not correspond to any quantity which would be measured in practice (one measures the degree of polarization of the emission at fixed ω and θ from a distribution of radiating electrons) it does provide an estimate of the expected degree of linear polarization. For non-relativistic particles ($p_\perp \ll mc$) (4.55) gives $r_l=\frac{1}{2}$, while for ultrarelativistic particles ($p_\perp \gg mc$) it gives $r_l=\frac{3}{4}$. Thus the maximum degree of linear polarization expected (from a source with a homogeneous magnetic field) would be around 75%.

The Rate Photons are Emitted

The foregoing method of evaluating $P^{\alpha\beta}$ may be extended to the evaluation of other quantities of the form, cf. (4.39),

$$2\pi\int_{-1}^{+1}\mathrm{d}\cos\theta\int_0^\infty \mathrm{d}\omega\,\omega^m\eta^{\alpha\beta}(s,\omega,\theta).$$

where m is an integer. (For $\alpha=\beta=1$ or 2, the sum over s can be performed for even values of m.) Of particular interest is the case $m=-1$. Let us write

$$R^{\alpha\beta}(s):=2\pi\int_{-1}^{+1}\mathrm{d}\cos\theta\int_0^\infty\frac{\mathrm{d}\omega}{\hbar\omega}\eta^{\alpha\beta}(s,\omega,\theta). \tag{4.56}$$

$R^{\alpha\beta}(s)$ may be interpreted as (the polarization tensor for) the rate photons are emitted at the sth harmonic. One finds

$$R^{11}(s)=\frac{P^{11}(s)}{\hbar s\Omega/(1-\beta_\parallel^2)},\quad R^{22}(s)=\frac{P^{22}(s)}{\hbar s\omega/(1-\beta_\parallel^2)},\quad R^{12}(s)=-R^{21}(s)=0, \tag{4.57}$$

where $P^{\alpha\beta}(s)$ is defined by writing (4.40) in the form

$$P^{\alpha\beta}=\sum_{s=1}^{\infty}P^{\alpha\beta}(s).$$

The sum over s cannot be performed explicitly to find the total rate photons are emitted. The rate is evaluated in the ultra-relativistic limit in (4.96c) below.

An interesting consequence of (4.57a, b) is that the mean frequency radiated at the sth harmonic, $\bar{\omega}_s$ say, is the same for the two linearly polarized components and is given by

$$\bar{\omega}_s = \frac{s\Omega}{1-\beta_\parallel^2} = \frac{s\Omega_0}{\gamma(1-\beta_\parallel^2)}, \tag{4.58}$$

with $\Omega_0 := |q|B/mc$. Comparison of this result with (4.38) (with $n(\omega)=1$) shows that for ultrarelativistic particles (when one has $1-\beta_\parallel^2 \approx \sin^2\alpha$, α=pitch angle) the emission must be concentrated at angles $\theta \approx \alpha$, as one expects on more general grounds.

The Angular Dependence

Now let us return to (4.40) and perform the sum over s first. The sum can be performed only for the terms $\alpha=\beta=1$ and $\alpha=\beta=2$. The relevant sums are

$$\sum_{s=1}^{\infty} s^2 J_s^2(sx) = \frac{x^2(4+x^2)}{16(1-x^2)^{7/2}} \qquad (0 \leqslant x < 1), \tag{4.59a}$$

$$\sum_{s=1}^{\infty} s^2 J_s'^2(sx) = \frac{4+3x^2}{16(1-x^2)^{5/2}} \qquad (0 \leqslant x < 1), \tag{4.59b}$$

which were also first evaluated by Schott (1912). The resulting expressions for P^{11} and P^{22} are

$$P^{11} = \frac{q^4B^2}{16m^2c^3}\,\beta_\perp^2(1-\beta^2)\int_{-1}^{+1} \mathrm{d}\cos\theta \times \frac{(\cos\theta-\beta_\parallel)^2\{4(1-\beta_\parallel\cos\theta)^2+\beta_\perp^2\sin^2\theta\}}{[(1-\beta_\parallel\cos\theta)^2-\beta_\perp^2\sin^2\theta]^{7/2}} \tag{4.60a}$$

and

$$P^{22} = \frac{q^4B^2}{16m^2c^3}\,\beta_\perp^2(1-\beta^2)\int_{-1}^{+1} \mathrm{d}\cos\theta\, \frac{\{4(1-\beta_\parallel\cos\theta)^2+3\beta_\perp^2\sin^2\theta\}}{[(1-\beta_\parallel\cos\theta)^2-\beta_\perp^2\sin^2\theta]^{5/2}}. \tag{4.60b}$$

For ultrarelativistic particles the angular dependence of the emission is dominated by the effect of the denominators in (4.60a, b). One has

$$(1-\beta_\parallel \cos\theta)^2-\beta_\perp^2 \sin^2\theta=\beta^2(\cos\alpha-\cos\theta)^2+(1-\beta^2)-2\beta(1-\beta)\cos\alpha\cos\theta$$
$$\approx\begin{cases}(\cos\alpha-\cos\theta)^2 & \text{for } |\alpha-\theta|>\gamma^{-1},\\ (\gamma^{-2}+(\alpha-\theta)^2)\sin^2\alpha & \text{for } |\alpha-\theta|\lesssim\gamma^{-1}.\end{cases} \qquad (4.61)$$

where α is the pitch angle of the particle. Consequently, most of the emission is confined to angles in the range $\alpha-\gamma^{-1}\lesssim\theta\lesssim\alpha+\gamma^{-1}$. This result is an example of the general result that emission by an ultrarelativistic particle is strongly concentrated in a cone with half angle $\approx\gamma^{-1}$ about the direction of the instantaneous velocity of the particle.

More generally, the result that all but a fraction γ^{-1} of the power emitted *in vacuo* by an ultrarelativistic particle is in a forward cone of half-angle γ^{-1} is proven as follows. Firstly, treat the emission in the instantaneous rest frame of the particle. The radiation is electric dipole radiation with the dipole axis along the instantaneous acceleration, and the power radiated is given by the Larmor formula (3.34). Next make a Lorentz transformation to the laboratory frame: one finds that this transforms the radiation pattern into one confined almost entirely to a forward cone with half angle γ^{-1} for $\gamma\gg 1$.

4.2 SYNCHROTRON EMISSIVITY

Qualitative Properties of Synchrotron Emission

Formulae which describe synchrotron emission may be derived from those which describe gyromagnetic emission in general by assuming that the particles are highly relativistic and by expanding in inverse powers of the Lorentz factor γ. (Actually, various expansions in γ^{-1} and ξ^{-1}, see (4.68) below, are involved.) There are two types of approximation made in order to perform such expansions. The first relates to the angular distribution of the emission from individual particles. As is implied by (4.60a, b) and (4.61), the emission *in vacuo* is confined to angles $|\theta-\alpha|\lesssim\gamma^{-1}$ and consequently an expansion in $\theta-\alpha$ is in effect an expansion in γ^{-1}. Thus to lowest order in γ^{-1} all the emission is regarded as being on the cone $\theta=\alpha$ (i.e. angle of emission equals pitch angle). The second type of approximation is to the Bessel functions, the conventional approximation being in terms of Airy integrals. The variety of different treatments of synchrotron emission which may be found in the literature, e.g. Schwinger (1949), Westfold (1959), le Roux (1961), Ginzburg and Syrovatskii (1965), differ only in the manner and the stage in which the approximations are introduced. The treatment given here involves leaving the approximation to the Bessel functions until last. (There is a practical reason for this: the Airy integral approximation is a relatively poor one in some respects and better approximations are available, see Section 4.3 below.) The treatment of synchrotron emission here is based on the derivation of the power radiated in gyromagnetic emission given in Section 4.1.

A general expression for the power per unit frequency range and per unit solid angle in gyromagnetic emission may be obtained by integrating the emissivity (4.33) over a distribution of particles and summing over harmonic numbers. Suppose the distribution $f(p, \alpha)$ is a separable function of p and α:

$$f(p, \alpha)=f_0(p)\phi(\alpha), \qquad \int_{-1}^{+1} d\cos\alpha\,\phi(\alpha)=2. \tag{4.62}$$

The integral over p may be set aside for the present, while we consider the emission at given ω and θ from particles with given momentum (i.e. given ε, p, γ or β) averaged over the pitch angle distribution $\phi(\alpha)$. To lowest order in γ^{-1} all the emission is at $\theta=\alpha$. Hence, if we averaged the emissivity over the pitch angle distribution $\phi(\alpha)$, then divide by $\phi(\theta)$, we may regard the resulting quantity as the emissivity in synchrotron radiation or, more simply, the *synchrotron emissivity*. This definition is artificial only in that, as it turns out, the circularly polarized component in the emission depends explicitly on the pitch angle distribution.

The Typical Frequency Emitted

There is one additional qualitative feature of the emission to which one must appeal when treating the theory in detail. This is that the emission occurs at high harmonics. It is instructive to estimate the characteristic frequency of emission, ω_c say, by the following semi-quantitative arguments. Once ω_c has been estimated one may compare it with the frequency $\langle\omega\rangle_s \approx s\Omega_e/\gamma\sin^2\alpha$ (which follows from (4.58) for $\gamma \gg 1$) to estimate the characteristic harmonic, s_c say, at which the emission occurs. That is s_c is to be estimated from

$$s_c=\frac{\omega_c}{\Omega_e}\gamma\sin^2\alpha. \tag{4.63}$$

Consider the emission from a particle with pitch angle $\alpha=\pi/2$. The emission is confined to a cone of angles with half-width $\approx\gamma^{-1}$, and which sweeps through 2π radians in a gyro-period $T_G=2\pi\gamma/\Omega_G$. The particle radiates in a given direction only for a time $\approx\gamma^{-1}T_G$. A fixed distant observer receives one pulse per gyroperiod. The particle is moving directly towards him when each pulse is emitted. Consequently the duration of each pulse is

$$\Delta t \approx (\gamma^{-1}T_G)(1-\beta) \simeq \tfrac{1}{2}T_G\gamma^{-3}.$$

A pulse of duration Δt contains frequencies $\omega \lesssim (\Delta t)^{-1}$. It follows that the typical frequency radiated is

$$\omega_c \approx (\Delta t)^{-1} \approx (\tfrac{1}{2}T_G\gamma^{-3})^{-1} \approx \frac{1}{\pi}\,\Omega_e\gamma^2.$$

The case of a particle with arbitrary pitch angle α may be obtained by applying a Lorentz transformation to relate ω and γ to the corresponding values ω' and γ' in the frame where the pitch angle is $\alpha' = \pi/2$. The Lorentz factor for this transformation is $(1-\beta_\parallel^2)^{-1/2} \approx (\sin\alpha)^{-1}$. Consequently, we have

$$\omega' = \omega \sin\alpha, \quad \gamma' = \gamma \sin\alpha,$$

and hence

$$\omega_c = \omega_c'/\sin\alpha \approx \frac{1}{\pi}\,\Omega_e\gamma'^2/\sin\alpha = \frac{1}{\pi}\,\Omega_e\gamma^2 \sin\alpha. \tag{4.64}$$

Thus the typical harmonic at which the emission occurs is

$$s_c \approx \frac{1}{\pi}\,(\gamma \sin\alpha)^3. \tag{4.65}$$

Calculation of the Emissivity

As defined above the emissivity in synchrotron radiation, $\bar{\eta}^{\alpha\beta}(\omega, \theta)$ say, is to be evaluated from

$$\bar{\eta}^{\alpha\beta}(\omega, \theta) = \sum_{s=1}^{\infty} \frac{1}{2\phi(\theta)} \int_{-1}^{+1} \mathrm{d}\cos\alpha\,\phi(\alpha)\eta^{\alpha\beta}(s, \omega, \theta), \tag{4.66}$$

with $\eta^{\alpha\beta}(s, \omega, \theta)$ given by (4.33). The following steps are made in evaluating this emissivity:–

1) Because high harmonics dominate in the emission the sum over s is replaced by an integral over s.

2) For $\gamma^2 \gg 1$ and $\omega^2 \gg \omega_p^2$ the following approximations are made

$$n(\omega) \approx 1 - \frac{\omega_p^2}{2\omega^2}, \quad \beta \approx 1 - \tfrac{1}{2}\gamma^{-2}, \quad n(\omega)\beta \approx 1 - \tfrac{1}{2}\xi^{-2}, \tag{4.67a, b, c}$$

with

$$\xi := \gamma\left(1 + \frac{\gamma^2\omega_p^2}{\omega^2}\right)^{-1/2}. \tag{4.68}$$

3) The relation $\alpha=\theta+(\alpha-\theta)$ together with the fact that the emission is confined to $|\theta-\alpha|\lesssim\xi^{-1}$ (this condition replaces the condition $|\theta-\alpha|\lesssim\gamma^{-1}$ which applies *in vacuo*) implies

$$\cos\alpha=\cos\theta+(\theta-\alpha)\sin\theta+0(\xi^{-2}). \tag{4.69}$$

4) The cos α-integral is performed by changing variables in a manner analogous to the way the cos θ-integral in (4.40) is performed:

$$\cos\alpha':=\frac{\cos\alpha-n(\omega)\beta\cos\theta}{1-n(\omega)\beta\cos\alpha\cos\theta}, \tag{4.70}$$

$$\beta':=\frac{n(\omega)\beta\sin\theta}{[1-n^2(\omega)\beta^2\cos^2\theta]^{1/2}}, \tag{4.71}$$

$$\beta'\sin\alpha'=\frac{n(\omega)\beta\sin\alpha\sin\theta}{1-n(\omega)\beta\cos\alpha\cos\theta}. \tag{4.72}$$

The limits $\cos\alpha=\pm1$ become $\cos\alpha'=\pm1$. The components of $F^{\alpha\beta}$ are given by (4.35), with

$$A=A^{(0)}+A^{(1)}, \tag{4.73a}$$

$$A^{(0)}=-\frac{\beta'}{n(\omega)\beta}\frac{\cos\alpha'}{\sin\alpha'}J_s(s\beta'\sin\alpha') \tag{4.73b}$$

$$A^{(1)}=\frac{\beta'(1-n^2(\omega)\beta^2)\cos\theta}{n^2(\omega)\beta^2}\frac{1}{\sin^2\theta\sin\alpha'}J_s(s\beta'\sin\alpha') \tag{4.73c}$$

$$C=-\epsilon J_s'(s\beta'\sin\alpha'). \tag{4.74}$$

5) The next step is to relate the expansion in $\theta-\alpha$ to an expansion in $\cos\alpha'$. This is achieved by noting that (4.67c), (4.69 and (4.70) imply

$$\theta-\alpha=\sin\theta\cos\alpha'+O(\xi^{-2}). \tag{4.75}$$

Consequently the expansion in $\theta-\alpha$ gives

$$\begin{aligned}\phi(\alpha)&=\phi(\theta)\left\{1+\frac{(\alpha-\theta)}{\phi(\theta)}\frac{d\phi(\theta)}{d\theta}+O(\xi^{-2})\right\}\\&=\phi(\theta)\{1-\cos\theta\cos\alpha' g(\theta)+O(\xi^{-2})\},\end{aligned} \tag{4.76}$$

with

$$g(\theta):=\frac{\tan\theta}{\phi(\theta)}\frac{d\phi(\theta)}{d\theta}, \tag{4.77}$$

and

$$1-n(\omega)\beta\cos\alpha\cos\theta=\sin^2\theta\{1-\cos\theta\cos\alpha'+O(\xi^{-2})\}. \tag{4.78}$$

We are concerned only with the terms up to first order in the small parameters $(\gamma\sin\theta)^{-1}$ or $(\xi\sin\theta)^{-1}$. To this order (4.66) is of the form

$$\int_0^\infty ds \int_{-1}^{+1} d\cos\alpha' f(s,\alpha')\delta\{\omega\sin^2\theta(1-\cos\theta\cos\alpha')-s\Omega\}$$

$$=\frac{1}{\Omega}\int_{-1}^{+1} d\cos\alpha' \left\{f(s_0,\alpha')-\cos\theta\cos\alpha' s_0\frac{\partial f(s_0,\alpha')}{\partial s_0}+O(\xi^{-2})\right\}, \tag{4.79}$$

with $\Omega=\Omega_e/\gamma$ and

$$s_0=\frac{\omega}{\Omega_e}\gamma\sin^2\theta. \tag{4.80}$$

The $\cos\alpha'$-integral can then be performed with the following result, cf. (4.48) and (4.49a, b, c):

$$\bar{\eta}^{\alpha\beta}(\omega,\theta)=\frac{e^2\Omega_e s_0^2}{2\pi c\gamma}\bar{G}^{\alpha\beta}(s_0) \tag{4.81}$$

with

$$\bar{G}^{11}(s)=\int_0^{\beta'}\frac{dy}{y}J_{2s}(2sy)-\frac{1}{\beta'}\int_0^{\beta'}dy\,J_{2s}(2sy), \tag{4.82a}$$

$$\bar{G}^{12}(s)=-\bar{G}^{21}(s)=i\epsilon\cos\theta\left[\left\{4+g(\theta)+s\frac{\partial}{\partial s}\right\}\frac{1}{2s}\int_0^{\beta'}dy\,J_{2s}(2sy)+\frac{1}{2s\xi^2\sin^2\theta}J_{2s}(2s\beta')\right], \tag{4.82b}$$

$$\bar{G}^{22}(s)=\frac{1}{s\beta'}J'_{2s}(2s\beta')-\frac{1}{\beta'^2}\int_0^{\beta'}\frac{dy}{y}J_{2s}(2sy)+\frac{1}{\beta'}\int_0^{\beta'}dy\,J_{2s}(2sy). \tag{4.82c}$$

with β' given by (4.71) and s_0 by (4.80).

From a formal viewpoint (4.81) with (4.82a, b, c) is the desired emissivity for synchrotron radiation to first order in small parameters. However s_0 is now a function of frequency and angle of emission and consequently the functional dependence on ω and θ involves a functional dependence on the order of the Bessel functions. This is impractical, but it can be overcome by introducing an appropriate approximation to the Bessel functions. The conventional approximation in this context is in terms of the Airy integral. Other approximations to the Bessel functions are mentioned in Section 4.3.

The Airy Integral Approximation

The Airy integral approximation to Bessel functions with argument z less than the order s (and with $z-s \ll s$, $s \gg 1$) is given by equation (A.30) of Appendix A. One has

$$J_{2s}(2s\beta') \approx \frac{1}{\pi\sqrt{3}\,\xi \sin\theta} K_{1/3}(R), \tag{4.83a}$$

$$J'_{2s}(2s\beta') \approx \frac{1}{\pi\sqrt{3}\,\xi^2 \sin^2\theta} K_{2/3}(R), \tag{4.83b}$$

$$\int_0^{\beta'} \mathrm{d}y\, J_{2s}(2sy) \approx \frac{1}{2s\pi\sqrt{3}} \int_R^{\infty} \mathrm{d}t\, K_{1/3}(t), \tag{4.83c}$$

$$\int_0^{\beta'} \mathrm{d}y \left(\frac{1}{y}-\frac{1}{\beta'}\right) J_{2s}(2sy) \approx \frac{1}{4s\pi\sqrt{3}\,\xi^2 \sin^2\theta} \left[\int_R^{\infty} \mathrm{d}t\, K_{5/3}(t) - K_{2/3}(R)\right], \tag{4.83d}$$

with

$$\beta' \approx 1 - \frac{1}{2\xi^2 \sin^2\theta}, \quad R := \frac{2s}{3\xi^3 \sin^3\theta}. \tag{4.84}$$

For $s = s_0$, with s_0 given by (4.80), the quantity R may be written

$$R = \frac{2s_0}{3\xi^3 \sin^3\theta} =: \frac{\omega}{\omega_c}, \tag{4.85}$$

which is equivalent to

$$\omega_c := \frac{3\Omega_e \xi^3 \sin\theta}{2\gamma}. \tag{4.86}$$

In vacuo, i.e. for $\xi = \gamma$, ω_c differs from the value of the characteristic frequency estimated semi-quantitatively in (4.64) only by a factor of order unity.

With the Airy integral approximation, the emissivity (4.81) reduces to

$$\bar{\eta}^{\alpha\beta}(\omega, \theta) = \frac{\sqrt{3}}{16\pi^2 c} \frac{e^2 \Omega_e \xi \sin\theta}{\gamma} H^{\alpha\beta}(\xi, \omega, \theta), \tag{4.87}$$

with

$$H^{11}(\xi, \omega, \theta) = \frac{\omega}{\omega_c}\left[\int_{\omega/\omega_c}^{\infty} \mathrm{d}t\, K_{5/3}(t) - K_{2/3}\left(\frac{\omega}{\omega_c}\right)\right], \tag{4.88a}$$

$$H^{12}(\xi, \omega, \theta) = -H^{21}(\xi, \omega, \theta) = \frac{2i\epsilon \cot\theta}{3\xi} \times \left[(2+g(\theta)) \int_{\omega/\omega_c}^{\infty} \mathrm{d}t\, K_{1/3}(t) + 2\frac{\omega}{\omega_c} K_{1/3}\left(\frac{\omega}{\omega_c}\right)\right], \tag{4.88b}$$

$$H^{22}(\xi, \omega, \theta)=\frac{\omega}{\omega_c}\left[\int_{\omega/\omega_c}^{\infty} \mathrm{d}t\, K_{5/3}(t)+K_{2/3}\left(\frac{\omega}{\omega_c}\right)\right]. \tag{4.88c}$$

For emission *in vacuo* ($\xi=\gamma$) it is an interesting exercise to calculate the total power radiated and compare it with the exact result derived in Section 4.1. Now $\bar{\eta}^{\alpha\beta}(\omega, \theta)$ is the power radiated per unit solid angle and per unit frequency range. Because, to a first approximation, all the power is emitted at $\theta=\alpha$, the total power radiated per particle could be identified as

$$\bar{P}^{\alpha\beta}:=4\pi \int_0^{\infty} \mathrm{d}\omega\, \bar{\eta}^{\alpha\beta}(\omega, \alpha). \tag{4.89}$$

The ω-integral can be reduced to one of the form

$$\int_0^{\infty} \mathrm{d}x\, x^{\mu} K_{\nu}(ax)=2^{\mu-1}a^{-\mu-1}\Gamma\left(\frac{1+\mu+\nu}{2}\right)\Gamma\left(\frac{1+\mu-\nu}{2}\right), \tag{4.90}$$

with the Γ-functions evaluated using

$$\Gamma(x+1)=x\Gamma(x), \quad \Gamma(1-x)\Gamma(x)=\frac{\pi}{\sin \pi x}. \tag{4.91}$$

One finds

$$\bar{P}^{11}=\tfrac{1}{8}\bar{P}, \quad \bar{P}^{22}=\tfrac{7}{8}\bar{P}, \tag{4.92a, b}$$

$$\bar{P}:=\bar{P}^{11}+\bar{P}^{22}=\frac{2}{3}\frac{q^4B^2}{m^2c^3}\gamma^2 \sin^2 \alpha, \tag{4.92c}$$

which is just the ultrarelativistic limit of the exact result (4.52a, b). In this case it is also possible to evaluate $\bar{P}^{12}$ explicitly:

$$\bar{P}^{12}=-\bar{P}^{21}=\frac{i\epsilon \cot \alpha}{\gamma}\frac{\sqrt{3}}{8}\{4+g(\alpha)\}\bar{P}. \tag{4.92d}$$

Discussion

The synchrotron emissivity (4.81) or (4.87) has been derived by expanding in γ^{-1} and retaining only the leading terms and the corrections of order γ^{-1}. The leading terms describe the linearly polarized component of the emission, and the terms of order γ^{-1} describe the circularly polarized emission. There is no correction of order γ^{-1} to the linearly polarized components. Also, the pitch-angle distribution appears in the synchrotron emissivity only in the form $g(\theta)$, cf. (4.77), as first shown by Legg and Westfold (1968).

Note that all corrections of order γ^{-2} are neglected in synchrotron theory. Thus one must set $\varepsilon=pc$ because the corrections to $\varepsilon/pc=1$ are of order γ^{-2}.

An important point is that one must neglect the difference between the refractive indices for the magnetoionic modes for the same reason. Specifically, it follows from (2.132) that the refractive indices for the two modes are equal to $(1-\omega_p^2/\omega^2)^{1/2}$ plus a correction of order (ω_p^2/ω^2) times Ω_e/ω. However, the typical frequency emitted is of order ω_c, i.e. of order $\Omega_e\gamma^2$ for $\omega \gg \gamma\omega_p$. Thus Ω_e/ω is of order γ^{-2} and is to be neglected. (For $\omega \lesssim \gamma\omega_p$ the emission is suppressed and although technically the difference between the refractive indices of the two modes could be significant in this case, the emission itself is of no interest.) Similarly, the longitudinal component of the polarization, described by K_σ is (2.132), is of the same order as the difference between the refractive indices, and it is to be neglected. Thus, in developing synchrotron theory it is appropriate to start from the emissivity (4.33) for transverse waves in an isotropic plasma, and not from the emissivity (4.24) for each of the magnetoionic modes. Although the difference between the two modes is to be neglected in treating the emission (and absorption), the difference between the two modes can be important in affecting the escaping radiation due to Faraday rotation (Section 6.2).

Before proceeding with a more detailed discussion of synchrotron emission, it is appropriate to make several comments of an historical nature. The historical development of ideas relating to synchrotron radiation has been discussed by Ginzburg and Syrovatskii (1965) from an astrophysical viewpoint and by Sokolov and Ternov (1968) from the viewpoint of laboratory accelerators. Besides the early work of Schott, summarized in his book (1912), the earliest quantitative treatments were by Ivanenko and Sokolov (1948) and Schwinger (1949). They derived a result (for the total power radiated) which is equivalent to the total emissivity $\bar{\eta}^{11}(\omega, \theta)+\bar{\eta}^{22}(\omega, \theta)$ implied by (4.87) with (4.88a, b). The suggestions that radio emission from cosmic sources might be synchrotron emission (Alfvén and Herlofson, 1950; Kiepenheuer, 1950) was pursued by Ginzburg (1951, 1953). Shklovskii (1953) suggested that part of the optical emission from the Crab Nebula might be synchrotron emission. The observation of polarized optical emission from the Crab Nebula (Oort and Walraven, 1956) confirmed Shklovskii's suggestion. A fully quantitative treatment of the linear polarization of synchrotron radiation was given by Westfold (1959), and le Roux (1961) obtained the same results using a slightly different approach. Westfold's (1959) result is equivalent to (4.87) with (4.88a) and (4.88c). Finally, the small circularly polarized component, described by (4.87) with (4.88b), was derived by Legg and Westfold (1968).

4.3 PROPERTIES OF SYNCHROTRON EMISSION

Now let us turn to the detailed properties of synchrotron emission. The discussion here covers three aspects of the emission (absorption is neglected entirely here and is treated separately in Section 6.3). The first is emission from individual particles. The properties of such emission may be deduced directly from the emissivity. The second is emission from power-law (energy spectra) distributions of particles, which is the case thought to be of most interest in practice. The third is a variety of special cases, including the low-frequency and small-pitch-angle limits. The effect of a medium (the Razin effect) is also discussed.

The Frequency Spectrum and the Polarization

The frequency spectrum of the radiation *in vacuo* is determined by the function $F(\omega/\omega_c)$ with, cf. (4.87) and (4.88a, c),

$$F(x) := x \int_x^\infty dt\, K_{5/3}(t). \tag{4.93}$$

Also, (4.86) gives $\omega_c = \frac{3}{2}\Omega_e \gamma^2 \sin\theta$ *in vacuo*. The function $F(x)$ is plotted in Fig. 4.1. It has a maximum at $x = x_m = 0.29$ with $F(x_m) = 0.92$, and it satisfies

$$F(x) = \begin{cases} \dfrac{4\pi}{\sqrt{3}\,\Gamma(\frac{1}{3})}\left(\dfrac{x}{2}\right)^{1/3}\left\{1 - \frac{1}{2}\Gamma(\frac{1}{3})\left(\dfrac{x}{2}\right)^{2/3} + \cdots\right\} & \text{for} \quad x \ll 1, \\ \left(\dfrac{\pi x}{2}\right)^{1/2} e^{-x}\left\{1 + \dfrac{55}{72x} + \cdots\right\} & \text{for} \quad x \gg 1, \end{cases} \tag{4.94}$$

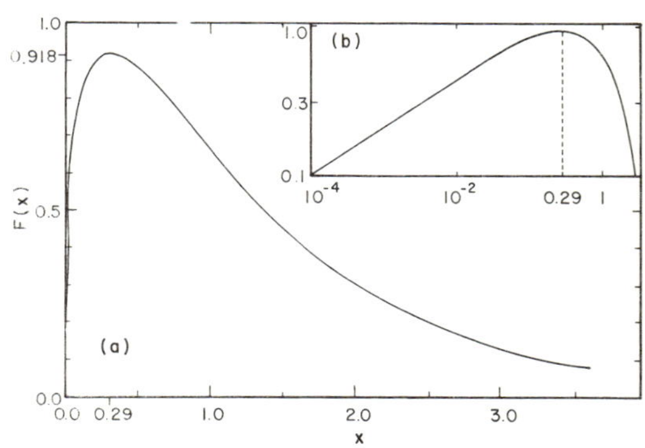

FIGURE 4.1 The function $F(x)$, defined by (4.93), is plotted (a) on a linear–linear scale, and (b) on a log–log scale.

with

$$F(x) \approx 1.8 x^{0\cdot 3}\, \mathrm{e}^{-x} \tag{4.94'}$$

between these two limits.

It is straightforward to determine the mean frequency $\langle\omega\rangle$ the mean square frequency $\langle\omega^2\rangle$, and so on, of the emitted radiation:

$$\langle\omega^n\rangle := \frac{\int_0^\infty \mathrm{d}\omega\, \omega^{n-1} F(\omega/\omega_c)}{\int_0^\infty \mathrm{d}\omega\, \omega^{-1} F(\omega/\omega_c)}. \tag{4.95}$$

The relevant integrals are of the form (4.90). The denominator in (4.95) is proportional to the rate photons are emitted, cf. the derivation of (4.58). One finds

$$\langle\omega\rangle = \frac{8}{15\sqrt{3}}\, \omega_c = 0.31\omega_c, \tag{4.96a}$$

$$\langle(\omega - \langle\omega\rangle)^2\rangle = \langle\omega^2\rangle - \langle\omega\rangle^2 = \frac{211}{64}\, \langle\omega\rangle^2 = (1.8\langle\omega\rangle)^2. \tag{4.96b}$$

The rate photons are emitted is given by

$$R = \frac{5}{2\sqrt{3}}\, \frac{e^2 \Omega_e \sin\alpha}{\hbar c}. \tag{4.96c}$$

(It has been pointed out by J. Romig (unpublished) that the total number of photons per pulse, cf. the discussion leading to (4.64), is of order $R/\Omega_e \approx e^2/\hbar c \approx \frac{1}{137}$, i.e. of order the fine structure constant; *inter alia* this implies that it is impossible in principle to detect a single such pulse). The mean frequency radiated is close to the frequency at which $F(\omega/\omega_c)$ is a maximum and (4.96c) implies that the emission from individual electrons is relatively broad band. In fact from Fig. 4.1 it can be seen that at a frequency about ten times that at which $F(\omega/\omega_c)$ is a maximum, the value of $F(\omega/\omega_c)$ has dropped to only about one tenth its maximum value.

The degree of linear polarization of the radiation is given by

$$r_l = \frac{|\bar{\eta}^{11}(\omega,\theta) - \bar{\eta}^{22}(\omega,\theta)|}{\bar{\eta}^{11}(\omega,\theta) + \bar{\eta}^{22}(\omega,\theta)} = \frac{K_{2/3}(\omega/\omega_c)}{\int_{\omega/\omega_c}^\infty \mathrm{d}t\, K_{5/3}(t)}$$

$$= \begin{cases} \dfrac{1}{2}\left[1 + \frac{1}{2}\Gamma(\frac{1}{3})\left(\dfrac{\omega}{2\omega_c}\right)^{2/3} + \cdots\right] & \text{for} \quad \omega \ll \omega_c \\[2ex] 1 - \dfrac{2\omega_c}{3\omega} + \cdots & \text{for} \quad \omega \gg \omega_c \end{cases} \tag{4.97}$$

The plane of polarization is along the direction orthogonal to both the direction of the magnetic field and the direction of wave propagation. At $x = x_m = 0.29$ one has $r_l = 0.645$.

The degree of circular polarization is given by

$$r_c = \frac{i\{\bar{\eta}^{12}(\omega, \theta) - \bar{\eta}^{21}(\omega, \theta)\}}{\bar{\eta}^{11}(\omega, \theta) + \bar{\eta}^{22}(\omega, \theta)}$$

$$= \frac{-\dfrac{2\epsilon \cot\theta}{3\xi}\left[(2+g(\theta))\displaystyle\int_{\omega/\omega_c}^{\infty} dt\, K_{1/3}(t) + 2\frac{\omega}{\omega_c} K_{1/3}\left(\frac{\omega}{\omega_c}\right)\right]}{\dfrac{\omega}{\omega_c}\displaystyle\int_{\omega/\omega_c}^{\infty} dt\, K_{5/3}(t)}$$

$$= -\frac{2\epsilon \cot\theta}{3\xi}\begin{cases} \frac{1}{4}\Gamma(\frac{1}{3})(2+g(\theta))(\omega/2\omega_c)^{-1/3} + \cdots & \text{for} \quad \omega \ll \omega_c, \\ 2 + (\frac{1}{3} + g(\theta))(\omega/\omega_c)^{-1} + \cdots & \text{for} \quad \omega \gg \omega_c. \end{cases} \tag{4.98}$$

For an isotropic distribution of electrons, i.e. for $\epsilon = -1$ and $g(\theta) = 0$, the degree of circular polarization is positive (i.e. RH polarization) for emission at angles $0 < \theta < \pi/2$ and is negative (i.e. LH polarization) for emission at angles $\pi/2 < \theta < \pi$.

Power-Law Spectrum

An advantage of the Airy integral approximation in the context of synchrotron theory is that it allows an analytic treatment of emission and absorption by particles with a power-law energy spectrum, e.g. one of the form

$$N(\varepsilon) = \begin{cases} K\varepsilon^{-a} & \text{for} \quad \varepsilon_1 < \varepsilon < \varepsilon_2, \\ 0 & \text{otherwise,} \end{cases} \tag{4.99}$$

where K, a, ε_1 and ε_2 are all constants. Actually the integrals can be performed in closed form only for $\varepsilon_1 = 0$ and $\varepsilon_2 = \infty$ (and for $a > \frac{1}{3}$) in (4.99). However, this is not a major restriction because semi-quantitative arguments enable one to infer the shape of the emitted spectrum for finite values of ε_1 and ε_2 (Section 9.1).

The emissivity per unit volume in synchrotron radiation is

$$\frac{d\eta^{\alpha\beta}(\omega, \theta)}{dV} = \int_0^{\infty} d\varepsilon\, N(\varepsilon)\bar{\eta}^{\alpha\beta}(\omega, \theta). \tag{4.100}$$

When the power-law spectrum (4.99) and the expression (4.87) for the emissivity are inserted in (4.100) the integral can be performed for the case of radiation *in vacuo* by replacing the variable of integration ε by, cf. (4.86),

$$\frac{\omega}{\omega_c}=\left(\frac{2}{3}\frac{\omega}{\Omega_e}\frac{1}{\sin\theta}\right)\left(\frac{mc^2}{\varepsilon}\right)^2. \qquad (4.101)$$

The integrals then reduce to ones of the form (4.90). There results

$$\frac{d\eta^{\alpha\beta}(\omega,\theta)}{dV}=\frac{K(mc^2)^{-a+1}}{16\pi^2 c}\sqrt{3}\,e^2\Omega_e\sin\theta J^{\alpha\beta}(a)\left(\frac{2\omega}{3\Omega_e\sin\theta}\right)^{-(a-1)}, \qquad (4.102)$$

with

$$J^{11}(a)=\frac{\frac{2}{3}}{a+1}\,2^{(a-3)/2}\,\Gamma\left(\frac{3a+7}{12}\right)\Gamma\left(\frac{3a-1}{12}\right), \qquad (4.103a)$$

$$J^{12}(a)=-j^{21}(a)=\frac{2i\epsilon\cot\theta}{3}\left(\frac{2\omega}{3\Omega_e\sin\theta}\right)^{-1/2}$$

$$\times 2^{(a-2)/2}\,\frac{(2+a+g(\theta))}{a}\,\Gamma\left(\frac{3a+8}{12}\right)\Gamma\left(\frac{3a+4}{12}\right), \qquad (4.103b)$$

$$J^{22}(a)=\frac{a+\frac{5}{3}}{a+1}\,2^{(a-3)/2}\,\Gamma\left(\frac{3a+7}{12}\right)\Gamma\left(\frac{3a-1}{12}\right). \qquad (4.103c)$$

Thus the frequency spectrum from an optically thin source is a power law $I(\omega)\propto\omega^{-\alpha}$ with

$$\alpha=\frac{a-1}{2}. \qquad (4.104)$$

The degrees of polarization are

$$r_l=-\frac{a+1}{a+\frac{7}{3}} \qquad (4.105)$$

and

$$r_c=-\frac{\epsilon\cot\theta}{3}\left(\frac{\omega}{3\Omega_e\sin\theta}\right)^{-1/2}\frac{(a+1)(2+a+g(\theta))}{a(a+\frac{7}{3})}\,\frac{\Gamma\left(\frac{3a+8}{12}\right)\Gamma\left(\frac{3a+4}{12}\right)}{\Gamma\left(\frac{3a+7}{12}\right)\Gamma\left(\frac{3a-1}{12}\right)}. \qquad (4.106)$$

The sign of (4.105) is negative; this corresponds to the major axis of the polarization ellipse lying along $\mathbf{e}^2$, i.e. along the direction orthogonal to both the line of sight and to the magnetic field in the source. The sign of (4.106) for electrons ($\epsilon=-1$) corresponds to RH polarization for $\cot\theta>0$, i.e. when the angle between the line of sight and the magnetic field in the source is acute, and to LH polarization for $\cot\theta<0$.

Limitations on the Airy Integral Approximation

The conventional theory of synchrotron theory, as developed above, breaks down in a variety of limiting cases, e.g. for particles with very small pitch angles or with relatively low energies, and at very low or very high frequencies. Firstly, consider the Airy integral approximation.

Wild and Hill (1971) made a numerical comparison of the Airy integral approximation in the form (4.83a, b) with the equivalent exact expressions for the Bessel functions with the following results. The Airy integral approximations to $J_s(sx)$ and $J'_s(sx)$ lead to large errors for

$$s \gtrsim (1-x^2)^{-5/2}. \tag{4.107}$$

In the particular case of synchrotron theory this condition, with (4.80) and (4.84), reduces to

$$\frac{\omega}{\Omega} \gtrsim \xi^5 \sin^3\theta, \quad \Omega := \frac{\Omega_e}{\gamma}. \tag{4.108}$$

In vacuo, with $\omega_c = \frac{3}{2}\Omega_e\gamma^2 \sin\theta$, (4.108) is equivalent to

$$\frac{\omega}{\omega_c} \gtrsim \tfrac{2}{3}(\gamma \sin\theta)^2. \tag{4.109}$$

Thus for $\gamma \sin\theta \gg 1$, the Airy integral approximation breaks down only at frequencies very much greater than the mean frequency radiated. One may conclude that the Airy integral approximation is adequate except for two cases: firstly, for $\gamma \sin\theta \lesssim 1$, which corresponds to particles with $p_\perp \lesssim mc$, and, secondly, for the very high frequency emission from individual particles.

Wild and Hill (1971) also constructed improved approximations to the Bessel functions. They showed that the following approximations lead to errors of no more than a few percent even for $x \ll 1$ and for low values of s:

$$J_s(sx) \approx \frac{e^{-s/2s_1}}{(2\pi s)^{1/2}} \left(\frac{3}{2s_2} + \frac{a}{s}\right)^{-1/6}, \tag{4.110a}$$

$$J'_s(sx) \approx \frac{e^{-s/2s_1}}{(2\pi s)^{1/2} x} \left(\frac{3}{2s_2} + \frac{b}{s}\right)^{1/6} \left(1 - \frac{1}{5s^{2/3}}\right), \tag{4.110b}$$

with

$$\frac{1}{2s_1} := \ln\{1 + (1-x^2)^{1/2}\} - \ln x - (1-x^2)^{1/2}, \tag{4.111a}$$

$$s_2 := \tfrac{3}{2}(1-x^2)^{-3/2}. \tag{4.111b}$$

and with

$$a = 0.50330, \quad b = 1.19300. \tag{4.111c, d}$$

These approximations are useful for numerical calculations of gyromagnetic emission from mildly relativistic electrons.

Trubnikov (1958) also developed an approximation to the Bessel functions for mildly relativistic electrons. His approximation is based on what Watson (1944) called the Carlini approximation, and is of a similar form to Wild and Hill's approximation. Trubnikov's approximation for mildly relativistic electrons is neither very simple nor very accurate; it does not seem to have been found useful in practice.

Emission at Low Frequencies

The use of the Airy integral approximation is inadequate at low frequencies as well as at high frequencies. The reasons are twofold. First, at low frequencies the emission from individual particles should not be regarded as a continuous function of frequency. Second, the Airy integral approximation requires $s \gg 1$ whereas emission at low frequencies involves low harmonics. These two points will now be discussed further.

Consider the frequency distribution in the emission from an individual particle at the sth harmonic. The frequency is a unique function of angle θ of emission; it is $\omega = \omega_s$ with

$$\omega_s := \frac{s\Omega}{1 - \beta \cos \alpha \cos \theta}. \tag{4.112}$$

According to (4.58) the mean frequency radiated at the sth harmonic is $\bar{\omega}_s$ with

$$\bar{\omega}_s = \frac{s\Omega}{1 - \beta^2 \cos^2 \alpha}. \tag{4.113}$$

For $\gamma \sin \alpha \gg 1$, (4.112) and (4.113) imply

$$\bar{\omega}_s(1 - |\cos \alpha|) \leqslant \omega_s \leqslant \bar{\omega}_s(1 + |\cos \alpha|). \tag{4.114}$$

Thus the emission at the sth harmonic falls in a restricted range. In practice the emission may be regarded as falling in a much narrower range than is implied by (4.114). This is because all but a fraction γ^{-1} of the power is radiated into a cone of angles $\Delta\theta \approx \gamma^{-1}$ about $\theta = \alpha$. (This result applies both to the total power and to the power radiated at each harmonic.) The effective bandwidth of the emission at the sth harmonic is therefore

$$\Delta\omega_s \approx \Delta\theta \left(\frac{d\omega_s}{d\theta}\right)_{\theta=\alpha} \approx \frac{\bar{\omega}_s}{\gamma \tan \alpha}. \tag{4.115}$$

The emission at high harmonics is a continuous function of frequencies in the sense that the bandwidth $\Delta\omega_s$ exceeds the separation between harmonics. However, this is not the case for

$$s \ll \gamma \tan \alpha. \tag{4.116}$$

When (4.116) is satisfied the emission should be regarded as line emission at $\bar{\omega}_s \approx s\Omega/\sin^2 \alpha$ with a bandwidth $\Delta\omega_s$. In place of the Airy integral approximation it is practical to use the exact expressions for the Bessel functions for low values of s. For example the Bessel functions in (4.49a, b, c) or (4.82a, b, c) appear in forms which are tabulated in standard references (e.g. Abramowitz and Stegun, 1964) or which may be reexpressed in such forms using

$$\int_0^{\beta'} \mathrm{d}y\, J_{2s}(2sy) = \int_0^{\beta'} \mathrm{d}y\, J_0(2sy) - \frac{1}{s}\sum_{n=0}^{s-1} J_{2n+1}(2s\beta'), \tag{4.117}$$

and

$$\int_0^{\beta'} \frac{\mathrm{d}y}{y} J_{2s}(2sy) = \frac{1}{2s}\left[1 - \frac{1}{s\beta'}\sum_{n=0}^{s-1} (2n+1) J_{2n+1}(2s\beta')\right]. \tag{4.118}$$

The polarization of the radiation at low frequencies needs to be discussed with some care. On the one hand, suppose the emission from individual particles is of interest. Then the emission at each harmonic $s \ll \gamma \tan \alpha$ may be treated separately, and at each harmonic the polarization is a unique function at frequency or angle (the frequency is a unique function of angle through (4.112)). Thus in principle the polarization is known but there is a rather complicated variation of the polarization with frequency and/or angle. On the other hand, suppose the emission from a distribution of particles is of interest and that the distribution covers a range of energies so that the emission may be regarded as continuous. For example if the energy range corresponds to $\gamma_1 < \gamma < \gamma_2$, then the emission at neighboring harmonics overlaps for

$$\frac{1}{\gamma_1} > \frac{s+1}{\gamma_2}. \tag{4.119}$$

In particular for $\gamma_2 \gg 2\gamma_1$, even emission at $s=1$ and 2 can be regarded as continuous in frequency. In this case the polarization follows from (4.82a, b, c). The derivative $s\partial/\partial s$ (i.e. with respect to harmonic numbers) in (4.82b) may be converted into an integral with respect to γ using the dependence of s_0 on γ through (4.80). By partial integration the derivative can be transformed into a derivative of the distribution function. It is interesting to note that the degree of circular polarization in the low frequency limit can become quite high. Indeed the Airy integral approximation (4.106) predicts that $|r_c|$ is of order unity for $\omega \approx \Omega_e$, i.e. for emission at $s \approx \gamma$, and predicts non-physical values $|r_c| > 1$ at even lower frequencies. Gyromagnetic emission from mildly relativistic electrons in the solar corona can be quite highly circularly polarized (Section 10.4).

However, it is unlikely that this low frequency emission would be observed in practice. The reason is as follows. Although little is known of the plasma frequency in synchrotron sources it would be reasonable to expect the plasma frequency to be less than or roughly comparable with the electron gyrofrequency Ω_e. Now the low frequency emission occurs at frequencies $\omega \approx s\Omega/\sin^2\alpha$, with $s \lesssim \gamma \tan\alpha$, i.e. at $\omega \lesssim \Omega_e/\sin\alpha\cos\alpha$. However, synchrotron emission is suppressed (the Razin effect) at $\omega \lesssim \gamma\omega_p$. Consequently, only for $\omega_p \ll \Omega_e/\gamma$ would this low frequency emission be possible. Such a low value of the plasma frequency seems implausible for the following reasons.

In practice radio observations of sources outside the solar system are restricted to frequencies above about several megahertz due to the ionosphere and to absorption in the interstellar medium. The low frequency synchrotron spectrum under discussion here could be observed only in sources where the electron gyrofrequency exceeds several megahertz, and this would require a magnetic field strength in excess of several gauss.

Emission from Particles with Small Pitch Angles

Emission from particles with $\gamma\sin\alpha \approx 1$ is dominated by low harmonics and the emission from particles with $\gamma\sin\alpha \ll 1$ is dominated by emission at the fundamental $s=1$. However, unlike the low frequency emission from particles with $\gamma\sin\alpha \gg 1$ this emission need not be at low frequencies. This follows from the form of the mean frequency (4.113) for $\gamma\sin\alpha \ll 1$, i.e. from

$$\bar{\omega}_s \approx \frac{s\Omega}{\sin^2\alpha + \gamma^{-2}} = \begin{cases} s\Omega/\sin^2\alpha & \text{for} \quad \gamma\sin\alpha \gg 1, \\ s\Omega_e\gamma & \text{for} \quad \gamma\sin\alpha \ll 1, \end{cases} \tag{4.120}$$

which can be much greater than the non-relativistic gyrofrequency.

It is relatively simple to treat emission in the case $\gamma\sin\alpha \ll 1$ because the quantity

$$x = \frac{n(\omega)\beta\sin\alpha\sin\theta}{1 - n(\omega)\beta\cos\alpha\cos\theta},$$

which appears in the argument of the Bessel in the emissivity, e.g. in (4.33), is much less than unity. This ensures that the power series expansion of each Bessel function converges rapidly. Thus in this limit the Bessel functions may be approximated by the leading term in its power series expansion:

$$J_s(sx) \approx \frac{1}{s!}\left(\frac{sx}{2}\right)^s, \quad J_s'(sx) \approx \frac{J_s(sx)}{sx}. \tag{4.121}$$

Although the usual formulae for synchrotron radiation cannot be used in the case of small pitch angles it is relatively simple to use the formulae for gyromagnetic emission directly. For example, for $\gamma \sin\alpha \ll 1$ only the term with $s=1$ contributes significantly to the emissivity (4.33). Consequently the emissivity may be approximated by

$$\eta^{\alpha\beta}(\omega, \theta) = \frac{n(\omega)\omega^2 e^2 \beta^2 \sin^2\alpha}{2\pi c(1-n(\omega)\beta\cos\alpha\cos\theta)} F^{\alpha\beta}(\theta)\delta\left(\omega - \frac{\Omega}{1-n(\omega)\beta\cos\alpha\cos\theta}\right), \tag{4.122}$$

with

$$F^{\alpha\beta}(\theta) := \frac{1}{4}\begin{pmatrix} z^2 & i\epsilon z \\ -i\epsilon z & 1 \end{pmatrix}, \tag{4.123}$$

and with

$$z := \frac{\cos\theta - n(\omega)\beta\cos\alpha}{1-n(\omega)\beta\cos\alpha\cos\theta}. \tag{4.124}$$

Furthermore, if we assume $\gamma\sin\alpha \ll 1$, $\gamma \gg 1$ and $\theta \ll 1$ (most of the power is emitted at angles $\theta \lesssim \gamma^{-1}$ for $\alpha \ll \gamma^{-1}$, or at $\pi - \theta \lesssim \gamma^{-1}$ for $\pi - \alpha \ll \gamma^{-1}$) then the following approximations to the total emissivity and to the degrees of polarization follow from (4.122) and (4.123) for $\omega^2 \gg \omega_p^2$:

$$\eta^{\alpha\alpha}(\omega, \theta) \approx \frac{2e^2\Omega_e^2\alpha^2}{\pi c\gamma^2} \frac{(1+\xi^4\theta^4)\xi^6}{(1+\xi^2\theta^2)^5} \delta\left\{\omega - \frac{2\Omega_e\xi^2}{\gamma(1+\xi^2\theta^2)}\right\}, \tag{4.125}$$

$$r_l = \frac{1-z^2}{1+z^2} \approx \frac{2\xi^2\theta^2}{1+\xi^4\theta^4}, \tag{4.126}$$

$$r_c = \frac{-2\epsilon z}{1+z^2} \approx -2\epsilon\frac{\cos\alpha}{|\cos\alpha|}\left(\frac{1-\xi^2\theta^2}{1+\xi^4\theta^4}\right), \tag{4.127}$$

with, as in (4.68),

$$\xi^2 := \gamma^2\left(1 + \frac{\gamma^2\omega_p^2}{\omega^2}\right)^{-1}.$$

(It may be inferred that the emission is strongly suppressed due to the presence of a medium except for $\omega_p \ll \Omega_e$. In other words emission from electrons with $\gamma\sin\alpha \ll 1$ would not occur for $\omega_p \gtrsim \Omega_e$.)

The emission from particles with $\gamma \sin\alpha \ll 1$ may be summarized as follows. For simplicity consider only the case $\alpha \approx 0$ (and not $\alpha \approx \pi$) and $\omega_p = 0$. The frequency $\omega = \Omega/(1-\beta\cos\alpha\cos\theta)$ varies monotonically with θ from $\omega \approx 2\gamma\Omega_e$ at $\theta = 0$ to $\omega \approx \Omega_e/2\gamma$ at $\theta = \pi$; most of the power is concentrated at $\theta \lesssim \gamma^{-1}$ with $\omega \gtrsim \gamma\Omega_e$. The degree of circular polarization varies montonically with θ from $r_c=1$ (for $\epsilon=-1$) at $\theta=0$ through $r_c=0$ at $\theta\approx\gamma^{-1}$ to $r_c=-1$ at $\theta=\pi$. The degree of linear polarization varies from $r_l=0$ at $\theta=0$ to $r_l=1$ at $\theta\approx\gamma^{-1}$ and back to $r_l=0$ at $\theta=\pi$.

Emission from particles with small pitch angles has been discussed in connection with the emission from pulsars and from other synchrotron sources, see for example the following articles for discussion and further reference to the literature:– Zheleznyakov and Shaposhnikov (1972); O'Dell and Sartori (1970); Cavaliere, Morrison and Pacini (1970); Epstein (1973); Epstein and Petrosian (1973); Westfold (1974).

The Razin Effect

The Razin effect is the following: in the presence of an ambient plasma the low frequency emission by relativistic particles is suppressed compared to the corresponding emission in vacuo. Of course, for $\omega<\omega_p$ emission of transverse waves is strictly forbidden. The Razin effect is the suppression of emission in the range $\omega_p<\omega\lesssim\gamma\omega_p$ where γ is the Lorentz factor of the radiating particle. It should be emphasized that this effect of the plasma is one of suppression and not one of absorption. Relativistic particles simply emit less power in the presence of a plasma.

A semi-quantitative treatment of the Razin effect is as follows. The emissivity in synchrotron radiation varies with frequency through $F(\omega/\omega_c)$ which, according to (4.94) say, falls off exponentially for

$$\omega\gg\omega_c=\frac{3\Omega_e\gamma^2\sin\theta}{2\left(1+\dfrac{\gamma^2\omega_p^2}{\omega^2}\right)^{3/2}}, \tag{4.128}$$

where (4.85) and (4.86) have been used. For sufficiently large γ (4.128) reduces to two inequalities:

$$\omega\gg\tfrac{3}{2}\Omega_e\gamma^2\sin\theta\gg\gamma\omega_p, \tag{4.129a}$$

$$\omega\ll\left(\frac{2\gamma\omega_p^3}{3\Omega_e\sin\theta}\right)^{1/2}\ll\gamma\omega_p. \tag{4.129b}$$

Thus there is an effective frequency range of emission with the emission falling off exponentially at both high *and low* frequencies. As γ is decreased the higher and lower limiting frequencies decrease and approach each other. For

$$\gamma=\gamma_R:=\frac{2\omega_p}{3\Omega_e\sin\theta} \tag{4.130}$$

the two limiting frequencies are equal and equal to $\gamma\omega_p$, and the effective range of emission has shrunk to zero. For smaller values of γ there is no range of frequency where $F(\omega/\omega_c)$ is of order unity, and one concludes that the emission is suppressed for particles with $\gamma \ll \gamma_R$. The frequency

$$\omega_{RT} := \gamma_R \omega_p = \frac{2\omega_p^2}{3\Omega_e \sin\theta}, \tag{4.131}$$

which is only defined to within a factor of order unity, is the characteristic frequency below which the emission is suppressed. This frequency is called the Razin frequency or the Razin-Tsytovich frequency.

Tsytovich's name became associated with this effect following the work of Hornby and Williams (1966). The history of the Razin effect is briefly as follows. Tsytovich (1951) treated gyromagnetic emission in the presence of a medium with $n(\omega) > 1$. His expression for the power radiated by a particle with speed βc moving around a circle of radius R in a medium with refractive index $n(\omega)$ is

$$P = \sum_{s=1}^{\infty} \frac{q^2\beta^3 cs}{R^2} \left[2J'_{2s}(2sn(\omega)\beta) + \left(1 - \frac{1}{n^2(\omega)\beta^2}\right) \int_0^{2sn(\omega)\beta} \mathrm{d}x\, J_{2s}(x) \right]. \tag{4.132}$$

Tsytovich (1951) showed that this reduces to the usual formula (3.54) for Cerenkov emission in the limit $R \to \infty$. Ginzburg (1957) alluded to the fact that synchrotron emission is suppressed in a plasma and Razin (1960) pointed this out explicitly. Scheuer (1965) first brought the existence of the effect to the attention of those not familiar with the Soviet literature. Hornby and Williams (1966) used the result of Tsytovich (1951) in working out the details of the suppression effect.

The Razin effect is a characteristic feature of the emission by relativistic particles in a plasma. The effect occurs for mildly relativistic particles, as was shown by the numerical calculations of Ramaty (1969) and also by the analytical method of Wild and Hill (1971) and Tarnstrom (1976). The Razin effect also occurs for bremsstrahlung from relativistic particles (Melrose, 1972a).

Scheuer (1965) gave a simple physical explanation of the Razin effect, but his explanation is unsatisfactory in that it fails to imply a corresponding suppression for bremsstrahlung. The suppression effect can be explained as follows. The rate at which an emission process proceeds depends, amongst other things, on the phase space (i.e. the integral over $\mathrm{d}^3\mathbf{k}/(2\pi)^3$) available to the emitted photon. For the emission at $\omega < \gamma\omega_p$ the phase space available is very much smaller than it would be for $\omega_p = 0$, as shown by Melrose (1972a).

4.4 INVERSE COMPTON SCATTERING

In this section the scattering of electromagnetic waves by relativistic electrons is treated. This scattering is closely analogous to synchrotron emission, and it is significant in theories of synchrotron sources. Rees' emission mechanism ("nonlinear inverse Compton emission") is also treated.

The terminology associated with this scattering can prove confusing. There are three names which describe essentially the same processes: Thomson scattering, Compton scattering and inverse Compton scattering. Thomson scattering is the classical theory of the scattering. Compton scattering is used both in a generic sense, meaning the quantum mechanical counterpart of Thomson scattering, and in a specific sense meaning the scattering of soft electrons by hard photons. Inverse Compton scattering is of soft photons by hard electrons. In astrophysical applications intrinsically quantum mechanical effects are unimportant, and the three names are used interchangeably. Here, "inverse Compton emission" is used to refer to the production of scattered photons by relativistic electrons. The theory is treated semi-classically using the theory of Thomson scattering as developed in Section 3.4. ("Semi-classical" here means a classical theory with a quantum mechanical notation, see Section 5.1 below.)

For transverse waves *in vacuo* the scattering probability (3.103) may be written as

$$w^{tt}(\mathbf{p}, \mathbf{k}, \mathbf{k}')=\frac{(2\pi)^3 e^4}{m_e^2\omega\omega'}\langle|a^{tt}(\mathbf{k}, \mathbf{k}', \mathbf{v})|^2\rangle\delta\{\omega(1-\boldsymbol{\kappa}\cdot\boldsymbol{\beta})-\omega'(1-\boldsymbol{\kappa}'\cdot\boldsymbol{\beta})\}, \quad (4.133)$$

with $\boldsymbol{\kappa}'$ a unit vector along $\mathbf{k}'$ and with $\boldsymbol{\beta}:=\mathbf{v}/c$. The remaining quantity in (4.133) is obtained from (3.104) after averaging over the initial states of polarization and summing over the final states of polarization, i.e.

$$\begin{aligned}\langle|a^{tt}|^2\rangle &:= \tfrac{1}{2}(\delta_{ij}-\kappa_i\kappa_j)(\delta_{rs}-\kappa'_r\kappa'_s)a_{ir}a_{js}\\ &=\frac{(1-\beta^2)}{2}\left[1+\left\{1-\frac{(1-\beta^2)(1-\boldsymbol{\kappa}\cdot\boldsymbol{\kappa}')}{(1-\boldsymbol{\kappa}\cdot\boldsymbol{\beta})(1-\boldsymbol{\kappa}'\cdot\boldsymbol{\beta})}\right\}^2\right],\end{aligned} \quad (4.134)$$

where arguments are omitted.

Let P, P' and P_p be the power gained by the scattered waves, the unscattered waves and the particle respectively. The unscattered waves and the particle lose energy so that P' and P_p will be negative. These are given by

$$P=\int\frac{d^3\mathbf{k}}{(2\pi)^3}\int\frac{d^3\mathbf{k}}{(2\pi)^3}\, w^{tt}(\mathbf{p}, \mathbf{k}, \mathbf{k}')\hbar\omega N(\mathbf{k}'), \quad (4.135a)$$

$$P' = -\int \frac{d^3\mathbf{k}}{(2\pi)^3} \int \frac{d^3\mathbf{k}'}{(2\pi)^3} w^{tt}(\mathbf{p}, \mathbf{k}, \mathbf{k}')\hbar\omega' N(\mathbf{k}'), \tag{4.135b}$$

$$P_p = -(P+P'). \tag{4.135c}$$

These three quantities can be calculated explicitly if the unscattered waves are isotropic and also in a number of other special cases.

Arbitrarily Distributed Unscattered Radiation

Suppose that $N(\mathbf{k}')$ is arbitrary and let us proceed as far as possible without introducing the assumption of isotropy. Let θ, ϕ and θ', ϕ' be polar angles for $\mathbf{k}$ and $\mathbf{k}'$, respectively, relative to the direction $\boldsymbol{\beta}$:

$$\boldsymbol{\kappa}\cdot\boldsymbol{\beta} =: \beta \cos\theta, \quad \boldsymbol{\kappa}'\cdot\boldsymbol{\beta} =: \beta\cos\theta'.$$

The angle between $\boldsymbol{\kappa}$ and $\boldsymbol{\kappa}'$ is then given by

$$\boldsymbol{\kappa}\cdot\boldsymbol{\kappa}' =: \cos\psi = \cos\theta\cos\theta' + \sin\theta\sin\theta'\cos(\phi-\phi'). \tag{4.136}$$

When the $\mathbf{k}$-integral is written in terms of the variable ω, θ and ϕ, i.e.

$$\int d^3\mathbf{k} \to \int_0^{2\pi} d\phi \int_{-1}^{+1} d\cos\theta \int_0^\infty \frac{d\omega\,\omega^2}{c^3},$$

the ω-integrals in (4.135a) and (4.135b) may be performed over the delta function in the probability (4.133) and the ϕ-integral and the $\cos\theta$-integral are then elementary (although the latter is somewhat lengthy). There results

$$P = \frac{8\pi}{3} r_0^2 c \int \frac{d^3\mathbf{k}'}{(2\pi)^3} \hbar\omega' N(\mathbf{k}') \frac{(1-\beta\cos\theta')^2}{1-\beta^2}, \tag{4.137a}$$

$$P' = -\frac{8\pi}{3} r_0^2 c \int \frac{d^3\mathbf{k}'}{(2\pi)^3} \hbar\omega' N(\mathbf{k}')(1-\beta\cos\theta') \tag{4.137b}$$

where $r_0 := e^2/m_e c^2$ is the classical radius of the electron. Other quantities of the form

$$F^{(n)} := \int \frac{d^3\mathbf{k}}{(2\pi)^3} \int \frac{d^3\mathbf{k}'}{(2\pi)^3} w^{tt}(\mathbf{p}, \mathbf{k}, \mathbf{k}')\omega^n N(\mathbf{k}'), \tag{4.138}$$

for $n > -1$, may be reduced to the form (4.137a, b) by the same procedure. Beside the case $n=1$ (one has $P = \hbar F^{(1)}$), the case $n=0$ is of interest. The rate R that photons are scattered is equal to $F^{(0)}$ and is given by

$$R = \frac{8\pi}{3} r_0^2 c \int \frac{d^3\mathbf{k}'}{(2\pi)^3} N(\mathbf{k}')(1-\beta\cos\theta'). \tag{4.139}$$

Isotropic Unscattered Radiation

For an isotropic distribution of unscattered waves (4.137a, b), (4.135c) and (4.139) give

$$P = \frac{8\pi}{9} r_0^2 c W' \frac{3+\beta^2}{1-\beta^2}, \tag{4.140a}$$

$$P' = -\frac{8\pi}{3} r_0^2 c W', \tag{4.140b}$$

$$P_p = -\frac{32\pi}{9} r_0^2 c W' \frac{\beta^2}{1-\beta^2}, \tag{4.140c}$$

$$R = \frac{8\pi}{3} r_0^2 c n'_{ph}, \tag{4.140d}$$

respectively, where W' and n'_{ph} are the energy density and the number density of photons in the unscattered waves:

$$W' := \int \frac{d^3\mathbf{k}'}{(2\pi)^3} \hbar\omega' N(\mathbf{k}'), \quad n'_{ph} := \int \frac{d^3\mathbf{k}'}{(2\pi)^3} N(\mathbf{k}') \tag{4.141a, b}$$

The mean frequency $\langle\omega\rangle$ of the scattered waves may be related to the mean frequency $\langle\omega'\rangle$ of the unscattered waves by writing $P = \hbar\langle\omega\rangle R$, and

$$\langle\omega'\rangle = \frac{W'}{\hbar n'_{ph}} = \frac{|P'|}{\hbar R}, \tag{4.142}$$

where (4.140b) and (4.140d) have been used.

Then (4.140a) and (4.140d) imply

$$\frac{\langle\omega\rangle}{\langle\omega'\rangle} = \frac{3+\beta^2}{3(1-\beta^2)} = \begin{cases} 1 & \text{for} \quad \beta^2 \ll 1, \\ \frac{4}{3}\gamma^2 & \text{for} \quad \gamma \gg 1. \end{cases} \tag{4.143}$$

In similar fashion $\langle\omega^2\rangle/\langle\omega'^2\rangle$ may be evaluated in terms of $F^{(2)}$. The result is

$$\frac{\langle\omega^2\rangle}{\langle\omega'^2\rangle} = \frac{1+\frac{4}{3}\beta^2+\frac{7}{15}\beta^4}{(1-\beta^2)^2} = \begin{cases} 1 & \text{for} \quad \beta^2 \ll 1, \\ \frac{14}{5}\gamma^4 & \text{for} \quad \gamma \gg 1. \end{cases} \tag{4.144}$$

In particular for *monochromatic* unscattered waves (4.143) and (4.144) may be used to estimate the effective bandwidth for inverse Compton radiation. This follows from

$$\langle(\omega - \langle\omega\rangle)^2\rangle = \frac{23}{40} \langle\omega\rangle^2 \quad \text{for} \quad \gamma \gg 1. \tag{4.145}$$

Uni-directional Unscattered Waves

Another special case which is of some interest is where the unscattered radiation is highly collimated. For example this would apply to unscattered radiation passing through the region in which the scattering occurs. Suppose the radiation is uni-directional and that θ' in (4.137a, b) is the angle between this direction and β. In this case (4.137a, b), (4.135c) and (4.139) give

$$P=\frac{8\pi}{3} r_0^2 c W' \frac{(1-\beta \cos \theta')^2}{1-\beta^2}, \tag{4.146a}$$

$$P'=-\frac{8\pi}{3} r_0^2 c W'(1-\beta \cos \theta'), \tag{4.146b}$$

$$P_p=-\frac{8\pi}{3} r_0^2 c W' \frac{\beta(1-\beta \cos \theta')(\beta-\cos \theta')}{1-\beta^2}, \tag{4.146c}$$

$$R=\frac{8\pi}{3} r_0^2 c n'_{ph}(1-\beta \cos \theta'), \tag{4.146d}$$

respectively. An interesting feature of this example is that for $\cos\theta'>\beta$ (e.g. for $\theta'\lesssim\gamma^{-1}$ for $\gamma\gg 1$) particles actually gain energy as a result of the scattering.

Consider the energy losses by a relativistic particle with pitch angle α due to synchrotron emission in comparison to the energy losses due to inverse Compton scattering. If the unscattered waves are isotropic the ratio of the energy losses is

$$\left(\frac{d\varepsilon}{dt}\right)_{\substack{\text{inverse}\\ \textit{Compton}}} : \left(\frac{d\varepsilon}{dt}\right)_{\textit{synchrotron}} = \tfrac{2}{3} W'_{ph} : \sin^2 \alpha\, W_m, \tag{4.147}$$

where W'_{ph} and $W_m := B^2/8\pi$ are the energy densities in the photons and in the magnetic field, respectively, and where (4.92c) and (4.140c) have been used. On the other hand, if the radiation is beamed along the direction of the magnetic field, then (4.146c), with $\theta'=\alpha$, in place of (4.140c) gives

$$\left(\frac{d\varepsilon}{dt}\right)_{\substack{\text{inverse}\\ \textit{Compton}}} : \left(\frac{d\varepsilon}{dt}\right)_{\textit{synchrotron}} = \tfrac{1}{2}(1-\cos \alpha)^2 W'_{ph} : \sin^2 \alpha\, W_m \tag{4.148}$$

for $\cos \alpha \ll \beta$. Woltjer (1966) appealed to this effect to alleviate a suggested theoretical difficulty, namely that existing models for quasars implied excessive Compton losses. Woltjer's idea was that if the particles are streaming outwards, and so are restricted to relatively small values of α, the ratio of Compton to synchrotron losses is reduced from $(2/3 \sin^2 \alpha)(W'_{ph}/W_m)$, for isotropic radiation, to $\frac{1}{2} \tan^2 (\alpha/2)(W'_{ph}/W_m)$, for highly collimated radiation. The reduction is by a factor $3 \sin^4 (\alpha/2)$.

Scattering of Langmuir Waves into Transverse Waves by Relativistic Electrons

Relativistic electrons can also scatter Langmuir waves into transverse waves. This process was first discussed by Gailitis and Tsytovich (1964) and independently by Colgate (1967), see also Colgate *et al.* (1975), and has been discussed further by Melrose (1971a, b), by authors cited by Kaplan and Tsytovich (1973, Section 13), by Windsor and Kellogg (1974) and by Chiuderi and Veltri (1974).

The probability for Thomson scattering of Langmuir waves into transverse waves may be obtained from (3.106),

$$w^{\mathrm{tl}}(\mathbf{p}, \mathbf{k}, \mathbf{k}')=\frac{(2\pi)^3 e^4}{m_e^2\omega_p\omega}\langle|a^{\mathrm{tl}}(\mathbf{k}, \mathbf{k}', \mathbf{v})|^2\rangle\,\delta\{\omega(1-\boldsymbol{\kappa}\cdot\boldsymbol{\beta})-\omega_p+\mathbf{k}'\cdot\mathbf{v}\}, \quad (4.149)$$

with (after summing over the two final states of polarization)

$$\begin{aligned}\langle|a^{\mathrm{tl}}|^2\rangle&=(\delta_{ij}-\kappa_i\kappa_j)\kappa'_r\kappa'_s a_{ir}a_{js}\\&=\frac{\omega_p^2}{\omega^2}\frac{(1-\beta^2)}{(1-\boldsymbol{\kappa}\cdot\boldsymbol{\beta})^4}[(1-\boldsymbol{\kappa}\cdot\boldsymbol{\beta})^2\{1-(\boldsymbol{\kappa}'\cdot\boldsymbol{\beta})^2\}-(1-\beta^2)(\boldsymbol{\kappa}\cdot\boldsymbol{\kappa}'-\boldsymbol{\kappa}'\cdot\boldsymbol{\beta})]. \quad (4.150)\end{aligned}$$

This scattering can lead to an emission analogous to inverse Compton emission.

Scattering of Langmuir waves by relativistic electrons is qualitatively different depending on whether the Langmuir waves have phase speeds $\beta_\phi c$ much less than the speed of light, or greater than about the speed of light. Omitting the details of the calculation, one finds (for $\gamma\gg 1$)

$$P\approx\frac{8\pi}{9}r_0^2 cW^{\mathrm{l}}\gamma^2\begin{cases}1 & \text{for } \beta_\phi\ll 1,\\ 2 & \text{for } \beta_\phi\gtrsim 1,\end{cases} \quad (4.151)$$

where W^{l} is the energy density in the Langmuir waves, and

$$\frac{\langle\omega\rangle}{\omega_p}\approx\begin{cases}\frac{2}{3}\gamma^2/\beta_\phi \ln(2\gamma^2/\beta_\phi) & \text{for } \beta_\phi\ll 1,\\ \gamma^2 & \text{for } \beta_\phi\gtrsim 1.\end{cases} \quad (4.152)$$

When considering possible astrophysical applications of this process the following point should be kept in mind. Apart from factors of order unity, the power radiated by an electron due to synchrotron radiation, inverse Compton radiation and scattering of Langmuir waves are in the ratio of the energy densities in the magnetic field, the unscattered photons and the Langmuir waves respectively. It would require exceptional circumstances for the energy density in Langmuir waves to exceed the energy density in the magnetic field, and it is highly implausible that this inequality could be satisfied other than in a very localized region of an extended source. Consequently, the mechanism should not be regarded as a realistic alternative to synchrotron radiation. Indeed the only suggested alternative to synchrotron radiation which seems at all plausible is Rees' mechanism.

Rees' Emission Mechanism (Nonlinear Inverse Compton)

Inverse Compton emission as treated so far presupposes that the field of the unscattered wave is weak in the sense that Ω_E, cf. (3.89), is much less than the frequency of the unscattered wave. In the opposite limit the effect of the wave on the motion of the particle cannot be treated as a perturbation. Rees (1971) proposed emission due to the motion of relativistic particles in the magnetic-dipole radiation field (Section 13.1) as an alternative to synchrotron emission for the Crab Nebula. Wider astrophysical applications of Rees' mechanism have been proposed by Blandford and Rees (1972). This emission mechanism has been treated in detail by Gunn and Ostriker (1971), Arons (1972), Blandford (1972) and Stewart (1972), see also Bonometto and Scrascia (1974) and Stewart (1974).

The most widely used name for this mechanism is "non-linear inverse Compton" emission. "Synchro-Compton" radiation has also been used. Firstly, it should be pointed out that most treatments of the emission are purely classical and that "Thomson" is more appropriate than "Compton". The point of departure from a conventional treatment of Thomson scattering is that the effect of the wave on the particle is treated exactly and not as a perturbation. Put another way, in the conventional treatment one regards Thomson scattering as consisting of a hierarchy of processes (single, double, and multiple scatterings) which have the quantum analogs of the scattering of one, two or more initial photons into one final photon; in the alternative ("nonlinear") approach one regards Thomson scattering as analogous to gyromagnetic emission in that the background field (now an electromagnetic wave) causes the motion of the particle to be qualitatively different from constant rectilinear motion. Thus one calculates the exact orbit of the particle in the field of the wave and then inserts this exact orbit in (3.29) to treat the emission.

The first step in deriving an explicit expression for the current is to find an explicit expression for the orbit of the particle. The relevant equation of motion is

$$\frac{d\mathbf{p}}{dt}=q\left[\mathbf{E}+\frac{\mathbf{v}\times\mathbf{B}}{c}\right], \tag{4.153}$$

with $\mathbf{B}=\boldsymbol{\kappa}'\times\mathbf{E}$ for an electromagnetic wave propagating in the direction $\boldsymbol{\kappa}'$ *in vacuo*. Explicit solutions for the orbit have been given by Batygin and Toptygin (1964), by Synge (1969) who also solved the equation with the radiation reaction terms included, and by Gunn and Ostriker (1971). For a monochromatic wave with frequency ω_0, the electric field is a function only of

$$\chi := \omega_0(t - \boldsymbol{\kappa} \cdot \mathbf{r}/c). \tag{4.154}$$

Noting that

$$\gamma(1 - \boldsymbol{\kappa} \cdot \boldsymbol{\beta}) =: a \tag{4.155}$$

is an integral of the motion, one may convert integrals and derivatives with respect to t into ones with respect to χ using

$$\gamma \frac{d\chi}{dt} = \omega_0 a. \tag{4.156}$$

Integrating the equation of motion once gives

$$\mathbf{p}_\perp(\chi) = \mathbf{p}_\perp(\chi_0) + \frac{q}{\omega_0} \int_{\chi_0}^{\chi} d\chi' \, \mathbf{E}(\chi'), \tag{4.157a}$$

$$p_\parallel(\chi) = p_\parallel(\chi_0) + \frac{1}{2amc} \{p_\perp^2(\chi) - p_\perp^2(\chi_0)\} = \frac{p_\perp^2(\chi)}{2amc} - \left(\frac{a^2 - 1}{2a}\right) mc, \tag{4.157b}$$

where $\perp$ and $\parallel$ refer to the direction $\boldsymbol{\kappa}$, and where χ_0 is an arbitrarily chosen value of χ. The second form of (4.157b) is implied by evaluating (4.155) at $\chi = \chi_0$. A further integration is straightforward:

$$\mathbf{r}_\perp(\chi) = \mathbf{r}_\perp(\chi_0) + \frac{1}{am\omega_0} \left[(\chi - \chi_0)\mathbf{p}_\perp(\chi_0) + \frac{q}{\omega_0} \int_{\chi_0}^{\chi} d\chi' \int_{\chi_0}^{\chi'} d\chi'' \mathbf{E}(\chi'') \right], \tag{4.158a}$$

$$r_\parallel(\chi) = r_\parallel(\chi_0) - \frac{c(a^2 - 1)}{\omega_0 2a^2} (\chi - \chi_0) + \frac{1}{2a^2 m^2 c\omega_0} \int_{\chi_0}^{\chi} d\chi' \, p_\perp^2(\chi'). \tag{4.158b}$$

Explicit evaluation of the orbit for an elliptically polarized wave, e.g. for

$$\mathbf{E}(\chi) = \frac{E_0}{(1 + T^2)^{1/2}} (T \cos \chi, \sin \chi, 0), \tag{4.159}$$

where $\boldsymbol{\kappa}$ is along the 3-axis and where T is the axial ratio, leads to an explicit expression for $\mathbf{r}(\chi)$ which depends on χ only through terms of the form χ, $\sin \chi$, $\cos \chi$, $\sin^2 \chi$ and $\cos^2 \chi$.

The next step is to evaluate the current (3.29). By using (4.156), (3.29) may be recast into the form

$$\mathbf{j}(\mathbf{k}, \omega) = q \int_{-\infty}^{\infty} d\chi \, \frac{\mathbf{p}(\chi)}{m\omega_0 a} \exp\left[-i \left\{ \left(\mathbf{k} - \frac{\omega}{c} \boldsymbol{\kappa} \right) \cdot \mathbf{r}(\chi) - \frac{\omega}{\omega_0} \chi \right\} \right]. \tag{4.160}$$

For the purpose of evaluating (4.160) it is appropriate to make further simplifications. It is possible to choose the initial conditions (at $\chi_0 = 0$) such that (4.157a, b) and (4.158a, b) with (4.159) simplify to

$$\mathbf{p}(\chi)=mc\gamma\boldsymbol{\beta}_0+mc$$
$$\times\left(\frac{vT\sin\chi}{(1+T^2)^{1/2}},\ \frac{-v\cos\chi}{(1+T^2)^{1/2}},\ \frac{v\gamma_0}{a(1+T^2)^{1/2}}(\beta_{ox}T\sin\chi-\beta_{oy}\cos\chi)\right.$$
$$\left.+\frac{v^2}{4a}\left\{1+\frac{(1-T^2)\cos 2\chi}{1+T^2}\right\}\right),\tag{4.161}$$

$$\mathbf{r}(\chi)=\frac{c\gamma_0}{a\omega_0}\boldsymbol{\beta}_0\chi+\frac{c}{a\omega_0}\left(\frac{-vT\cos\chi}{(1+T^2)^{1/2}},\ \frac{-v\sin\chi}{(1+T^2)^{1/2}},\ \frac{-v\gamma_0}{a(1+T^2)^{1/2}}\right.$$
$$\left.\times(\beta_{ox}T\cos\chi+\beta_{oy}\sin\chi)+\frac{v^2}{4a}\left\{\chi+\frac{(1-T^2)\sin 2\chi}{2(1+T^2)}\right\}\right)\tag{4.162}$$

where $\boldsymbol{\beta}_0$ is a constant initial vector, with $\gamma_0:=(1-\beta_0^2)^{-1/2}$, and with

$$a=\gamma_0(1-\boldsymbol{\kappa}'\cdot\boldsymbol{\beta}_0),\quad v:=\frac{qE_0}{mc\omega_0}.\tag{4.163}$$

On inserting (4.162) in the exponent in (4.160) with $\mathbf{k}=:(\omega/c)\boldsymbol{\kappa}$ one has

$$-i\left\{\left(\mathbf{k}-\frac{\omega}{c}\boldsymbol{\kappa}'\right)\cdot\mathbf{r}(\chi)-\frac{\omega}{\omega_0}\chi\right\}=-if_0+i(p\chi+R\cos(\chi-\rho)+A\sin 2\chi),\tag{4.164}$$

where f_0 is a constant and with

$$p=\frac{\omega}{\omega_0}\left\{1-\frac{\gamma_0}{a}(\boldsymbol{\kappa}-\boldsymbol{\kappa}')\cdot\boldsymbol{\beta}_0+\frac{v^2}{4a^2}(1-\boldsymbol{\kappa}\cdot\boldsymbol{\kappa}')\right\},$$
$$R:=(q^2+r^2)^{1/2},\quad \tan\rho:=q/r,$$
$$q:=\frac{\omega}{\omega_0}\frac{v}{a(1+T^2)^{1/2}}\left\{\kappa_y-(1-\kappa_z)\frac{\gamma_0\beta_{oy}}{a}\right\},$$
$$r:=\frac{\omega}{\omega_0}\frac{vT}{a(1+T^2)^{1/2}}\left\{\kappa_x-(1-\kappa_z)\frac{\gamma_0\beta_{ox}}{a}\right\},$$
$$A:=\frac{\omega}{\omega_0}\frac{v^2}{8a^2}\left(\frac{1-T^2}{1+T^2}\right)(1-\boldsymbol{\kappa}'\cdot\boldsymbol{\kappa}).\tag{4.165}$$

The evaluation of the current (4.160) would now be straightforward except for the term $A\sin 2\chi$ in (4.164). If this term were absent, as it is for circularly polarized waves ($T^2=1$), the evaluation of the current could be achieved by expanding in Bessel functions as in the derivation of (4.12). Arons (1972) and Stewart (1972) considered only the case of a circularly polarized wave for which this difficulty does not arise. Blandford (1972) considered an elliptically polarized wave and by-passed this difficulty by proceeding directly to the Airy integral approximation. It is possible to evaluate the current explicitly by making a double expansion in Bessel functions, i.e. by expanding both $\exp[i(\omega/\omega_0)R\cos(\chi-\rho)]$ and $\exp[-i(\omega/\omega_0)A\sin 2\chi]$ independently. However, before doing so it is convenient to rewrite the solution for the orbit in yet another form.

Following Arons (1972) it is convenient to introduce a "drift velocity"

$$\boldsymbol{\beta}_D := \frac{\boldsymbol{\beta}_0 + (v^2/4a\gamma_0)\boldsymbol{\kappa}'}{1+v^2/4a\gamma_0}. \tag{4.166}$$

The quantities $\boldsymbol{\beta}_0$ and γ_0 are of no particular physical significance themselves, but the drift velocity may be interpreted as the time-averaged velocity of the particle. The physical significance of $\boldsymbol{\beta}_D$, and the associated momentum $\mathbf{p}_D$ was explained by Brown and Kibble (1964), cf. (5.101) below. One has

$$\gamma_0 = \gamma_D(1+v^2/2)^{1/2} - v^2/4a, \quad \gamma_D = \frac{\gamma_0(1+v^2/4a\gamma_0)}{(1+v^2/2)^{1/2}},$$

$$a = \gamma_D(1-\boldsymbol{\kappa}'\cdot\boldsymbol{\beta}_D)(1+v^2/2)^{1/2} = \gamma_0(1-\boldsymbol{\kappa}'\cdot\boldsymbol{\beta}_0),$$

$$\gamma_0\boldsymbol{\beta}_0 = \gamma_D\boldsymbol{\beta}_D(1+v^2/2)^{1/2} - (v^2/4a)\boldsymbol{\kappa}'. \tag{4.167}$$

It is also convenient to introduce the polarization vector for the electromagnetic wave (4.159),

$$\mathbf{e}' := \frac{(T, i, 0)}{(T^2+1)^{1/2}}. \tag{4.168}$$

In place of (4.165) one has

$$\mathrm{p} = \frac{\omega(1-\boldsymbol{\kappa}\cdot\boldsymbol{\beta}_D)}{\omega_0(1-\boldsymbol{\kappa}'\cdot\boldsymbol{\beta}_D)},$$

$$Re^{i\rho} = \frac{\omega}{\omega_0}\frac{v\mathbf{e}'\cdot\{(1-\boldsymbol{\kappa}'\cdot\boldsymbol{\beta}_D)\boldsymbol{\kappa} - (1-\boldsymbol{\kappa}\cdot\boldsymbol{\kappa}')\boldsymbol{\beta}_D\}}{\gamma_D(1-\boldsymbol{\kappa}'\cdot\boldsymbol{\beta}_D)^2(1+v^2/2)^{1/2}},$$

$$A = \frac{\omega}{\omega_0}\frac{v^2(1-\boldsymbol{\kappa}\cdot\boldsymbol{\kappa}')}{8\gamma_D^2(1-\boldsymbol{\kappa}'\cdot\boldsymbol{\beta}_D)^2(1+v^2/2)}\left(\frac{1-T^2}{1+T^2}\right). \tag{4.169}$$

Now expanding in Bessel functions and writing the result in a form analogous to (4.12), one has (recall that we have assumed $\mathbf{k} = (\omega/c)\boldsymbol{\kappa}$)

$$\mathbf{j}(\mathbf{k}, \omega) = 2\pi q \sum_{s=-\infty}^{\infty} \mathbf{V}^{EM}(s)\,\delta\{\omega(1-\boldsymbol{\kappa}\cdot\boldsymbol{\beta}_D) - s\omega_0(1-\boldsymbol{\kappa}'\cdot\boldsymbol{\beta}_D)\}, \tag{4.170}$$

where the dependence of $\mathbf{V}^{EM}$ on all variables apart from harmonic number

s is not shown explicitly. Detailed calculations give

$$\mathbf{V}^{EM}(s)=e^{-ifo}\sum_{t=-\infty}^{\infty}e^{i(s-2t)(\rho-\pi/2)}$$
$$\times\left[\boldsymbol{\beta}_D J_{s-2t}(R)-\frac{v}{2}\frac{\{(1-\boldsymbol{\kappa}'\cdot\boldsymbol{\beta}_D)\mathbf{E}_{s-2t}(R)+\boldsymbol{\kappa}'\boldsymbol{\beta}_D\cdot\mathbf{E}_{s-2t}(R)\}}{(1-\boldsymbol{\kappa}'\cdot\boldsymbol{\beta}_D)\gamma_D(1+v^2/2)^{1/2}}\right.$$
$$\left.+\frac{v^2(1-T^2)\{e^{-2i\rho}J_{s-2t-2}(R)+e^{2i\rho}J_{s-2t+2}(R)\}}{8(1+T^2)\gamma_D^2(1-\boldsymbol{\kappa}'\cdot\boldsymbol{\beta}_D)(1+v^2/2)}\right]J_t(A), \quad (4.171)$$

with

$$\mathbf{E}_s(R):=\mathbf{e}'\,e^{-i\rho}\,J_{s-1}(R)+\mathbf{e}'^*\,e^{i\rho}\,J_{s+1}(R). \quad (4.172)$$

With the current in the form (4.170), the analysis of "nonlinear inverse Compton emission" is directly analogous to that of gyromagnetic emission, as given in Section 4.1, although the algebra is considerably more tedious.

"Linear" and "Nonlinear" Thomson Scattering

It should be possible to derive the current (3.96) associated with Thomson scattering from (4.170) with (4.171). Thomson scattering is first order in the electric field E_0 of the unscattered waves, and this suggests that we should expand (4.171) for $v\ll 1$. The lowest order term is for $R=0$, $A=0$ which imply $s=0$ and $t=0$. In this limit the current is actually independent of v, and may be identified as the current associated with Cerenkov emission (which is forbidden here). To first order in v only $s=1$ and $s=-1$ contribute. By assuming the electric field to be of the form (4.159), we have included both positive and negative frequencies in the unscattered wave, and both $s=\pm 1$ do need to be considered. However, the two cases $s=\pm 1$ are closely analogous and only $s=+1$ will be considered explicitly. In this case the δ-function in (4.170) implies $\omega(1-\boldsymbol{\kappa}\cdot\boldsymbol{\beta})=\omega_0(1-\boldsymbol{\kappa}'\cdot\boldsymbol{\beta})$, which we may use to write the leading term in (4.171) as

$$e^{-i(\rho-\pi/2)}\,\boldsymbol{\beta}_D J_1(R)\approx-\frac{iv}{2\gamma_D}\frac{\mathbf{e}'\cdot\{(1-\boldsymbol{\kappa}'\cdot\boldsymbol{\beta}_D)\boldsymbol{\kappa}-(1-\boldsymbol{\kappa}\cdot\boldsymbol{\kappa}')\boldsymbol{\beta}_D\}}{(1-\boldsymbol{\kappa}'\cdot\boldsymbol{\beta})(1-\boldsymbol{\kappa}\cdot\boldsymbol{\beta})},$$

where $J_1(R)\approx\frac{1}{2}R$ and the expression (4.169) (with $v\ll 1$) for $R\,e^{i\rho}$ have been used. Combining this approximation to the leading term with the $R=0$ limit of the second term in (4.171) gives

$$V_i^{EM}(s=1)=\frac{iv}{2}\,e^{-ifo}\,e'_j a_{ij}((\omega/c)\boldsymbol{\kappa},\omega;(\omega_0/c)\boldsymbol{\kappa}',\omega_0;\boldsymbol{\beta}_D c), \quad (4.173)$$

where a_{ij} is given by (3.99). The final term in (4.171) does not contribute to first order in v. There is an analogous contribution from $s=-1$. With v given by (4.163), it is now straightforward to establish that for the electric field (4.159), the two results, namely (3.96) and (4.170) for $v \ll 1$, are the same apart from an unimportant phase factor (which incorporates the choice of initial conditions). Thus (4.170) with (4.171) is a generalization of the current associated with Thomson scattering for the case where the parameter v is arbitrary. The next order terms in v should give a current related to that for double Thomson scattering (see remarks at the end of Section 3.4).

The opposite limit $v \gg 1$ has been treated in detail by the authors cited above. Qualitatively, the dominant contribution for R and s comes from the region where the argument R of the Bessel functions in (4.171) is comparable with their order s. From the δ-function in (4.170) one has $\omega/\omega_0 = s(1-\boldsymbol{\kappa}'\cdot\boldsymbol{\beta}_D)/(1-\boldsymbol{\kappa}\cdot\boldsymbol{\beta}_D)$ and, apart from exceptional initial conditions, $1-\boldsymbol{\kappa}'\cdot\boldsymbol{\beta}_D$ should be of order unity. Consequently, in order to have $R \approx s$, cf. (4.169), one must have $1-\boldsymbol{\kappa}\cdot\boldsymbol{\beta}_D$ of order γ_D^{-1}. From the relation (4.167) between γ_D and γ_0, one finds that for $\gamma_0 \gg v \gg 1$, γ_D is of order γ_0/v. Hence the typical frequency radiated is $\omega \approx s\omega_0/\gamma_D \approx s\omega_0 v/\gamma_D$. As with synchrotron emission the dominant contribution comes from harmonics $s \approx \gamma_0^3$, implying a typical frequency of emission $\omega \approx |qE_0/mc|\gamma_0^2$.

For detailed analysis of the case $v \gg 1$ the reader is referred to the original papers: Arons (1972), Blandford (1972) and Stewart (1972, 1974). For a semi-quantitative analysis it should suffice to note that the typical frequency emitted is $\omega \approx |qE_0/mc|\gamma_0^2$, and that the power radiated per particle is $\sigma_T c\gamma_0^2$ times the energy density in the electromagnetic waves. (The formula (power radiated)$=(\sigma_T c\gamma^2)\times$(energy density) applies to synchrotron emission, and to inverse Compton emission from transverse waves and Langmuir waves as well as to the present case.) The details of the polarization of the emission have been discussed in detail by Blandford (1972). Stewart (1974) included the effect of a static background magnetic field (for the case of the large-amplitude electromagnetic wave propagating along this field) and found that the significant modification was to the "nonlinearity parameter" v. The static magnetic field $\mathbf{B}_0$ may be taken into account by replacing the definition (4.163) of v by

$$v := \frac{qE_0}{mc\omega_0}\left(1+\frac{qB_0}{mc\omega_0}\right)^{-1}. \tag{4.174}$$

PROBLEM SET 4

Answers to relevant questions are given at the end of the text. More difficult questions are marked with an asterisk.

4.1 (a) Find an approximate expression for the emissivity at the s-th harmonic in transverse waves in an isotropic plasma for a non-relativistic electron by retaining the leading term in the power series expansion of the Bessel functions, cf. Appendix A.
(b) Integrate this approximate emissivity over a Maxwellian distribution of electrons to find the emissivity at the s-th harmonic from a thermal plasma.
(c) On summing over s the emissivity in part (b) diverges! Identify the source of this error.

4.2 Find the angle θ at which the handedness of gyromagnetic emission in an isotropic plasma changes sign.

4.3 Suppose that the emissivity in synchroton radiation *in vacuo* is approximated by

$$\eta^{\alpha\beta}(\omega, \theta)=\frac{P^{\alpha\beta}}{2\pi \sin \alpha}\,\delta(\omega-\langle\omega\rangle)\delta(\theta-\alpha), \tag{P4-1}$$

with $P^{\alpha\beta}$ given by (4.52) to (4.54) for $\alpha=\beta=1$ or 2, and with $\langle\omega\rangle$ given by, cf. (4.96a) with (4.86),

$$\langle\omega\rangle=\frac{4}{5\sqrt{3}}\,\Omega_e\gamma^2 \sin \theta. \tag{P4-2}$$

Evaluate the corresponding emissivity (for $\alpha=\beta=1$ or 2) per unit volume from a power-law distribution (4.99) with $\varepsilon_1=0$ and $\varepsilon_2=\infty$ and pitch-angle distribution $\phi(\alpha)$.

4.4 Show that the effective bandwidth for gyromagnetic emission at the s-th harmonic *in vacuo* is

$$\Delta\omega_s=\frac{\Delta\theta s\Omega_e|\cos \theta|}{\gamma \sin^3 \theta}, \qquad \Delta\theta \approx \frac{1}{\gamma}. \tag{P4-3}$$

Hence show that synchrotron emission can be regarded as continuous in frequency (because the bandwidth at a given harmonic is greater than the separation between harmonics) for $s \gg \gamma \tan \theta$.

Remark: Conversely, at low harmonics, $s \lesssim \gamma \tan \theta$, synchrotron emission would consist of discrete lines. However, such emission occurs at $\omega \lesssim \Omega_e/\sin \theta|\cos \theta|$ and is unlikely to be of interest in practice.

4.5 Make the non-relativistic approximation and find the power radiated per unit wave-number k due to gyromagnetic emission by thermal electrons into Langmuir waves. Assume $\Omega_e \ll \omega_p$ so that the dispersion relation of the Langmuir waves may be approximated by $\omega^2=\omega_p^2+3k^2V_e^2$.

4.6 (a) Derive probabilities for gyromagnetic emission at $s=0, \pm 1$, into any mode for a particle with $k_\perp v_\perp \ll \Omega$.
(b) Apply the result of (a) to Alfvén waves, magnetoacoustic waves, and whistlers, cf. (2.135), (2.136), and (2.133) respectively.

Remark: These probabilities are important in the theory of resonant scattering (Section 7.2).

4.7 Derive Tsytovich's (1951) result (4.132).

4.8 Using Wild and Hill's (1971) formulae, cf. (4.110) and (4.111), show that the power radiated in a plasma differs from the power radiated *in vacuo* by a factor

$$\exp\left[-\frac{\lambda}{\sin^2\alpha}\frac{(1-x^2)^{1/2}}{y}\frac{\omega_p^2}{\omega\Omega_e}\right], \tag{P4-4}$$

with

$$x := \frac{\beta\sin\alpha\sin\theta}{1-\beta\cos\alpha\cos\theta}, \quad y := \frac{(1-\beta^2)^{1/2}}{1-\beta\cos\alpha\cos\theta}, \tag{P4-5}$$

and where λ is approximately unity.

4.9 Derive (4.134) and (4.150).

4.10* Find the current associated with double Thomson scattering by a charge at rest, and hence derive the probability for double Thomson scattering.

4.11* Develop the following explanation of the Razin effect: emission at decreasing $\omega < \gamma\omega_p$ in confined to a rapidly decreasing range of angles.
(a) Show that $\omega^2 = \omega_p^2 + k^2c^2$ in the laboratory frame implies $\omega'^2 = \omega_p^2 + k'^2c^2$ in any other inertial frame.
(b) Show that generalizations (for $\omega_p \neq 0$) of the usual formulae for the aberration of light are

$$\omega = \frac{\omega'}{\gamma(1-\beta\cos\theta)}\left[\frac{1+\beta\cos\theta\{1-(\gamma\omega_p/\omega')^2(1-\beta^2\cos^2\theta)\}^{1/2}}{1+\beta\cos\theta}\right] \tag{P4-6}$$

and

$$\tan\theta = \frac{\sin\theta'[\cos\theta'(1-(\omega_p^2/\omega^2)+\beta\{1-(\gamma\omega_p/\omega)^2(1-\beta^2\cos^2\theta')\}^{1/2}]}{\gamma\{\cos^2\theta'-\beta^2-(\omega_p/\gamma\omega)^2\cos^2\theta'\}}$$

$$\approx \frac{\sin\theta'}{\gamma\{\cos\theta'-\beta\}}\frac{\{\cos\theta'+\beta[1-(\gamma\omega_p/\omega)^2(1-\beta^2\cos^2\theta')]^{1/2}\}}{\cos\theta'+\beta} \tag{P4-7}$$

where the approximate expression applies for $\omega^2 \gg \omega_p^2$.
(c) Emission in the instantaneous rest frame of the particle is roughly isotropic (a dipolar pattern in fact). Show that emission in the range $\omega_p \ll \omega \ll \gamma\omega_p$ arises from emission in the instantaneous rest frame at $\omega_p < \omega' < \gamma\omega_p$ and at $\theta' < \omega/\gamma\omega_p$. Moreover, show that the stated emission is restricted to $\theta \lesssim \omega/\gamma\omega_p$.

Remarks: (i) The angles θ and θ' are the angles between **k** and **v** and **k**$'$ and **v** in the laboratory and rest frames.
(ii) The Razin effect may be interpreted as the shrinking of the phase space available for emission due to the restriction $\theta \lesssim \omega/\gamma\omega_p$.

CHAPTER 5

The Induced Processes and Quasi-Linear Theory

In this chapter a semi-classical theory is introduced and used to relate absorption to emission and to determine the secular changes in the distribution of particles due to their emission and absorption of waves. This so-called quasi-linear theory of particle-wave interactions is then extended to treat scattering of waves by particles and other nonlinear wave-particle interactions.

5.1 SEMI-CLASSICAL FORMALISM

For practical reasons it is convenient to introduce a quantum mechanical formalism at this stage. There are two practical benefits which result from the introduction of quantum mechanical ideas. Both are related to the treatment of emission (and absorption) by individual particles. From a classical viewpoint it is straightforward to treat emission by a single particle, e.g. as is done in Section 3.2. However, it is considerably more difficult to take account of the effect of such emission on the particles. One of the benefits of a quantum mechanical formalism is that it ensures that energy (and, where relevant, momentum) is conserved on a microscopic level. One says that a particle emits a photon and that the photon carries off known energy and momentum. The other benefit is that a quantum mechanical formalism allows one to relate absorption to emission on a microscopic level. The general concept is the so-called principle of detailed balancing (or microscopic reversibility) which, for emission and absorption, may be expressed through the Einstein coefficients. (The Einstein coefficients were first used in the connection by Smerd and Westfold (1949), and the quantum approach itself originated with Ginzburg (1939).) Thus the benefit of a quantum mechanical formalism is that it allows one to develop the theory of wave-particle interactions from a single-particle viewpoint (Pines and Schrieffer 1962, Melrose 1968a, Harris 1969), whereas within a purely classical approach the wave-particle interaction is treated as a collective one.

A theory in which the particles are treated using quantum mechanics but the radiation field is treated classically is said to be a *semi-classical* theory. In plasma physics the term "semi-classical" is often used in the more restricted sense that even the particles are treated classically and only the notation is quantum mechanical. That is, in plasma physics it is usually only the formalism which remains semi-classical, with all the calculations being classical. It is interesting that the scope of the semi-classical theory turns out to be wider than has often been thought. For example multi-photon emission and absorption, the photo-electric effect and the polarization of the vacuum can be treated semi-classically, despite the fact that they have often been regarded as evidence for the quantization of the radiation field. In other words, these effects may be treated using a classical theory of the electromagnetic field together with a quantum theory of particles. Thus one may treat the photo-electric emission due to plasmons, or quanta of other plasma modes, without quantizing the electromagnetic field, e.g. Melrose (1974a, b).

The introduction of a quantum mechanical formalism has been viewed with suspicion by some authors. In a formal sense criticism of the introduction of quantum mechanical ideas is unjustified; *the most general theory is quantum electrodynamics*. (A relativistic quantum treatment of plasma processes is developed in Section 13.4 below.) However, there is a limitation in practice in that the usual semi-classical theory presupposes the random phase approximation, see the discussion following (5.3) below. (A quantum theory of coherence does exist, e.g. Glauber (1963), Klauder and Sudarshan (1968), but it has not been used in plasma theory.)

Two specific points have led to controversy concerning the use of a semi-classical formalism. The first controversy was between Twiss (1963) and Ginzburg and Zheleznyakov (1965) on whether or not the quantum approach and the conventional purely classical approach led to equivalent results for the growth rate of Langmuir waves in the two-stream instability. The situation is now clear. The two approaches are equivalent when treating one form of the instability—called the quasi-linear version in Section 10.1 below—but the quantum approach cannot be used to treat another form of the instability—called the reactive version in Section 10.1. One reason is that reactive instabilities involve phase-coherent interactions. The other controversy concerned the conservation of energy and momentum. The disagreement that existed seemed to reduce to a matter of definition of the energy and momentum, e.g. Fukai and Harris (1972). With the definition of the total energy adopted in Section 2.3 above, no difficulty arises. However, the question of the definition of the energy and momentum remains an interesting one, cf. Dewar (1973, 1976).

Photons

The essential idea in the semi-classical formalism is as follows. When a particle in an initial state with energy ε makes a transition to a final state with energy ε' due to the emission of radiation, then the frequency of the emitted wave is determined by

$$\omega = \frac{\varepsilon - \varepsilon'}{\hbar}, \tag{5.1}$$

with $\hbar := h/2\pi$, where h is Planck's constant. Likewise for free particles, if the initial momentum is $\mathbf{p}$ and the final momentum is $\mathbf{p}'$ then the wave vector of the emitted waves is determined by

$$\mathbf{k} = \frac{\mathbf{p} - \mathbf{p}'}{\hbar}. \tag{5.2}$$

One says that a wave quantum with energy $\hbar\omega$ and momentum $\hbar\mathbf{k}$ is emitted. In the presence of an external force, momentum (of the particle plus the radiation field) need not be conserved, and so (5.2) need not apply.

In general we shall call any wave quantum a *photon*. There is a widely used specific name for the quanta of Langmuir waves: these are called *plasmons*. Although other specific names are sometimes used, e.g. phonon for a quantum of a sound wave, only "photon" and "plasmon" will be used here.

Consider the quantity $W^\sigma(\mathbf{k})$ introduced in Section 3.1. By definition

$$\frac{d^3\mathbf{k}}{(2\pi)^3} W^\sigma(\mathbf{k})$$

is the energy density in the waves in the mode σ in the unit elemental range of $\mathbf{k}$-space, denoted $d^3\mathbf{k}/(2\pi)^3$. Thus $W^\sigma(\mathbf{k})$ has the dimensions of an energy. Now each photon has an energy $\hbar\omega^\sigma(\mathbf{k})$. Consequently the quantity

$$N^\sigma(\mathbf{k}) := \frac{W^\sigma(\mathbf{k})}{\hbar\omega^\sigma(\mathbf{k})} \tag{5.3}$$

is a dimensionless function which describes the distribution of the waves (or, as one says, of the photons). The quantity $N^\sigma(\mathbf{k})$ is the *occupation number*. The classical limit applies only for $N^\sigma(\mathbf{k}) \gg 1$.

As already stated, it is sometimes said that the use of a semi-classical description introduces restrictions on the theory. Although this cannot be the case in principle, there is an important practical restriction implicit in the assumption that the occupation number $N^\sigma(\mathbf{k})$ is well-defined. In quantum mechanics the uncertainty relation between conjugate variables, such as momentum and position or energy and time, implies that the uncertainty in occupation number times the uncertainty in phase is of order unity. Consequently, by assuming $N^\sigma(\mathbf{k})$ to be well-defined one is in effect making the *random phase approximation*. This does not mean that the waves must have random phases for the theory to apply. Rather it means that the theory applies provided the phase of the waves is unimportant. Now there are many situations in laboratory plasma physics, and there is one specific context in magnetospheric physics (Helliwell's (1967) theory of whistler emissions) where the phase of the wave is important. The semi-classical theory cannot be used to describe such phase-coherent processes. In astrophysical applications we have no direct information on the phases and, in the absence of any evidence to the contrary, it is appropriate to make the random phase approximation.

The Doppler Condition

According to (3.46) a free particle emits radiation only when the Cerenkov condition

$$\omega - \mathbf{k}\cdot\mathbf{v} = 0 \tag{5.4}$$

is satisfied. For a particle spiralling in a magnetic field the corresponding condition, which follows from (4.15), is the Doppler condition

$$\omega - s\Omega - k_\parallel v_\parallel = 0 \quad \text{with} \quad s = 0, \pm 1, \pm 2, \ldots . \tag{5.5}$$

From a classical viewpoint these conditions can be interpreted as resonance conditions: (5.4) implies that the frequency of the wave is zero in the rest frame of the particle, and (5.5) implies that the frequency of the wave is an integral (including zero and negative integers) multiple of the gyrofrequency of the particle in the frame in which the particle has zero velocity along the field lines. In a semi-classical formalism (5.4) and (5.5) are implied by conservation of energy (in the classical limit).

In fact (5.4) follows from (5.1), (5.2) and

$$\varepsilon = [m^2c^4 + p^2c^2]^{1/2}, \quad \varepsilon' = [m^2c^4 + p'^2c^2]^{1/2}. \tag{5.6}$$

These imply

$$\varepsilon - \hbar\omega = [m^2c^4 + (\mathbf{p} - \hbar\mathbf{k})^2c^2]^{1/2} = [\varepsilon^2 - 2\hbar c^2\mathbf{p}\cdot\mathbf{k} + \hbar^2c^2k^2]^{1/2}. \tag{5.7}$$

On expanding in powers of $\hbar$ (that is assuming $\hbar\omega \ll \varepsilon$ and $\hbar k \ll p$, one has

$$\omega = \mathbf{k}\cdot\mathbf{v} + \frac{\hbar}{2\varepsilon}\{(\mathbf{k}\cdot\mathbf{v})^2 - k^2c^2\} + \cdots. \tag{5.8}$$

The expansion (5.8) is equivalent to a Taylor series expansion:

$$\varepsilon - \hbar\omega = \left\{1 - \hbar\mathbf{k}\cdot\frac{\partial}{\partial\mathbf{p}} + \frac{1}{2}\left(\hbar\mathbf{k}\cdot\frac{\partial}{\partial\mathbf{p}}\right)^2 + \cdots\right\}\varepsilon,$$

i.e.

$$\omega = \frac{1}{\hbar}\left\{\hbar\mathbf{k}\cdot\frac{\partial}{\partial\mathbf{p}} - \frac{1}{2}\left(\hbar\mathbf{k}\cdot\frac{\partial}{\partial\mathbf{p}}\right)^2 + \cdots\right\}\varepsilon. \tag{5.9}$$

In the classical limit ($\hbar \to 0$), (5.8) reduces to (5.4). The next term on the right hand side of (5.8) is the first quantum correction and is said to describe the recoil of the particle due to the emission of the photon. This recoil term is important in that it allows one to take account of the back-reaction of the emission on the particle.

A Spiralling Charge

Now consider a charge spiralling in a magnetic field. According to non-relativistic quantum theory, the energy eigenvalues are

$$\hbar(n + \tfrac{1}{2} + \tfrac{1}{2}\mu_s)\frac{|q|B}{mc} + \frac{p_\parallel{}^2}{2m}.$$

This result arises as follows. The motion perpendicular to the magnetic field is circular motion with a frequency $|q|\mathrm{B}/mc$. Circular motion is simple harmonic motion and the energy eigenvalues are those of a simple harmonic oscillator, i.e. $\hbar(n+\frac{1}{2})|q|B/mc$, with $n = 0, 1, 2, \ldots$. (The quantum number n should not be confused with the refractive index, also denoted by n.) For a spin-$\frac{1}{2}$ particle the energy associated with the spin is $\frac{1}{2}\mu_s\hbar|q|B/mc$, with $\mu_s = \pm 1$ describing the projection of the spin along $\mathbf{B}$. Finally the component of the momentum along $\mathbf{B}$ is described by a continuous quantum number $p_\parallel$, and the associated contribution to the energy is $p_\parallel{}^2/2m$ as in the classical (i.e. non-quantum) case.

The non-relativistic case is too restrictive for our purposes here. The energy of a particle in a magnetic field in the relativistic case for a spin-$\frac{1}{2}$ particle is given by the energy eigenvalues of the Dirac equation (Section 13.4):

$$\varepsilon = \{m^2c^4 + (2n + 1 + \mu_s)|q|B\hbar c + p_\parallel{}^2c^2\}^{1/2}. \tag{5.10}$$

We shall be concerned ultimately only with the classical limit in which case the spin is of no significance. Consequently, it is convenient to ignore the spin at the outset, e.g. by regarding the value of μ_s as fixed at $\mu_s=-1$.

Now consider emission of a photon by a particle with energy ε given by (5.10). The final energy is ε' which, according to (5.1), is equal to $\varepsilon-\hbar\omega$. Once again ε' must be of the form (5.10) with n and $p_\parallel$ replaced by n' and $p_\parallel'$, respectively, say. The value of $p_\parallel'$ is determined by conservation of the component of momentum along **B**, i.e. $p_\parallel'=p_\parallel-\hbar k_\parallel$. The value of n' is not pre-determined, but n' must be a non-negative integer, $n'=n-s$ say, with $s\leqslant n$. With this choice, ε' is given by

$$\varepsilon'=\{m^2c^4+2n'|q|B\hbar c+(p_\parallel-\hbar k_\parallel)^2c^2\}^{1/2}. \tag{5.11}$$

An expansion analogous to (5.9) involves an expansion in s/n and in $\hbar k_\parallel/p_\parallel$. This expansion gives

$$\omega=\frac{1}{\hbar}\left\{s\frac{\partial}{\partial n}+\hbar k_\parallel\frac{\partial}{\partial p_\parallel}-\frac{1}{2}\left(s\frac{\partial}{\partial n}+\hbar k_\parallel\frac{\partial}{\partial p_\parallel}\right)^2+\cdots\right\}\varepsilon \tag{5.12}$$

In this case the classical limit corresponds to $\hbar\to 0$, $n\to\infty$ such that $\hbar n\to p_\perp^2c/2|q|B$. With $\Omega:=|q|B/\gamma mc$ and $p_\perp=:\gamma mv_\perp$, $p_\parallel=:\gamma mv_\parallel$, (5.12) reduces to

$$\omega=s\Omega+k_\parallel v_\parallel+\frac{\hbar}{2\varepsilon}\{(s\Omega+k_\parallel v_\parallel)^2-k_\parallel{}^2c^2\}+\cdots. \tag{5.12'}$$

In the limit $\hbar=0$, (5.12) reproduces the Doppler condition (5.5). The next term on the right hand side of (5.12), i.e. the first quantum correction, describes the effect of the recoil of the particle due to the emission of the photon.

For a spiralling charge there is an additional quantum number. Suppose the axes are oriented such that the wave vector of the emitted photon is in the x-z plane. It is then convenient to describe the ambient magnetic field ($\mathbf{B}=\text{curl}\,\mathbf{A}$) in terms of the vector potential

$$\mathbf{A}=(-By,0,0). \tag{5.13}$$

Then because this **A** is not a function of x, the component of the kinetic momentum (i.e. of $\gamma m\mathbf{v}$) along the x-axis is a constant of the motion. This component, p_x say, determines the y-component of the center of gyration of the particle through

$$y=\frac{\epsilon p_xc}{|q|B},\quad \epsilon:=\frac{q}{|q|}. \tag{5.14}$$

On emission of a photon p_x changes to p_x', say, given by

$$p_x'=p_x-\hbar k_x. \tag{5.15}$$

Thus, according to (5.14), the y-component of the center of gyration shifts as a result of the emission of a photon with $\mathbf{k}$ in the x-z plane. This effect is of practical interest because it leads to diffusion of particles across mangetic field lines (Section 7.4). Of course, not all photons are emitted in the x-z plane. In fact one expects the distributions of particles and waves to be axially symmetric about the magnetic field lines. In this case emission of a photon with given k_x, k_y, k_z will cause the center of gyration to move by $\Delta y = -\epsilon\hbar k_x c/|q|B$ for $k_y=0$, as implied by (5.14) and (5.15), and also by $\Delta x = \epsilon\hbar k_y c/|q|B$ for $k_y \neq 0$.

As an aside, it might be mentioned that although the inclusion of the changes in the position of the center of gyration is probably not very important in astrophysical applications of quasi-linear theory (Section 7.4), their inclusion is essential in a quasi-linear theory of drift waves. Drift waves (or at least some kinds of drift waves) exist only in the presence of spatial gradients in the plasma parameters, e.g. Mikhailovskii (1967), Krall (1968). The spatial dependence of the distribution function and the way it changes as the waves are emitted and absorbed needs to be taken into account for drift waves.

The Normal and Anomalous Doppler Conditions

The semi-classical formalism leads to the following interesting implications of emission at $s>0$, $s=0$ and $s<0$: emission at $s>0$ (the normal Doppler effect) involves a decrease in $p_\perp$, that at $s=0$ (the Cerenkov effect) involves no change in $p_\perp$, and that at $s<0$ (the anomalous Doppler effect) involves an increase in $p_\perp$. The total energy, and hence $p_\perp^2+p_\parallel^2$, must always decrease however. Furthermore for $s\neq 0$ the Doppler condition (5.5) implies

$$\omega = \frac{s\Omega}{1-n\beta\cos\alpha\cos\theta}. \tag{5.16}$$

In (5.16) and henceforth $n := n(\omega, \theta)$ has its usual meaning of refractive index. It follows from (5.16), and from $|\cos\alpha\ \cos|\leqslant 1$, that emission at $s<0$ is possible only for $n\beta<1$, and *a forteriori* only for $n<1$. Thus *emission via the anomalous Doppler effect ($s<0$), and also emission via the Cerenkov effect ($s=0$), is possible only for waves with refractive index greater than unity.*

Emission at $s=0$ at a given frequency occurs on a cone, $\theta=\theta_c$ say, where θ_c is the solution of $n(\omega, \theta)\beta\cos\alpha\cos\theta=1$ for θ as a function of ω, β and α. In cases of interest, whenever emission at a given frequency can occur due to resonances at $s>0$, $s=0$ and $s<0$, the emission at $s>0$ occurs outside this cone ($\theta>\theta_c$) and emission at $s<0$ occurs inside the cone ($\theta<\theta_c$). (Here it is implicit that α and θ_c are acute; for α and θ_c obtuse, one replaces θ by $\pi-\theta$

and θ_c by $\pi-\theta_c$ in the foregoing statement.) A detailed proof of this statement is not available, and it is possible that exceptions to it exist. Briefly, a sufficient condition for the stated result to hold is that $n\cos\theta$ be a decreasing function of θ over the range where the emission of interest occurs, and $n\cos\theta$ is usually a decreasing function of θ.

Changes due to Emission

When a photon is emitted $p_{\parallel}$ and ε decrease by $\hbar k_{\parallel}$ and $\hbar\omega$ respectively. The change in $p_{\perp}$ may be determined from the change in $p_{\perp}^2$ as follows. According to (5.10) and (5.11), $p_{\perp}^2$ decreases by $2\hbar s|q|B/c$, and hence $p_{\perp}$ decreases by $1/2p_{\perp}$ times $2\hbar s|q|B/c$ in the classical limit. Thus the changes $\Delta p_{\parallel}$, $\Delta p_{\perp}$ and ε in $p_{\parallel}$, $p_{\perp}$ and ε during emission (or absorption) are given by

$$\Delta p_{\parallel}=\hbar k_{\parallel},\quad \Delta p_{\perp}=\frac{\hbar s|q|B}{p_{\perp}c}=\frac{\hbar s\Omega}{v_{\perp}}, \tag{5.17a, b}$$

$$\Delta\varepsilon=\hbar\omega, \tag{5.17c}$$

respectively. The Doppler condition (5.5) then follows from the limit $\hbar\to 0$ of

$$\Delta\varepsilon^2=\Delta(m^2c^4+p_{\perp}^2c^2+p_{\parallel}^2c^2), \tag{5.18}$$

i.e.

$$\Delta\varepsilon=v_{\perp}\Delta p_{\perp}+v_{\parallel}\Delta p_{\parallel}. \tag{5.18$'$}$$

More generally the change in any quantity which is a function of $p_{\perp}$ and $p_{\parallel}$ may be obtained using (5.17a, b) and partial differentiation, cf. (5.70) and (5.71) below.

One major simplification which results from the semi-classical approach is that one can take account of the changes in the properties of the particles at a microscopic level. However, in a purely classical approach, the changes $\Delta p_{\parallel}$ and $\Delta p_{\perp}$ are zero because $\hbar$ is zero. An alternative approach must then be used in which the changes in a distribution of particles is found directly. Put another way, in the semi-classical approach, one finds the way a distribution of particles evolves due to the effects of emission and absorption by summing over the elemental changes due to each emission and absorption event, but in a strictly classical approach the concept of elemental changes in specific emission and absorption events does not exist. In the classical approach conservation of energy and momentum can be imposed only on a macroscopic level and not at the microscopic level.

5.2 THE INDUCED PROCESSES

To every emission process there must be a corresponding absorption process. That this is so can be seen as follows. Suppose, on the contrary, that there exists an emission process which has no associated absorption process. The power radiated would then be proportional to the number of radiating particles. Consider a thermal distribution of particles at a given temperature in a fixed volume. By increasing the number of thermal particles the power radiated could be increased without limit. In particular the power radiated could be made to exceed that of a black body with the given surface area and temperature. This would violate the second law of thermodynamics (e.g. another body absorbing the radiation could be heated to higher than the given temperature).

By reversing this argument it is possible to relate absorption to emission for a thermal distribution of particles radiating by any given process. The basic result used is that, for any thermal distribution of radiating particles, emission and absorption must be related in just the way required to produce thermal radiation when the source is optically thick. This is a statement of *Kirchhoff's law*. A simple but general version of Kirchhoff's law is as follows. Let $\mathrm{d}P^\sigma(\mathbf{k})/\mathrm{d}V$ be the power radiated per unit volume in waves in the mode σ in the range $\mathrm{d}^3\mathbf{k}/(2\pi)^3$ due to spontaneous emission by particles at a temperature T, and let $\gamma^\sigma(\mathbf{k})$ be the absorption coefficient, as defined by (2.86), for this thermal distribution of particles. Then Kirchhoff's law requires

$$\frac{\mathrm{d}P^\sigma(\mathbf{k})}{\mathrm{d}V}=\gamma^\sigma(\mathbf{k})T, \tag{5.19}$$

where, as always here, Boltzmann's constant is chosen to be unity. In the classical case, called the Rayleigh–Jeans limit, a thermal distribution of waves corresponds to $W^\sigma(\mathbf{k})=T$ or, in accord with (5.3), $N^\sigma(\mathbf{k})=T/\hbar\omega^\sigma(\mathbf{k})$.

However Kirchhoff's law applies only to thermal distributions and, although it is possible to derive results which continue to apply when the requirement of thermal equilibrium is relaxed, this is an inconvenient way to proceed. It is much more convenient to impose the requirements of the second law of thermodynamics at a microscopic level. This was first done by Einstein and the resulting relations may be expressed in terms of the Einstein coefficients.

"*Induced*", "*Stimulated*" *and* "*Coherent*"

Before proceeding it is appropriate to discuss the usage of three words: induced, stimulated and coherent. These are used with several different meanings which overlap.

Firstly, on a microscopic level emission may be separated into two parts, one of which is called *spontaneous emission*, which is independent of the ambient radiation field, and the other is called *stimulated emission*. The inverse process is called *true absorption*, stimulated absorption or simply absorption. Collectively, stimulated emission and true absorption are called the *induced processes* and their net effect is called *absorption* or net absorption. There is little room for confusion at this stage although "stimulated" and "induced" are sometimes used interchangeably. Secondly, when a collection of particles is considered, the total effect of spontaneous emission is called *incoherent emission*, and the total effect of the induced processes is called absorption or net absorption. It is possible for absorption to be netative, i.e. for the total effect of stimulated emission to dominate over that of true absorption. The consequence of negative absorption is called induced emission or stimulated emission. "Stimulated" emission is used in laser physics. To avoid confusion the term *maser action* is often used in astrophysical applications to imply amplification due to induced or stimulated emission.

Thirdly, the three words are used in connection with any scattering process, e.g. Thomson scattering. The scattering by individual particles, as discussed in Section 3.4, is the counterpart for scattering of spontaneous or incoherent emission. The counterpart of absorption is called *induced scattering* or stimulated scattering, the former usage being the accepted one in plasma physics and the latter in laser physics. The term *coherent scattering* is used but it has an entirely different meaning: coherent scattering refers to several particles scattering in phase with each other.

The Einstein Coefficients

The Einstein coefficients (Einstein, 1917) may be derived as follows. Suppose we are given a system (an atom for example) each of whose states is described by a set of quantum numbers $\{q\}$, some or all of which may be continuous. Let the energy of the state $\{q\}$ be ε_q. Suppose that a radiative transition from $\{q\}$ to another state $\{q'\}$ with energy $\varepsilon_{q'}<\varepsilon_q$ is possible. Let $p^{\sigma}_{qq'}(\mathbf{k})$ be the probability that a particle in the state $\{q\}$ emits a photon in the mode σ in the range $d^3\mathbf{k}/(2\pi)^3$ with transition to the state $\{q'\}$. (We need not be concerned whether $p^{\sigma}_{qq'}(\mathbf{k})$ is a probability or a probability per unit time; the argument is the same in either case.) Of course, this probability is non-zero only for $\hbar\omega^{\sigma}(\mathbf{k})=\varepsilon_q-\varepsilon_{q'}$.

Now consider a thermal distribution of particles in the states $\{q\}$ and $\{q'\}$. Let N_q and $N_{q'}$ be the number of particles in the two states. In thermal equilibrium at a temperature T one has

$$\frac{N_q}{N_{q'}} = \exp\left[-\frac{\varepsilon_q - \varepsilon_{q'}}{T}\right]. \tag{5.20}$$

(The degeneracy factors for the two states have been assumed equal for simplicity.) Furthermore, suppose that this distribution of particles is in equilibrium with a thermal distribution of waves in the mode σ at the same temperature T. A thermal distribution is the Planck distribution

$$N^\sigma(\mathbf{k}) = [\exp\{\hbar\omega^\sigma(\mathbf{k})/T\} - 1]^{-1}. \tag{5.21}$$

The requirement which we now impose is that the total probability of a transition from $\{q\}$ to $\{q'\}$ due to emission of a photon must be equal to the total probability of a transition from $\{q'\}$ to $\{q\}$ due to absorption of a photon. Furthermore this equality must obtain for any temperature. This requirement is essentially the same as the requirement imposed by Kirchhoff's law, but differs from the latter in that it is imposed at a microscopic level.

By hypothesis the probability of a spontaneous transition from $\{q\}$ to $\{q'\}$ by a single particle is $p^\sigma_{qq'}(\mathbf{k})$, and so the probability for N_q particles is N_q times $p^\sigma_{qq'}(\mathbf{k})$. Likewise, the probability of a transition due to stimulated emission is proportional to N_q, and the probability of true absorption with transition from $\{q'\}$ to $\{q\}$ is proportional to $N_{q'}$. The probabilities of stimulated emission and of true absorption are also proportional to the level of excitation of the waves, e.g. to $N^\sigma(\mathbf{k})$. Let us write the probabilities per particle of stimulated emission $\{q\}\to\{q'\}$ and of true absorption $\{q'\}\to\{q\}$ as

$$\underset{s}{p^\sigma_{qq'}}(\mathbf{k})N^\sigma(\mathbf{k}), \quad \underset{a}{p^\sigma_{q'q}}(\mathbf{k})N^\sigma(\mathbf{k})$$

respectively. The equality of the total probability of transitions $\{q\}\to\{q'\}$ and $\{q'\}\to\{q\}$ requires

$$\{p^\sigma_{qq'}(\mathbf{k}) + \underset{s}{p^\sigma_{qq'}}(\mathbf{k})N^\sigma(\mathbf{k})\}N_q = \underset{a}{p^\sigma}_{q'q}(\mathbf{k})N^\sigma(\mathbf{k})N_{q'}. \tag{5.22}$$

Then (5.21) and (5.22), with $\hbar\omega^\sigma(\mathbf{k}) = \varepsilon_q - \varepsilon_{q'}$, together with the fact that (5.22) must hold for all T, imply

$$\underset{s}{p^\sigma_{qq'}}(\mathbf{k}) = \underset{a}{p^\sigma_{q'q}}(\mathbf{k}) = p^\sigma_{qq'}(\mathbf{k}). \tag{5.23}$$

Thus in the present notation the three basic probabilities of spontaneous emission, stimulated emission and true absorption are all equal. This is an expression of the Einstein relations. Note that according to (5.22) and (5.23) the total probability of emission is proportional to $\{1 + N^\sigma(\mathbf{k})\}$ while the probability of true absorption is proportional to $N^\sigma(\mathbf{k})$.

Although the relations (5.23) have been derived by appealing to thermodynamic arguments, the relations themselves apply irrespective of whether or not thermodynamic equilibrium obtains.

Kinetic Equations

Now let us relax the assumptions that the distributions of particles and of waves are thermal and consider the way these distributions evolve due to the emission and absorption of the waves by the particles. For continuous emission processes, the probability per unit time $w^{\sigma}_{qq'}(\mathbf{k})$ is a more appropriate quantity than the probability $p^{\sigma}_{qq'}(\mathbf{k})$ itself.

Consider transitions from $\{q\}$ to $\{q'\}$. In each such transition N_q decreases by unity, $N_{q'}$ increases by unity and, because a photon is emitted, $N^{\sigma}(\mathbf{k})$ increases by unity. The total probability per unit time of such transitions is

$$w^{\sigma}_{qq'}(\mathbf{k})\{1+N^{\sigma}(\mathbf{k})\}N_q.$$

Now consider transitions from $\{q'\}$ to $\{q\}$. In each such transition N_q increases by unity, $N_{q'}$ decreases by unity and $N^{\sigma}(\mathbf{k})$ decreases by unity. The total probability per unit time of such transitions is

$$w^{\sigma}_{qq'}(\mathbf{k})N^{\sigma}(\mathbf{k})N_{q'}.$$

The rate of change of $N^{\sigma}(\mathbf{k})$ due to transitions between $\{q\}$ and $\{q'\}$ may be found by subtracting the rate photons are abosrbed from the rate they are emitted. After summing over all initial states $\{q\}$ the rate of change of $N^{\sigma}(\mathbf{k})$ is given by

$$\begin{aligned}\frac{\mathrm{d}N^{\sigma}(\mathbf{k})}{\mathrm{d}t} &= V^{-1}\sum_q w^{\sigma}_{qq'}(\mathbf{k})[\{1+N^{\sigma}(\mathbf{k})\}N_q - N^{\sigma}(\mathbf{k})N_{q'}] \\ &= V^{-1}\sum_q w^{\sigma}_{qq'}(\mathbf{k})[N_q + N^{\sigma}(\mathbf{k})(N_q - N_{q'})] \qquad (5.24)\end{aligned}$$

where, for simplicity, it has been assumed that for given $\{q\}$ and $\mathbf{k}$ there is a unique final state $\{q'\}$. The term independent of $N^{\sigma}(\mathbf{k})$ on the right hand side pf (5.24) describes the effect of spontaneous emission, and the term proportional to $N^{\sigma}(\mathbf{k})$ describes the effects of the induced processes.

The appearance of a power of the volume V in (5.24) requires comment. The probability has been defined per elemental range of continuous $\mathbf{k}$-values. The discussion leading to (5.24) presupposes that all states including those of the photons are discrete. Discrete $\mathbf{k}$-values, i.e. eigenmodes, may be obtained by imposing appropriate boundary conditions. If the probability is redefined per eigenmode, then in the limit of continuous eigenmodes this redefined probability is V^{-1} times the probability defined for continuous **k**-values.

In a similar fashion by considering transitions between three states q'', q and q', with $\varepsilon_{q''}-\varepsilon_q=\varepsilon_q-\varepsilon_{q'}=\hbar\omega^\sigma(\mathbf{k})$, the rate of change of the number of particles in the state $\{q\}$ may be found:

$$\frac{\mathrm{d}N_q}{\mathrm{d}t}=V^{-1}\sum_{\mathbf{k}}[w^\sigma_{q''q}(\mathbf{k})\{N_{q''}+(N_{q''}-N_q)N^\sigma(\mathbf{k})]$$
$$-w^\sigma_{qq'}(\mathbf{k})\{N_q+(N_q-N_{q'})N^\sigma(\mathbf{k})\}]. \tag{5.25}$$

For continuous **k**-values $V^{-1}\sum_{\mathbf{k}}$ is to be interpreted as $\int\frac{\mathrm{d}\mathbf{k}}{(2\pi)^3}$, while for discrete **k**-values V^{-1} is to be incorporated into $w^\sigma_{qq'}(\mathbf{k})$, which is reinterpreted as the probability per eigenmode.

Equations (5.24) and (5.25) are the kinetic equations describing the effects of emission and absorption on the distributions of particles and of waves.

Specific Probabilities

In the following $p^\sigma(\mathbf{k})$ denotes a probability and $w^\sigma(\mathbf{k})$ denotes a probability per unit time, both for emission of a photon in the mode σ in the elemental range $\mathrm{d}^3\mathbf{k}/(2\pi)^3$.

The classical expressions derived in Section 3.1 for spontaneous emission may be used to derive explicit expressions for the probabilities. Consider the energy radiated in the mode σ in the range $\mathrm{d}^3\mathbf{k}/(2\pi)^3$, i.e. $U^\sigma(\mathbf{k})$ as given by (3.18). On the other hand, this energy radiated is just the energy per photon times the probability of emission, i.e. one has

$$U^\sigma(\mathbf{k})=\hbar\omega^\sigma(\mathbf{k})p^\sigma(\mathbf{k}). \tag{5.26}$$

The resulting expression for $p^\sigma(\mathbf{k})$ is

$$p^\sigma(\mathbf{k})=\frac{4\pi R^\sigma_E(\mathbf{k})}{\hbar\omega^\sigma(\mathbf{k})}\left|\mathbf{e}^{\sigma *}(\mathbf{k})\cdot\mathbf{j}^{\text{ext}}(\mathbf{k},\omega^\sigma(\mathbf{k}))\right|^2. \tag{5.27}$$

With this identification we may proceed to rewrite the classical theory in terms of the semi-classical formalism.

Specific probabilities per unit time for Cerenkov emission and for gyromagnetic emission may be derived in an analogous manner from the results of Section 3.1 and Section 4.1 respectively. The probability for Cerenkov emission follows from (3.47):

$$w^\sigma(\mathbf{p},\mathbf{k})=\frac{8\pi^2q^2R^\sigma_E(\mathbf{k})}{\hbar\omega^\sigma(\mathbf{k})}\left|\mathbf{e}^{\sigma *}(\mathbf{k})\cdot\mathbf{v}\right|^2\delta\{\omega^\sigma(\mathbf{k})-\mathbf{k}\cdot\mathbf{v}\}. \tag{5.28}$$

The probability for gyromagnetic emission follows from (4.18) and may be separated into contributions from each harmonics:

$$w^{\sigma}(\mathbf{p}, \mathbf{k}) =: \sum_{s=-\infty}^{\infty} w^{\sigma}(s, \mathbf{p}, \mathbf{k}), \tag{5.29}$$

$$w^{\sigma}(s, \mathbf{p}, \mathbf{k}) = \frac{8\pi^2 q^2 R_E^{\sigma}(\mathbf{k})}{\hbar\omega^{\sigma}(\mathbf{k})} |\mathbf{e}^{\sigma *}(\mathbf{k}) \cdot \mathbf{V}(s, \mathbf{p}, \mathbf{k})|^2 \delta\{\omega^{\sigma}(\mathbf{k}) - s\Omega - k_{\parallel} v_{\parallel}\}. \tag{5.30}$$

The probability (5.28) is that for emission of a photon with wave vector $\mathbf{k}$ by a particle with momentum $\mathbf{p}$. In taking the classical limit of the kinetic equation (5.25) we shall require an explicit expression for the probability of emission by a particle with momentum $\mathbf{p}+\hbar\mathbf{k}$. It is reasonable to suppose that the required result in the limit $\hbar \to 0$ is given by

$$w^{\sigma}(\mathbf{p}+\hbar\mathbf{k}, \mathbf{k}) = \left(1 + \hbar\mathbf{k} \cdot \frac{\partial}{\partial \mathbf{p}} + \cdots \right) w^{\sigma}(\mathbf{p}, \mathbf{k}). \tag{5.31}$$

Comparison of the exact quantum mechanical expression (see (13.175) below) with that derived by the Taylor series expansion (5.31) shows that the two agree only up to the first quantum correction. This means that starting from the classical result, the first quantum correction can be derived using semi-classical arguments, but that higher order quantum corrections cannot be so obtained.

The result corresponding to (5.31) for gyromagnetic emission is

$$w^{\sigma}(s, \mathbf{p}+\Delta\mathbf{p}, \mathbf{k}) = \left\{1 + \hbar \left(\frac{s\Omega}{v_{\perp}} \frac{\partial}{\partial p_{\perp}} + k_{\parallel} \frac{\partial}{\partial p_{\parallel}}\right) + \cdots \right\} w^{\sigma}(s, \mathbf{p}, \mathbf{k}), \tag{5.32}$$

where $\mathbf{p}+\Delta\mathbf{p}$ is defined to have components $p_{\perp} + \hbar s\Omega/v_{\perp}$ and $p_{\parallel} + \hbar k_{\parallel}$.

5.3 THE QUASI-LINEAR EQUATIONS

The quasi-linear equations (Vedenov *et al.* 1962, Drummond and Pines 1962) are equivalent to the kinetic equations (5.24) and (5.25) in the classical limit. Thus the quasi-linear equations describe the evolution of distributions of waves and of particles due to the emission and absorption of the waves by the particles.

The essential assumption in quasi-linear theory is that the effects of the particle-wave interaction on the orbit of the particle can be treated as a perturbation. This assumption is most easily explained in connection with the specific example of gyromagnetic emission and absorption. In this case the unperturbed orbit is a spiralling motion along the field lines. The emission and absorption of waves tends to scatter the particles, and it is legitimate to regard this scattering as the net effect of many small impulses. Specifically, each time a photon is emitted or absorbed by a particle, the particle recoils. It is important to note that the recoil can be treated semi-classically, and in effect this has been done in Section 5.1. Now suppose that the rate at which photons are emitted and absorbed is increased gradually. Eventually the random walk (as the scattering due to impulsive changes is called) would cause the particle to move by more than a gyroradius in a gyroperiod. In this case the motion of the particle cannot be regarded as spiralling motion to any approximation; it is simply a random walk. Thus for gyromagnetic emission and absorption, the essential condition required for quasi-linear theory to apply is as follows: the randomness in the motion of individual particles due to their emission and absorption of waves must be such that a typical particle deviates from its unperturbed orbit by much less than a gyroradius in a gyroperiod. In practice this assumption is not very restrictive.

Semi-Classical Form of the Quasi-Linear Equations

A semi-classical form of the quasi-linear equations may be obtained from (5.24) and (5.25) by taking the classical limit.

Consider first the case where there is no background magnetic field. The quantum numbers of a particle are then just the components of the momentum $\mathbf{p}$. Consequently, the sets of quantum numbers $\{q''\}$, $\{q\}$ and $\{q'\}$ are to be replaced by $\mathbf{p}+\hbar\mathbf{k}$, $\mathbf{p}$ and $\mathbf{p}-\hbar\mathbf{k}$ respectively. The sum in (5.24) is to be replaced according to

$$V^{-1}\sum_{q} N_q \rightarrow \int \frac{\mathrm{d}^3\mathbf{p}}{(2\pi\hbar)^3}\, n(\mathbf{p}) \rightarrow \int \mathrm{d}^3\mathbf{p} f(\mathbf{p}), \tag{5.33}$$

where $n(\mathbf{p})$ is the occupation number for the particles. In the classical case $n(\mathbf{p})/(2\pi\hbar)^3$ will be rewritten as $f(\mathbf{p})$ and will be called the single-particle distribution function. (In Section 13.4, where the quantum case is treated explicitly, there is an additional factor of 2 in the relation between $n(\mathbf{p})$ and $f(\mathbf{p})$ due to the two spin states, cf. (3.160) below.) The sum in (5.25) is rewritten according to

$$V^{-1}\sum_{\mathbf{k}} N^{\sigma}(\mathbf{k}) \rightarrow \int \frac{\mathrm{d}^3\mathbf{k}}{(2\pi)^3}\, N^{\sigma}(\mathbf{k}). \tag{5.34}$$

The only remaining steps are to use the expansion (5.31) for $w^{\sigma}(\mathbf{p}+\hbar\mathbf{k}, \mathbf{k})$, and to use the Taylor series expansions

$$f(\mathbf{p}\pm\hbar\mathbf{k})=\left\{1\pm\hbar\mathbf{k}\cdot\frac{\partial}{\partial\mathbf{p}}+\frac{1}{2}\left(\hbar\mathbf{k}\cdot\frac{\partial}{\partial\mathbf{p}}\right)^2+\cdots\right\}f(\mathbf{p}). \tag{5.35}$$

When only those terms which remain meaningful in the limit $\hbar\to 0$ are retained, (5.24) reduces to

$$\frac{\mathrm{d}N^\sigma(\mathbf{k})}{\mathrm{d}t}=\int \mathrm{d}^3\mathbf{p}\,w^\sigma(\mathbf{p},\mathbf{k})\left\{f(\mathbf{p})+N^\sigma(\mathbf{k})\hbar\mathbf{k}\cdot\frac{\partial f(\mathbf{p})}{\partial\mathbf{p}}\right\}. \tag{5.36}$$

Similarly (5.25) reduces to

$$\frac{\mathrm{d}f(\mathbf{p})}{\mathrm{d}t}=\int\frac{\mathrm{d}^3\mathbf{k}}{(2\pi)^3}\hbar\mathbf{k}\cdot\frac{\partial}{\partial\mathbf{p}}\left[w^\sigma(\mathbf{p},\mathbf{k})\left\{f(\mathbf{p})+N^\sigma(\mathbf{k})\hbar\mathbf{k}\cdot\frac{\partial f(\mathbf{p})}{\partial\mathbf{p}}\right\}\right]. \tag{5.37}$$

The corresponding equations for the case of gyromagnetic emission may be derived by first identifying $\{q''\}$, $\{q\}$ and $\{q'\}$ as $\{n+s,\ p_\parallel+\hbar k_\parallel\}$, $\{n,\ p_\parallel\}$ and $\{n-s,\ p_\parallel-\hbar k_\parallel\}$ respectively. In place of (5.36) and (5.37) one finds

$$\frac{\mathrm{d}N^\sigma(\mathbf{k})}{\mathrm{d}t}=\sum_{s=-\infty}^{\infty}\int \mathrm{d}^3\mathbf{p}\,w^\sigma(s,\mathbf{p},\mathbf{k})\left\{f(\mathbf{p})+N^\sigma(\mathbf{k})\hbar\left(\frac{s\Omega}{v_\perp}\frac{\partial}{\partial p_\perp}+k_\parallel\frac{\partial}{\partial p_\parallel}\right)f(\mathbf{p})\right\} \tag{5.38}$$

and

$$\frac{\mathrm{d}f(\mathbf{p})}{\mathrm{d}t}=\sum_{s=-\infty}^{\infty}\int\frac{\mathrm{d}^3\mathbf{k}}{(2\pi)^3}\hbar\left(\frac{s\Omega}{v_\perp}\frac{\partial}{\partial p_\perp}+k_\parallel\frac{\partial}{\partial p_\parallel}\right)\left[w^\sigma(s,\mathbf{p},\mathbf{k})\left\{f(\mathbf{p})+N^\sigma(\mathbf{k})\hbar\left(\frac{s\Omega}{v_\perp}\frac{\partial}{\partial p_\perp}+k_\parallel\frac{\partial}{\partial p_\parallel}\right)f(\mathbf{p})\right\}\right] \tag{5.39}$$

respectively. The change in the coordinate y, which describes the center of gyration of particles, may be taken into account by replacing the differential operator in (5.38) and (5.39) according to

$$\frac{s\Omega}{v_\perp}\frac{\partial}{\partial p_\perp}+k_\parallel\frac{\partial}{\partial p_\parallel}\to\frac{s\Omega}{v_\perp}\frac{\partial}{\partial p_\perp}+k_\parallel\frac{\partial}{\partial p_\parallel}-\frac{ck_x}{qB}\frac{\partial}{\partial y}. \tag{5.40}$$

This generalization follows by including $p_x+\hbar k_x$, p_x and $p_x-\hbar k_x$ in $\{q''\}$, $\{q\}$ and $\{q'\}$ respectively.

Conservation of Energy and Momentum

One of the important properties of the quasi-linear equations is that they ensure conservation of energy and, where relevant, of momentum. As a result of emission, energy is transferred from the particles to the waves and, as a result of absorption, energy is transferred from the waves to the particles.

Conservation of energy can be expressed by the requirement that the sum of the energy in particles and that in waves remains constant, i.e. by

$$\frac{\mathrm{d}}{\mathrm{d}t}\left[\int \mathrm{d}^3\mathbf{p}\varepsilon f(\mathbf{p})+\int \frac{\mathrm{d}^3\mathbf{k}}{(2\pi)^3}\,\hbar\omega^\sigma(\mathbf{k})N^\sigma(\mathbf{k})\right]=0. \tag{5.41}$$

The derivation of (5.41) is as follows. Starting with (5.37) one finds

$$\int \mathrm{d}^3\mathbf{p}\varepsilon\frac{\mathrm{d}f(\mathbf{p})}{\mathrm{d}t}=-\int \mathrm{d}^3\mathbf{p}\int\frac{\mathrm{d}^3\mathbf{k}}{(2\pi)^3}\,\hbar\mathbf{k}\cdot\mathbf{v}w^\sigma(\mathbf{p},\mathbf{k})\left\{f(\mathbf{p})+N^\sigma(\mathbf{k})\hbar\mathbf{k}\cdot\frac{\partial f(\mathbf{p})}{\partial\mathbf{p}}\right\} \tag{5.42}$$

where a partial integration has been performed (and $\partial\varepsilon/\partial\mathbf{p}=\mathbf{v}$ has been used). The probability includes a δ-function which implies $\mathbf{k}\cdot\mathbf{v}=\omega^\sigma(\mathbf{k})$. In view of (5.36), the right hand member of (5.42) can be rewritten as

$$-\int\frac{\mathrm{d}^3\mathbf{k}}{(2\pi)^3}\,\hbar\omega^\sigma(\mathbf{k})\frac{\mathrm{d}N^\sigma(\mathbf{k})}{\mathrm{d}t},$$

which establishes (5.41). The corresponding proof for gyromagnetic emission is analogous, with (5.38), (5.39) and (5.30) used in place of (5.36), (5.37) and (5.28) respectively.

Conservation of momentum may be expressed by

$$\frac{\mathrm{d}}{\mathrm{d}t}\left[\int \mathrm{d}^3\mathbf{p}\,\mathbf{p}f(\mathbf{p})+\int\frac{\mathrm{d}^3\mathbf{k}}{(2\pi)^3}\,\hbar\mathbf{k}N^\sigma(\mathbf{k})\right]=0. \tag{5.43}$$

The derivation of (5.43) is closely analogous to that of (5.41). For gyromagnetic emission conservation of momentum is relevant only for the component along the field lines. (Conservation of momentum in the directions perpendicular to the field lines is not relevant because cylindrical symmetry about the field lines has been assumed throughout.)

Incoherent Emission

The terms on the right hand sides of (5.36) to (5.39) may be divided into those which are independent of $N^\sigma(\mathbf{k})$ and those linear in $N^\sigma(\mathbf{k})$. The former describe the effects of spontaneous emission and the latter the effects of the induced processes.

The collective effect of spontaneous emission by the particles on the waves is referred to as incoherent emission: each particle radiates independently of the other particles. The power radiated per unit volume through incoherent

Cerenkov emission is, from (5.36),

$$\int \frac{d^3\mathbf{k}}{(2\pi)^3}\,\hbar\omega^\sigma(\mathbf{k}) \int d^3\mathbf{p}\,w^\sigma(\mathbf{p},\mathbf{k})f(\mathbf{p}) = \int d^3\mathbf{p}\,f(\mathbf{p}) \int \frac{d^3\mathbf{k}}{(2\pi)^3}\,\hbar\omega^\sigma(\mathbf{k})w^\sigma(\mathbf{p},\mathbf{k}),$$

which is just the power radiated per particle integrated over the distribution of particles. This result could have been written down in Section 3.2, e.g. by integrating

$$\int \frac{d^3\mathbf{k}}{(2\pi)^3}\,P^\sigma(\mathbf{k})$$

over the distribution of particles, with $P^\sigma(\mathbf{k})$ given by (3.47).

On the other hand, the term in (5.37) which describes the effect of incoherent emission on the distribution of particles is new in the sense that it is not implicit in the results of Section 3.2. Likewise the corresponding term in (5.39) is not implicit in the results of Section 4.1. These terms may be used to determine how a spectrum of energetic electrons evolves due to radiation losses (Section 9.3 and Section 9.4 below).

The Absorption Coefficient

The terms which describe the effects of the induced processes on the distribution of waves may be written in the form

$$\left[\frac{dN^\sigma(\mathbf{k})}{dt}\right]^{ind} = -\gamma^\sigma(\mathbf{k})N^\sigma(\mathbf{k}), \tag{5.44}$$

with, from (5.36),

$$\gamma^\sigma(\mathbf{k}) = -\int d^3\mathbf{p}\,w^\sigma(\mathbf{p},\mathbf{k})\hbar\mathbf{k}\cdot\frac{\partial}{\partial\mathbf{p}}f(\mathbf{p}), \tag{5.45}$$

or with, from (5.38),

$$\gamma^\sigma(\mathbf{k}) = -\sum_{s=-\infty}^{\infty}\int d^3\mathbf{p}\,w^\sigma(s,\mathbf{p},\mathbf{k})\hbar\left(\frac{s\Omega}{v_\perp}\frac{\partial}{\partial p_\perp} + k_\parallel\frac{\partial}{\partial p_\parallel}\right)f(\mathbf{p}). \tag{5.46}$$

The quantity $\gamma^\sigma(\mathbf{k})$ is the *absorption coefficient*.

In (5.36), (5.38) and (5.44) the variation of $N^\sigma(\mathbf{k})$ has been written as a temporal one. In practice one is more often concerned with a spatial variation, e.g. with the build-up due to incoherent emission and the attenuation due to absorption along a ray path. To treat spatial variations, or any combination of temporal and spatial variations, one simply rewrites the total time derivative d/dt as a partial time-derivative plus the group speed times the spatial derivative along the ray path. This is done in Section 6.1, where the transfer of radiation through a slowly varying medium is considered.

The absorption coefficient has already been defined in Section 2.3, where the damping of the waves appeared due to the anti-hermitian part of the dielectric tensor being non-zero. It is appropriate to ask how the two definitions of the absorption coefficient are related. To be specific let us consider the case where there is no ambient magnetic field and compare (5.45) with (2.86). The result implied by (5.45) is

$$\gamma^\sigma(\mathbf{k}) = -\frac{8\pi^2 q^2 R_E^\sigma(\mathbf{k})}{\omega^\sigma(\mathbf{k})} \int \mathrm{d}^3\mathbf{p}\, |\mathbf{e}^{\sigma *}(\mathbf{k})\cdot\mathbf{v}|^2 \delta\{\omega^\sigma(\mathbf{k}) - \mathbf{k}\cdot\mathbf{v}\}\mathbf{k}\cdot\frac{\partial f(\mathbf{p})}{\partial \mathbf{p}}, \quad (5.47)$$

where the explicit expression (5.28) for $w^\sigma(\mathbf{p}, \mathbf{k})$ has been inserted. The alternative expression is (2.86), viz.

$$\gamma^\sigma(\mathbf{k}) = -2i\omega^\sigma(\mathbf{k}) R_E^\sigma(\mathbf{k})\, e_i^{\sigma *}(\mathbf{k}) e_j^\sigma(\mathbf{k}) \varepsilon_{ij}^a(\mathbf{k}, \omega^\sigma(\mathbf{k})). \quad (5.48)$$

In (5.47) only the contribution to the damping from one species of particle has been included. We should sum over the contributions from all species before comparing (5.47) and (5.48) because the contributions to $\varepsilon_{ij}^a(\mathbf{k}, \omega)$ from all species is included in (5.48).

The relation between (5.47) (summed over all species) and (5.48) is as follows: the two expressions are equivalent if $\varepsilon_{ij}^a(\mathbf{k}, \omega)$ is identified as the anti-hermitian part of the dielectric tensor for a collisionless plasma, i.e. with the anti-hermitian part of (2.62). That this is so can be seen directly by inserting the relevant expression for $\varepsilon_{ij}^a(\mathbf{k}, \omega)$ in (5.48). Similarly, for a collisionless plasma with an ambient magnetic field, (5.46) with (5.29) and (5.30) (and summed over all species) is equivalent to (5.48) with $\varepsilon_{ij}^a(\mathbf{k}, \omega)$ given by the anti-hermitian part of (2.63). Thus the two expressions for the absorption coefficient are equivalent for collisionless plasmas. In other words, they are equivalent when used to treat collisionless damping.

Diffusion in Momentum Space ($\mathbf{B}=0$)

In (5.37) and (5.39) the terms which describe the effects of the induced processes have the form of a diffusion equation in momentum space. The relevant terms in (5.37) may be written as

$$\frac{\mathrm{d}f(\mathbf{p})}{\mathrm{d}t} = \frac{\partial}{\partial p_i}\left\{D_{ij}(\mathbf{p})\frac{\partial f(\mathbf{p})}{\partial p_i}\right\} \quad (5.49)$$

with

$$D_{ij}(\mathbf{p}) = \int \frac{\mathrm{d}^3\mathbf{k}}{(2\pi)^3}\hbar^2 k_i k_j w^\sigma(\mathbf{p}, \mathbf{k}) N^\sigma(\mathbf{k}). \quad (5.50)$$

Here it is implicit that the particles interact with waves in only one mode. The generalization to several modes involves merely summing over the modes in (5.50). The left hand member of (5.49) is the total time-derivative of f. In general, f may be a function of **p**, **r** and **t**, and then the total time-derivative is to be interpreted according to

$$\frac{\mathrm{d}}{\mathrm{d}t} \rightarrow \frac{\partial}{\partial t} + \mathbf{v} \cdot \frac{\partial}{\partial \mathbf{r}} + \mathbf{F} \cdot \frac{\partial}{\partial \mathbf{p}}, \tag{5.51}$$

where $\mathbf{v}(=\mathrm{d}\mathbf{r}/\mathrm{d}t)$ is the particle velocity and $\mathbf{F}(=\mathrm{d}\mathbf{p}/\mathrm{d}t)$ is the force acting on the particle.

It is often convenient to have the diffusion equation in terms of spherical polar or cylindrical polar coordinates in momentum space. Let us rewrite (5.49) in terms of an arbitrary orthogonal coordinate system with coordinates α, β, γ and with the corresponding set of orthonormal vectors $\hat{\alpha}$, $\hat{\boldsymbol{\beta}}$ and $\hat{\gamma}$ say. Let the quantities h_α, h_β, h_γ be defined by

$$|\mathrm{d}^3\mathbf{p}|^2 =: (h_\alpha \mathrm{d}\alpha)^2 + (h_\beta \mathrm{d}\beta)^2 + (h_\gamma \mathrm{d}\gamma)^2. \tag{5.52}$$

Then the desired general form of the diffusion equation is

$$\frac{\mathrm{d}f}{\mathrm{d}t} = \sum_{\lambda,\mu} \frac{1}{\sqrt{g}} \frac{\partial}{\partial \lambda} \left\{ \frac{\sqrt{g}}{h_\lambda h_\mu} D_{\lambda\mu} \frac{\partial f}{\partial \mu} \right\}, \tag{5.53}$$

with

$$\sqrt{g} := h_\alpha h_\beta h_\gamma, \tag{5.54}$$

where the sums are over $\lambda = \alpha, \beta, \gamma$ and $\mu = \alpha, \beta, \gamma$, and where the arguments of f has been omitted. In particular, (a) for Cartesian coordinates p_1, p_2, p_3 one has $h_1 = h_2 = h_3 = 1$, and (5.54) reproduces (5.49), (b) for spherical polar coordinates p, θ, ϕ one has

$$h_p = 1, \quad h_\theta = p, \quad h_\phi = p \sin\theta, \tag{5.55}$$

and (c) for cylindrical polar coordinates $p_\perp$, ϕ, $p_\parallel$ one has

$$h_\perp = 1, \quad h_\phi = p_\perp, \quad h = 1. \tag{5.56}$$

Finally, the components $D_{\alpha\beta}$ follow from

$$D_{\alpha\beta} = \int \frac{\mathrm{d}^3\mathbf{k}}{(2\pi)^3} \Delta\alpha \Delta\beta w^\sigma(\mathbf{p}, \mathbf{k}) N^\sigma(\mathbf{k}), \tag{5.57}$$

where the change $\Delta\alpha$ in α may be determined in terms of the change $\Delta\mathbf{p}$ in $\mathbf{p}$ by

$$\Delta\alpha = \Delta\mathbf{p} \cdot \frac{\partial\alpha}{\partial\mathbf{p}}, \quad \Delta\mathbf{p} := \hbar\mathbf{k}. \tag{5.58}$$

As a specific example suppose that both the distributions of waves and of particles are isotropic. Then, if the spherical polar coordinates (5.55) are chosen, there is only one non-zero term on the right hand side of (5.53):

$$\frac{\mathrm{d}f(p)}{\mathrm{d}t}=\frac{1}{p^2}\frac{\partial}{\partial p}\left\{p^2 D_{pp}(p)\frac{\partial f(p)}{\partial p}\right\}. \tag{5.59}$$

Furthermore $D_{pp}(p)$ follows from (5.57),

$$D_{pp}(p)=\int\frac{\mathrm{d}^3\mathbf{k}}{(2\pi)^3}(\Delta p)^2 w^\sigma(\mathbf{p},\mathbf{k})N^\sigma(\mathbf{k}), \tag{5.60}$$

with Δp determined from (5.58):

$$\Delta p=\hbar\mathbf{k}\cdot\frac{\partial p}{\partial\mathbf{p}}=\frac{\hbar\mathbf{k}\cdot\mathbf{v}}{v}$$

(The δ-function in the probability $w^\sigma(\mathbf{p},\mathbf{k})$ implies $\omega^\sigma(\mathbf{k})=\mathbf{k}\cdot\mathbf{v}$.) Of course one could derive (5.59) with (5.60) directly from (5.49) with (5.50) by assuming isotropy, but the derivation is laborious.

For a later purpose let us note that the prescription (5.58) for determining the change in any quantity $Q(\mathbf{p})$ due to emission of a photon may be written in terms of the differential operator

$$\hat{D}:=\hbar\mathbf{k}\cdot\frac{\partial}{\partial\mathbf{p}}. \tag{5.62}$$

Thus one has, cf. (5.58),

$$\Delta Q(\mathbf{p}):=Q(\mathbf{p})-Q(\mathbf{p}-\hbar\mathbf{k})=(\hat{D}-\tfrac{1}{2}\hat{D}^2+\cdots)Q(\mathbf{p}). \tag{5.63}$$

Diffusion in Momentum Space ($\mathbf{B}\neq 0$)

When an ambient magnetic field is present the relevant diffusion equation follows from (5.39) which is already in terms of the cylindrical polar coordinates (5.56). The diffusion equation is

$$\frac{\mathrm{d}f(\mathbf{p})}{\mathrm{d}t}=\frac{\partial}{\partial\mathbf{p}_\parallel}\left[\left\{D_{\parallel\parallel}\frac{\partial}{\partial\mathbf{p}_\parallel}+D_{\parallel\perp}\frac{\partial}{\partial\mathbf{p}_\perp}\right\}f(\mathbf{p})\right]$$
$$+\frac{1}{p_\perp}\frac{\partial}{\partial p_\perp}\left[p_\perp\left\{D_{\perp\parallel}\frac{\partial}{\partial p_\parallel}+D_{\perp\perp}\frac{\partial}{\partial p_\perp}\right\}f(\mathbf{p})\right]. \tag{5.64}$$

The diffusion coefficients are

$$D_{AB}=\sum_{s=-\infty}^{\infty}\int\frac{\mathrm{d}^3\mathbf{k}}{(2\pi)^3}\Delta p_A\Delta p_B w^\sigma(s,\mathbf{p},\mathbf{k})N^\sigma(\mathbf{k}), \tag{5.65}$$

where A and B denote $\|$ or $\perp$ with, from (5.17),

$$\Delta p_{\|}=\hbar k_{\|},\quad \Delta p_{\perp}=\frac{\hbar s\Omega}{v_{\perp}}. \tag{5.66}$$

There is no term involving the azimuthal angle ϕ because cylindrical symmetry about **B** is implicit. Any attempt to relax the assumption of cylindrical symmetry leads to the complication that any ϕ-dependence of f implies an associated time-dependence due to the spiralling motion of the particles. More generally, the distribution function must be a function only of constants of the motions, where "the motion" is implicit in the definition of the system. Here "the motion" refers to spiralling in a magnetic field and ϕ is not a constant of the motion. This point is discussed further in connection with (5.76) below.

It is often more convenient to express (5.64) in terms of spherical polar coordinates p, α and ϕ (ϕ will not appear), where α is the pitch angle. The relevant form of the diffusion equation is

$$\frac{\mathrm{d}f(p,\alpha)}{\mathrm{d}t}=\frac{1}{\sin\alpha}\frac{\partial}{\partial\alpha}\left[\sin\alpha\left\{D_{\alpha\alpha}\frac{\partial}{\partial\alpha}+D_{\alpha p}\frac{\partial}{\partial p}\right\}f(p,\alpha)\right]$$
$$+\frac{1}{p^2}\frac{\partial}{\partial p}\left[p^2\left\{D_{p\alpha}\frac{\partial}{\partial\alpha}+D_{pp}\frac{\partial}{\partial p}\right\}f(p,\alpha)\right], \tag{5.67}$$

with

$$\begin{pmatrix} D_{\alpha\alpha} \\ D_{\alpha p}=D_{p\alpha} \\ D_{pp}\end{pmatrix}=\sum_{s=-\infty}^{\infty}\int\frac{\mathrm{d}^3\mathbf{k}}{(2\pi)^3}\,w^{\sigma}(s,\mathbf{p},\mathbf{k})N^{\sigma}(\mathbf{k})\begin{pmatrix}(\Delta\alpha)^2 \\ (\Delta\alpha)(\Delta p) \\ (\Delta p)^2\end{pmatrix}, \tag{5.68}$$

The changes Δp and $\Delta\alpha$ are given by

$$\Delta p=\hbar\left(\frac{\omega-k_{\|}v_{\|}}{v_{\perp}}\frac{\partial}{\partial p_{\perp}}+k_{\|}\frac{\partial}{\partial p_{\|}}\right)p=\frac{\hbar\omega}{v}, \tag{5.69a}$$

and

$$\Delta\alpha=\hbar\left(\frac{\omega-k_{\|}v_{\|}}{v_{\perp}}\frac{\partial}{\partial p_{\perp}}+k_{\|}\frac{\partial}{\partial p_{\|}}\right)\arctan\frac{p_{\perp}}{p_{\|}}=\frac{\hbar(\omega\cos\alpha-k_{\|}v)}{pv\sin\alpha}. \tag{5.69b}$$

Quite generally, the differential operator used in (5.69a, b) may be used to find the change ΔQ in any function Q of $p_{\perp}$ and $p_{\|}$. That is, one may use (5.63) with the differential operator replaced by

$$\hat{D}=\hbar\left(\frac{\omega-k_{\|}v_{\|}}{v_{\perp}}\frac{\partial}{\partial p_{\perp}}+k_{\|}\frac{\partial}{\partial p_{\|}}\right). \tag{5.70}$$

Alternatively, if Q is expressed as a function of p and α one may use (5.69a, b) to rewrite (5.70) in the form

$$\hat{D}=\Delta p\,\frac{\partial}{\partial p}+\Delta\alpha\,\frac{\partial}{\partial\alpha}=\frac{\hbar\omega}{v}\left\{\frac{\partial}{\partial p}+\frac{\cos\alpha-(kv/\omega)\cos\theta}{p\sin\alpha}\frac{\partial}{\partial\alpha}\right\}, \tag{5.71}$$

where $\cos\theta:=k_{\parallel}/k$ has been introduced.

Diffusion Across the Field Lines

The change in the position of the center of gyration of the particle may be included in (5.70) or (5.71) by adding the terms, cf. (5.40),

$$\hat{D}':=\Delta x\,\frac{\partial}{\partial x}+\Delta y\,\frac{\partial}{\partial y}, \tag{5.72}$$

with, cf. (5.14) and (5.15),

$$\Delta y=\frac{\hbar c}{qB}\,k_y, \quad \Delta x=-\frac{\hbar c}{qB}\,k_x. \tag{5.73a, b}$$

In (5.40) the photon was assumed to be in the x-z plane. The generalization to (5.72) requires comment, which is given below in a more general discussion of quasi-linear diffusion. For the present it is appropriate to note that when the distribution of waves is either assumed or known to by cylindrically symmetric, the perpendicular diffusion may be described by

$$\left[\frac{\mathrm{d}f}{\mathrm{d}t}\right]_{\text{perp}}=\frac{\partial}{\partial x}\left(D_{\text{perp}}\,\frac{\partial f}{\partial x}\right)+\frac{\partial}{\partial y}\left(D_{\text{perp}}\,\frac{\partial f}{\partial y}\right), \tag{5.74}$$

with

$$D_{\text{perp}}=\sum_{s=-\infty}^{\infty}\int\frac{\mathrm{d}^3\mathbf{k}}{(2\pi)^3}\,\frac{\hbar^2k_\perp^2c^2}{2q^2B^2}\,w^\sigma(s,\mathbf{p},\mathbf{k})N^\sigma(\mathbf{k}). \tag{5.75}$$

For the terms on the right hand of (5.74) to be non-zero, f must be a function of x or y, and at least one of B, the direction $\mathbf{b}$ of $\mathbf{B}$, $\omega^\sigma(\mathbf{k})$ or $N^\sigma(\mathbf{k})$ must be a function of the same coordinate. Alternatively, if $\rho:=(x^2+y^2)^{1/2}$ is the radial distance from an arbitrary field line, (5.74) with (5.75) may be written as

$$\left[\frac{\mathrm{d}f}{\mathrm{d}t}\right]_{\text{perp}}=\frac{1}{\rho}\,\frac{\partial}{\partial\rho}\left[\rho D_{\text{perp}}\,\frac{\partial f}{\partial\rho}\right]. \tag{5.74}$$

Quasi-Linear Diffusion Under General Conditions

So far we have written the diffusion equation in terms of Cartesian components (5.49), cylindrical components (5.64), spherical polar components

(5.67) and in terms of arbitrary orthogonal components (5.53). There is an even more general form of the diffusion equation in which diffusion is described in terms of a more general set of variables describing the particle. However, before writing down this general form it is necessary to answer the following questions: in how many independent particle-variables can diffusion occur simultaneously and what restrictions are there on these variables? The answers are: diffusion can occur in at most three independent particle-variables simultaneously and these variables must be constants of the unperturbed motion. The reasoning behind these answers is as follows. The distribution function is a function in a 6-dimensional single-particle phase space (the six coordinates may be chosen to be x, y, z, p_x, p_y, p_z for example). The unperturbed distribution function must be independent of time, and hence it can be expressed as a function of constants of the unperturbed motion. A general form of the solution of the equation(s) of motion for any particle consists of three momenta which are constants of the motion and the three conjugate coordinates which are explicit functions of time. These could be the variables involved in the Hamilton-Jacobi equation, e.g. Goldstein (1959). Consequently, the distribution function can depend on at most three independent constants of the motion, and diffusion can occur simultaneously in at most three such constants of the unperturbed motion.

An example of three non-trivial constants of the unperturbed motion would be for a particle trapped in the Earth's radiation belts. In this case there are three periodicities in the motion, and these allow one to define three adiabatic invariants (Section 7.1) which are action-angle variables (Goldstein, 1959). Another example relates directly to (5.74) with (5.75). As pointed out in Section 5.1, when the vector potential is chosen to be independent of x, cf. (5.13), the x-component of $\mathbf{p}$ is a constant of the motion and this constant is reinterpreted in terms of the y-component of the center of gyration. Consequently, if photons were emitted only in the x-z plane, diffusion in y would occur. Alternatively, in the axially symmetric case, photons are emitted uniformly in azimuthal angle, and it is the azimuthal component of momentum p_ϕ which is the relevant constant of the motion. Now p_ϕ may be reexpressed in terms of the radial position of the center of gyration, and diffusion occurs radially with respect of the field lines. Such diffusion may be described by (5.74′) with (5.75).

Let $\Gamma_1, \Gamma_2, \Gamma_3$ be three independent constants of the unperturbed motion, and let J be a relevant Jacobian. The general form of the diffusion equation is

$$\frac{\partial f}{\partial t}=\sum_a \frac{1}{J}\frac{\partial}{\partial\Gamma_a}\{JA_a f\}+\sum_{a,b}\frac{1}{J}\frac{\partial}{\partial\Gamma_a}\left\{JD_{ab}\frac{\partial f}{\partial\Gamma_b}\right\}, \tag{5.76}$$

where the sums are over a, $b=1, 2, 3$. The first term on the right hand side describes the effect of spontaneous emission and the other term describes the effect of the induced processes. The coefficients in (5.76) are given by, cf. (5.65),

$$\begin{bmatrix} A_a \\ D_{ab} \end{bmatrix} = \sum_{s=-\infty}^{\infty} \int \frac{d^3\mathbf{k}}{(2\pi)^3}\, w^\sigma(s, \mathbf{p}, \mathbf{k}) \begin{bmatrix} \Delta\Gamma_a \\ \Delta\Gamma_a \Delta\Gamma_b N^\sigma(\mathbf{k}) \end{bmatrix}, \tag{5.77}$$

with

$$\Delta\Gamma_a = \hat{D}\Gamma_a \tag{5.78}$$

where $\hat{D}$ is given by (5.70) or (5.71). For the choice $p_\parallel$, $p_\perp$ and ρ of variables, cf. (5.74′), one has $J = p_\perp \rho$, and for the choice p, α and ρ one has $J = \rho p^2 \sin\alpha$. For any other choice J is to be obtained by multiplying say $p_\perp \rho$ by the Jacobian of the transformation from $p_\parallel$, $p_\perp$ and ρ to Γ_1, Γ_2 and Γ_3. For the choice of adiabatic invariants mentioned above (assuming the action-angle variables themselves are chosen) one has $J = 1$.

Quasi-linear Equations for Bremsstrahlung

Quasi-linear theory cannot be applied directly to bremsstrahlung because the momentum of the individual electrons is not conserved during collisions. However, the energy is conserved, and provided the distributions of both particles and waves are isotropic one can apply the isotropic form of the quasi-linear equations to bremsstrahlung.

By way of illustration consider bremsstrahlung due to electrons of arbitrary energy interacting with ions at rest. The relevant current is given by (3.86). The probability (per collision) of emission of a photon in the mode σ in the elemental range $d^3\mathbf{k}/(2\pi)^3$ may be obtained by dividing the energy radiated (3.18) by $\hbar\omega^\sigma(\mathbf{k})$. After making the approximations (3.74a, b), which applies for $|\omega^\sigma(\mathbf{k}) - \mathbf{k}\cdot\mathbf{v}| \ll v/b$, one finds

$$p^\sigma(\mathbf{k}) = \frac{16\pi Z_i^2 e^6 R_E^\sigma(\mathbf{k})}{\hbar\omega^\sigma(\mathbf{k}) m^2 \gamma^2 v^2 b^4} \left| \frac{\mathbf{e}^{\sigma *}(\mathbf{k})\cdot\mathbf{b}}{\omega^\sigma(\mathbf{k}) - \mathbf{k}\cdot\mathbf{v}} + \frac{\mathbf{k}\cdot\mathbf{b}\mathbf{e}^{\sigma *}(\mathbf{k})\cdot\mathbf{v}}{\{\omega^\sigma(\mathbf{k}) - \mathbf{k}\cdot\mathbf{v}\}^2} \right|^2. \tag{5.79}$$

The corresponding probability under the more restrictive assumptions of Section 3.3 follows from (5.79) by setting $\gamma = 1$ and $\mathbf{k}\cdot\mathbf{v} = 0 = \mathbf{k}\cdot\mathbf{b}$.

For present purposes we wish to define an average probability per unit time for the emission of bremsstrahlung. As in (3.76) an electron has $2\pi n_i v b\, db$ encounters per unit time with ions of species i and with impact parameters in the range db at b. However, one cannot integrate (5.79) over b because the emission also depends on the direction of $\mathbf{b}$ which is different for different encounters. Nevertheless, it is possible to define a probability averaged over the direction of $\mathbf{v}$, and to interpret this as the average probability of emission per unit time per member of an isotropic distribution of electrons. Let us perform this average over angles for the case where the dipole approximation applies. (It is relatively simple to calculate the averaged quadrupole correction by expanding in $\mathbf{k}\cdot\mathbf{v}/\omega$ in (5.79), retaining terms linear in $\mathbf{k}$ inside the modulus sign and then performing the average.) For $\mathbf{k}\cdot\mathbf{v}=0=\mathbf{k}\cdot\mathbf{b}$ in (5.79), the average probability per unit time, after summing over all species and integrating over b, is

$$\overline{w^\sigma(\mathbf{p},\mathbf{k})}=\sum_i \frac{32\pi^2 Z_i^2 n_i e^6 R_E^\sigma(\mathbf{k}) \ln(b_2/b_1)}{3\hbar m^2\gamma^2 v\{\omega^\sigma(\mathbf{k})\}^3}, \tag{5.80}$$

where the bar denotes the average over directions. The logarithmic factor in (5.80) arises from cutting off the b-integral for $b>b_2$ amd $b<b_1$. It is reasonable to set

$$b_2=\min\left[\lambda_D, v/\omega^\sigma(\mathbf{k})\right],\quad b_1=\max\left[\frac{Z_i e^2}{mv^2},\frac{\hbar}{mv}\right]. \tag{5.81a, b}$$

Although formally (5.80) applies for relativistic particles, the dipole approximation $\omega^\sigma(\mathbf{k})\gg|\mathbf{k}\cdot\mathbf{v}|$ restricts (5.80) to $v\ll c$ $(\gamma\approx 1)$ except for waves, e.g. Langmuir waves, with phase speeds much less than the speed of light. An alternative expression may be obtained for transverse waves (with

$$k=(\omega/c)n(\omega),\quad R_E=\tfrac{1}{2}n(\omega)\,\mathrm{d}\{\omega n(\omega)\}/\mathrm{d}\omega).$$

In this case the average over angles may be performed without further approximation to (5.79). The relevant result is (an extra factor of two arises from summing over the two states of polarization)

$$\overline{w^t(\mathbf{p},\mathbf{k})}=\sum_i \frac{32\pi^2 Z_i^2 n_i e^6 \ln(b_2/b_1)}{3\hbar m^2\gamma^2 v\omega^3(\mathrm{d}/\mathrm{d}\omega)\{\omega n(\omega)\}}\left\{1-\frac{v^2}{c^2}n^2(\omega)\right\}^{-1}, \tag{5.82}$$

with

$$b_2=\min\left[\lambda_D,\frac{\gamma^2 v}{\omega}\right] \tag{5.83}$$

in place of (5.81a).

If the derivation of the quasi-linear equations is now repeated under the assumption that the distributions of waves and particles are isotropic and that the probability is averaged over directions of emission, then in place of (5.36) and (5.37) one finds

$$\frac{\mathrm{d}N^\sigma(k)}{\mathrm{d}t}=\int \mathrm{d}^3\mathbf{p}\,\overline{w^\sigma(\mathbf{p},\mathbf{k})}\left\{f(p)+N^\sigma(k)\frac{\hbar\omega^\sigma(\mathbf{k})}{v}\frac{\partial f(p)}{\partial p}\right\}, \tag{5.84}$$

and

$$\frac{\mathrm{d}f(p)}{\mathrm{d}t}=\int\frac{\mathrm{d}^3\mathbf{k}}{(2\pi)^3}\,\hbar\omega^\sigma(\mathbf{k})\,\frac{1}{p^2}\frac{\partial}{\partial p}\left[\!\left[p^2\left[\overline{w^\sigma(\mathbf{p},\mathbf{k})}\left\{f(p)+N^\sigma(k)\,\frac{\hbar\omega^\sigma(\mathbf{k})}{v}\,\frac{\partial f(p)}{\partial p}\right\}\right]\right]\!\right] \tag{5.85}$$

respectively. The result (5.85) is of little interest in practice because the energy changes resulting from emission and absorption of radiation during collisions are usually unimportant in comparison with the energy changes which result from the collisions themselves.

The result (5.84) leads to the following expression for the absorption coefficient for collisional damping

$$\gamma_c^\sigma(\mathbf{k})=-\int\mathrm{d}^3\mathbf{p}\,\overline{w^\sigma(\mathbf{p},\mathbf{k})}\,\hbar\omega^\sigma(\mathbf{k})\,\frac{\partial f(p)}{\partial\varepsilon}, \tag{5.86}$$

with $\overline{w^\sigma(\mathbf{p},\mathbf{k})}$ given by (5.80). For transverse waves (5.86) is replaced by

$$\gamma_c^t(\omega)=-\frac{1}{2}\int\mathrm{d}^3\mathbf{p}\,\overline{w^t(\mathbf{p},\mathbf{k})}\hbar\omega\,\frac{f(p)}{\partial\varepsilon} \tag{5.87}$$

with $\overline{w^t(\mathbf{p},\mathbf{k})}$ given by (5.82). The extra factor of $\frac{1}{2}$ appears in (5.87) because when treating absorption one is to average rather than sum over the two states of polarization.

An interesting and important result may be established by partially integrating (5.87) and inserting the probability (5.82). One finds that $\gamma_c^t(\omega)$ cannot be negative. Thus, as with synchrotron emission *in vacuo* (Section 6.3) and curvative emission (Section 13.3), no maser action is possible for bremsstrahlung in a plasma. This conclusion refers specifically to a plasma, and does not preclude negative collisional damping in other media, cf. Bekefi (1966, Section 9.3).

5.4 KINETIC EQUATIONS FOR NONLINEAR PROCESSES

Now let us use the semi-classical approach to derive analogs of the quasilinear equations for various nonlinear processes. Of the hierarchy of nonlinear wave-particle and wave-wave interactions, we shall be concerned in detail with only the simplest. These are the scattering of waves (e.g. Thomson scattering) and the coalescence of two waves into a third. Thomson scattering is of interest in itself (Section 3.4, Section 4.4, Section 6.4) and a related "nonlinear" scattering is involved in the plasma emission processes in the solar corona (Section 10.4). The coalescence (and the inverse decay) is also involved in plasma emission.) Kinetic equations for other nonlinear processes (e.g. double emission, photon-photon scattering, nonlinear inverse Compton scattering) are written down at the end of this section.

Kinetic equations for nonlinear processes may be derived by setting up a semi-classical formalism and using the appropriate generalizations of the Einstein relations. Given the basic probability (the counterpart of the probability of spontaneous emission) the total probability of a transition in any direction is obtained by multiplying by the number of particles in the initial state by the occupation number for each absorbed photon and by one plus the occupation number for each emitted photon. The kinetic equations are then obtained by considering the net effect of the various transitions, as in the derivation of the quasi-linear equations.

Scattering Processes

Consider firstly the case of scattering of waves from the mode σ' into the mode σ say. If the scattering particle has initial momentum $\mathbf{p}$ and energy ε, then after the scattering its final momentum and energy are

$$\mathbf{p}'=\mathbf{p}+\hbar\mathbf{k}'-\hbar\mathbf{k}, \quad \varepsilon'=\varepsilon+\hbar\omega^{\sigma'}(\mathbf{k}')-\hbar\omega^{\sigma}(\mathbf{k}).$$

Let $w^{\sigma'\sigma}(\mathbf{p}, \mathbf{k}', \mathbf{k})$ be the basic scattering probability, e.g. (3.106) for Thomson scattering. Then the rate of transitions $\sigma'\rightarrow\sigma$ is proportional to

$$w^{\sigma'\sigma}(\mathbf{p}, \mathbf{k}', \mathbf{k})f(\mathbf{p})N^{\sigma'}(\mathbf{k}')\{1+N^{\sigma}(\mathbf{k})\}.$$

The rate of transitions in the opposite direction $\sigma\rightarrow\sigma'$ is proportional to

$$w^{\sigma'\sigma}(\mathbf{p}, \mathbf{k}', \mathbf{k})f(\mathbf{p}+\hbar\mathbf{k}'-\hbar\mathbf{k})N^{\sigma}(\mathbf{k})\{1+N^{\sigma'}(\mathbf{k}')\}.$$

The probabilities of the direct and inverse processes are the same, i.e.

$$w^{\sigma\sigma'}(\mathbf{p}, \mathbf{k}, \mathbf{k}')=w^{\sigma'\sigma}(\mathbf{p}, \mathbf{k}', \mathbf{k}), \tag{5.88}$$

as may be argued on general grounds (by generalizing the Einstein relations in fact).

Each transition $\sigma'\rightarrow\sigma$ increases N^{σ} and decreases $N^{\sigma'}$ by unity, and each transition $\sigma\rightarrow\sigma'$ decreases N^{σ} and increases $N^{\sigma'}$ by unity. Hence, in the classical limit, one finds

$$\frac{\mathrm{d}N^{\sigma}(\mathbf{k})}{\mathrm{d}t}=\int\frac{\mathrm{d}^3\mathbf{k}'}{(2\pi)^3}\int \mathrm{d}^3\mathbf{p}\, w^{\sigma'\sigma}(\mathbf{p}, \mathbf{k}', \mathbf{k})$$
$$\times\left[f(\mathbf{p})\{N^{\sigma'}(\mathbf{k}')-N^{\sigma}(\mathbf{k})\}+N^{\sigma'}(\mathbf{k}')N^{\sigma}(\mathbf{k})\hbar(\mathbf{k}-\mathbf{k}')\cdot\frac{\partial f(\mathbf{p})}{\partial \mathbf{p}}\right], \tag{5.89a}$$

$$\frac{\mathrm{d}N^{\sigma'}(\mathbf{k}')}{\mathrm{d}t} = -\int \frac{\mathrm{d}^3\mathbf{k}}{(2\pi)^3} \int \mathrm{d}^3\mathbf{p}\, w^{\sigma'\sigma}(\mathbf{p}, \mathbf{k}', \mathbf{k})$$

$$\times \left[f(\mathbf{p})\{N^{\sigma'}(\mathbf{k}) - N^{\sigma}(\mathbf{k})\} + N^{\sigma'}(\mathbf{k}')N^{\sigma}(\mathbf{k})\hbar(\mathbf{k}-\mathbf{k}') \cdot \frac{\partial f(\mathbf{p})}{\partial \mathbf{p}} \right]. \quad (5.89b)$$

The kinetic equation for the particles may be derived by considering three states, as in the derivation of (5.35),

$$\frac{\mathrm{d}f(\mathbf{p})}{\mathrm{d}t} = \int \frac{\mathrm{d}^3\mathbf{k}}{(2\pi)^3} \int \frac{\mathrm{d}^3\mathbf{k}'}{(2\pi)^3} \hbar(\mathbf{k}-\mathbf{k}') \cdot \frac{\partial}{\partial \mathbf{p}}$$

$$\left[\!\left[w^{\sigma'\sigma}(\mathbf{p}, \mathbf{k}', \mathbf{k}) \left[f(\mathbf{p})\{N^{\sigma'}(\mathbf{k}') - N^{\sigma}(\mathbf{k})\} + N^{\sigma'}(\mathbf{k}')N^{\sigma}(\mathbf{k})\hbar(\mathbf{k}-\mathbf{k}') \cdot \frac{\partial f(\mathbf{p})}{\partial \mathbf{p}} \right] \right]\!\right]. \quad (5.89c)$$

The first written terms on the RH side of (5.49a, b, c) describe the effect of *spontaneous scattering*, and the other terms describe the effects of *induced scattering*. The relation between induced scattering and spontaneous scattering is analogous to the relation between absorption and spontaneous emission.

An important feature of induced scattering by isotropic particles is that it involves a net transfer of photons from higher to lower frequencies. The associated damping of the higher frequency waves is called *nonlinear Landau damping*.

The generalization of (5.89a, c, b) to the case of scattering by a particle spiralling in magnetic field may be found as follows. Firstly, write the differential operator as

$$\hat{D} := \hbar(\mathbf{k}-\mathbf{k}') \cdot \frac{\partial}{\partial \mathbf{p}}. \quad (5.90)$$

Next replace the scattering probability by the probability (3.113) for scattering by a spiralling charge, and sum over harmonic number s. Finally replace the differential operator (5.90) by

$$\hat{D} := \hbar \left[\frac{\omega - \omega' - (k_\parallel - k'_\parallel)v_\parallel}{v_\perp} \frac{\partial}{\partial p_\perp} + (k_\parallel - k'_\parallel) \frac{\partial}{\partial p_\parallel} \right]. \quad (5.91)$$

It is straightforward to show that the kinetic equations (5.89a, b, c) imply conservation of energy in the sense that the sum of the energy densities in the unscattered and scattered waves and in the particles is constant. Both spontaneous and induced scattering satisfy this conservation law independently. Momentum is also conserved in the absence of a magnetic field, but only the component of momentum along the field is conserved in the magnetized case.

Wave-wave Interactions

Consider the coalescence process $\sigma'+\sigma''\rightarrow\sigma$ in which two initial photons coalesce into a third with

$$\omega^{\sigma'}(\mathbf{k}')+\omega^{\sigma''}(\mathbf{k}'')=\omega^{\sigma}(\mathbf{k}),\quad \mathbf{k}'+\mathbf{k}''=\mathbf{k}. \tag{5.92}$$

Let the basic probability for this process be

$$u^{\sigma\sigma'\sigma''}(\mathbf{k},\mathbf{k}',\mathbf{k}'').$$

The inverse process is the decay process $\sigma\rightarrow\sigma'+\sigma''$ and it is described by the same probability.

The rate of transitions $\sigma'+\sigma''\rightarrow\sigma$ is

$$u^{\sigma\sigma'\sigma''}(\mathbf{k},\mathbf{k}',\mathbf{k}'')\{1+N^{\sigma}(\mathbf{k})\}N^{\sigma'}(\mathbf{k}')N^{\sigma''}(\mathbf{k}''),$$

and the rate of transitions $\sigma\rightarrow\sigma'+\sigma''$ is

$$u^{\sigma\sigma'\sigma''}(\mathbf{k},\mathbf{k}',\mathbf{k}'')N^{\sigma}(\mathbf{k})\{1+N^{\sigma'}(\mathbf{k}')\}\{1+N^{\sigma''}(\mathbf{k}'')\}.$$

The former transitions decrease $N^{\sigma'}$ and $N^{\sigma''}$ and increase N^{σ} by unity, and the latter transitions increase $N^{\sigma'}$ and $N^{\sigma''}$ and decrease N^{σ} by unity. Hence, in the classical limit, one finds

$$\frac{\mathrm{d}N^{\sigma}(\mathbf{k})}{\mathrm{d}t}=\int\frac{\mathrm{d}^3\mathbf{k}'}{(2\pi)^3}\int\frac{\mathrm{d}^3\mathbf{k}''}{(2\pi)^3}\,u^{\sigma\sigma'\sigma''}(\mathbf{k},\mathbf{k}',\mathbf{k}'')$$
$$\times[N^{\sigma'}(\mathbf{k}')N^{\sigma''}(\mathbf{k}'')-N^{\sigma}(\mathbf{k})\{N^{\sigma'}(\mathbf{k}')+N^{\sigma''}(\mathbf{k}'')\}], \tag{5.93a}$$

$$\frac{\mathrm{d}N^{\sigma'}(\mathbf{k}')}{\mathrm{d}t}=-\int\frac{\mathrm{d}^3\mathbf{k}}{(2\pi)^3}\int\frac{\mathrm{d}^3\mathbf{k}''}{(2\pi)^3}\,u^{\sigma\sigma'\sigma''}(\mathbf{k},\mathbf{k}',\mathbf{k}'')$$
$$\times[N^{\sigma'}(\mathbf{k}')N^{\sigma''}(\mathbf{k}'')-N^{\sigma}(\mathbf{k})\{N^{\sigma'}(\mathbf{k}')+N^{\sigma''}(\mathbf{k}'')\}], \tag{5.93b}$$

$$\frac{\mathrm{d}N^{\sigma''}(\mathbf{k}'')}{\mathrm{d}t}=-\int\frac{\mathrm{d}^3\mathbf{k}}{(2\pi)^3}\int\frac{\mathrm{d}^3\mathbf{k}'}{(2\pi)^3}\,u^{\sigma\sigma'\sigma''}(\mathbf{k},\mathbf{k}',\mathbf{k}'')$$
$$\times[N^{\sigma'}(\mathbf{k}')N^{\sigma''}(\mathbf{k}'')-N^{\sigma}(\mathbf{k})\{N^{\sigma'}(\mathbf{k}')+N^{\sigma''}(\mathbf{k}'')\}]. \tag{5.93c}$$

In this case taking the classical limit consists simply of neglecting a term involving N^{σ} on the RH sides of (5.93a, b, c); this term is intrinsically quantum mechanical (there is no way of describing the associated physical process without using Planck's constant explicitly).

No modification of the kinetic equations is required when a magnetic field is included. However, the magnetic field must be taken into account in deriving an explicit expression for the probability (Section 12.2).

"Emission and Absorption" of Waves by Waves

An interesting limiting case of the coalescence and decay processes is when one high frequency wave emits a low frequency wave, i.e. the limit

$$\omega^{\sigma} \approx \omega^{\sigma''} \gg \omega^{\sigma'}, \quad k \approx k'' \gg k'.$$

This could apply to a higher frequency transverse wave emitting an ion sound wave, a Langmuir wave or some other low frequency wave. Another example is for a magnetized vacuum where a high frequency transverse wave could emit a low frequency transverse wave ("photon splitting").

The probability for such emission of a low frequency wave in the mode σ' by a high frequency wave in the mode σ may be obtained from the probability $u^{\sigma\sigma'\sigma''}(\mathbf{k}, \mathbf{k}', \mathbf{k}'')$ as follows. Firstly note that the conditions (5.92) must be included in δ-functions in $u^{\sigma\sigma'\sigma''}$, cf. Section 10.4 below,

$$u^{\sigma\sigma'\sigma''}(\mathbf{k}, \mathbf{k}', \mathbf{k}'') \propto \delta^3(\mathbf{k}-\mathbf{k}'-\mathbf{k}'')\delta\{\omega^{\sigma}(\mathbf{k})-\omega^{\sigma'}(\mathbf{k}')-\omega^{\sigma''}(\mathbf{k}'')\}.$$

One expands in $\omega^{\sigma'}(\mathbf{k}')/\omega^{\sigma}(\mathbf{k})$ and k'/k, and this involves replacing the final δ-function by

$$\delta(\omega^{\sigma}(\mathbf{k})-\omega^{\sigma'}(\mathbf{k}')-\omega^{\sigma''}(\mathbf{k}'')\} \to \delta\{\omega^{\sigma'}(\mathbf{k}')-\mathbf{k}'\cdot\mathbf{v}_g^{\sigma}(\mathbf{k})\}, \tag{5.94}$$

with $\mathbf{v}_g^{\sigma}(\mathbf{k}) = \partial\omega^{\sigma}(\mathbf{k})/\partial\mathbf{k}$. The probability is then defined by

$$w_{\sigma}^{\sigma'}(\mathbf{k}, \mathbf{k}') := \int \frac{d^3\mathbf{k}''}{(2\pi)^3}\, u^{\sigma\sigma'\sigma}(\mathbf{k}, \mathbf{k}', \mathbf{k}''). \tag{5.95}$$

The kinetic equations (5.93b) and (5.93a) reduce to

$$\frac{dN^{\sigma'}(\mathbf{k}')}{dt} = \int \frac{d^3\mathbf{k}}{(2\pi)^3}\, w_{\sigma}^{\sigma'}(\mathbf{k}, \mathbf{k}')\mathbf{k}' \cdot \frac{\partial N^{\sigma}(\mathbf{k})}{\partial \mathbf{k}}\, N^{\sigma'}(\mathbf{k}') \tag{5.96a}$$

and

$$\frac{dN^{\sigma}(\mathbf{k})}{dt} = \frac{\partial}{\partial k_i}\left[D_{ij}^{\sigma\sigma'}(\mathbf{k})\,\frac{\partial N^{\sigma'}(\mathbf{k})}{\partial k_j}\right], \tag{5.96b}$$

with

$$D_{ij}^{\sigma\sigma'}(\mathbf{k}) = \int \frac{d^3\mathbf{k}'}{(2\pi)^3}\, k_i' k_j' w_{\sigma}^{\sigma'}(\mathbf{k}, \mathbf{k}')N^{\sigma'}(\mathbf{k}'). \tag{5.97}$$

The equations (5.96a, b) are analogous to the quasi-linear equation (5.36) and (5.37) apart from the fact that the spontaneous term is absent. The spontaneous term would be intrinsically quantum mechanical, and could be included by retaining the term N^{σ} neglected in (5.93a, b, c).

For example the equation (5.96b) could be used to treat scattering of high-frequency photons by low-frequency plasma turbulence in Colgate's (1967) theory of quasars.

Kinetic Equations for Other Nonlinear Processes

The approach used above may be extended to derive kinetic equations for any other nonlinear processes of interest. The following are examples.

1) *Double Emission.* The probability for double emission (of two photons simultaneously) is $w^{\sigma'\sigma}(\mathbf{p}, -\mathbf{k}', \mathbf{k})$ where $\omega^{\sigma'}(-\mathbf{k}') = -\omega^{\sigma'}(\mathbf{k}')$ is implied. The rate of emission is proportional to $f(\mathbf{p})(1+N^{\sigma})(1+N^{\sigma'})$, and the rate of absorption is proportional to $f(\mathbf{p}-\hbar\mathbf{k}-\hbar\mathbf{k}')N^{\sigma}N^{\sigma'}$. The kinetic equation for $N^{\sigma}(\mathbf{k})$, for example, is

$$\frac{\mathrm{d}N^{\sigma}(\mathbf{k})}{\mathrm{d}t} = \int \mathrm{d}^3\mathbf{p} \int \frac{\mathrm{d}^3\mathbf{k}'}{(2\pi)^3} w^{\sigma'\sigma}(\mathbf{p}, -\mathbf{k}', \mathbf{k}) \times \left[\{1 + N^{\sigma}(\mathbf{k}) + N^{\sigma'}(\mathbf{k}')\} f(\mathbf{p}) + \hbar(\mathbf{k}+\mathbf{k}') \cdot \frac{\partial f(\mathbf{p})}{\partial \mathbf{p}} N^{\sigma}(\mathbf{k}) N^{\sigma'}(\mathbf{k}') \right]. \quad (5.98)$$

The unit term is intrinsically quantum mechanical and is to be omitted in a classical theory.

2) *Double Scattering.* Let $w^{\sigma\sigma'\sigma''}(\mathbf{p}, \mathbf{k}, \mathbf{k}', \mathbf{k}'')$ be the probability that a photon in the mode σ is scattered into two photons in the modes σ' and σ''. The kinetic equation for $N^{\sigma}(\mathbf{k})$ is

$$\frac{\mathrm{d}N^{\sigma}(\mathbf{k})}{\mathrm{d}t} = \int \mathrm{d}^3\mathbf{p} \int \frac{\mathrm{d}^3\mathbf{k}'}{(2\pi)^3} \int \frac{\mathrm{d}^3\mathbf{k}''}{(2\pi)^3} w^{\sigma\sigma'\sigma''}(\mathbf{p}, \mathbf{k}, \mathbf{k}', \mathbf{k}'') \times \Big[\{N^{\sigma}(\mathbf{k})(1 + N^{\sigma'}(\mathbf{k}') + N^{\sigma''}(\mathbf{k}'')) - N^{\sigma'}(\mathbf{k}')N^{\sigma''}(\mathbf{k}'')\} f(\mathbf{p}) + \hbar(\mathbf{k}-\mathbf{k}'-\mathbf{k}'') \cdot \frac{\partial f(\mathbf{p})}{\partial \mathbf{p}} N^{\sigma}(\mathbf{k}) N^{\sigma'}(\mathbf{k}') N^{\sigma''}(\mathbf{k}'') \Big]. \quad (5.99)$$

The term proportional to $N^{\sigma}(\mathbf{k})$ is intrinsically quantum mechanical.

3) *Wave-Wave Scattering.* Let $u^{\sigma_1\sigma_2\sigma_3\sigma_4}(\mathbf{k}_1, \mathbf{k}_2, \mathbf{k}_3, \mathbf{k}_4)$ be the probability that waves in the modes σ_1 and σ_2 scatter into waves in the modes σ_3 and σ_4 ("photon-photon" scattering). The kinetic equation for $N^{\sigma'}(\mathbf{k}_1)$ is

$$\frac{\mathrm{d}N^{\sigma_1}(\mathbf{k}_1)}{\mathrm{d}t}=\int\frac{\mathrm{d}^3\mathbf{k}_2}{(2\pi)^3}\int\frac{\mathrm{d}^3\mathbf{k}_3}{(2\pi)^3}\int\frac{\mathrm{d}^3\mathbf{k}_4}{(2\pi)^3}u^{\sigma_1\sigma_2\sigma_3\sigma_4}(\mathbf{k}_1,\mathbf{k}_2,\mathbf{k}_3,\mathbf{k}_4)$$
$$\times[N^{\sigma_3}(\mathbf{k}_3)N^{\sigma_4}(\mathbf{k}_4)\{N^{\sigma_1}(\mathbf{k}_1)+N^{\sigma_2}(\mathbf{k}_2)\}$$
$$-N^{\sigma_1}(\mathbf{k}_1)N^{\sigma_2}(\mathbf{k}_2)\{N^{\sigma_3}(\mathbf{k}_3)+N^{\sigma_4}(\mathbf{k}_4)],\quad(5.100)$$

where intrinsically quantum mechanical terms are neglected. ("Photon-photon" scattering *in vacuo* is quantum mechanical only to the extent that one must use a quantum theory to calculate the second order nonlinear response tensor, cf. Section 10.3, for the vacuum; this response tensor is required in the calculation of the probability.)

4) *Nonlinear Inverse Compton.* Finally let us consider the emission mechanism discussed at the end of Section 4.4, i.e. nonlinear inverse Compton emission. The energy-momentum conservation law may be written (Brown and Kibble, 1964)

$$\mathbf{p}'_D+\mathbf{k}=\mathbf{p}_D+s(\omega_0/c)\boldsymbol{\kappa}',\quad \varepsilon'_D+\omega=\varepsilon_D+s\omega_0.\quad(5.101)$$

The classical limit of (5.101) is implied by the δ-function (4.170). The probability of emission at the sth harmonic is

$$w^{EM}(s,\mathbf{p}_D,\mathbf{k})=\frac{4\pi^2q^2}{\hbar\omega}|\boldsymbol{\kappa}\times\mathbf{V}^{EM}(s)|^2\delta\{\omega(1-\boldsymbol{\kappa}\cdot\boldsymbol{\beta}_D)-s\omega_0(1-\boldsymbol{\kappa}'\cdot\boldsymbol{\beta}_D)\}\quad(5.102)$$

with $\mathbf{V}^{EM}(s)$ given by (4.171). In this case the kinetic equation for $N(\mathbf{k})$ is

$$\frac{\mathrm{d}N(\mathbf{k})}{\mathrm{d}t}=\sum_{s=-\infty}^{\infty}\int\mathrm{d}^3\mathbf{p}_D w^{EM}(s,\mathbf{p}_D,\mathbf{k})\left[f(\mathbf{p}_D)+\hbar N(\mathbf{k})(\mathbf{k}-s(\omega_0/c)\boldsymbol{\kappa}')\cdot\frac{\partial f(\mathbf{p}_D)}{\partial\mathbf{p}_D}\right],$$

with $\mathbf{k}=(\omega/c)\boldsymbol{\kappa}$ and where $\mathbf{p}_0$ is reexpressed in terms of $\mathbf{p}_D$ using (4.166) and (4.167). In the limit $v\ll 1$ only $|s|=1$ contributes and by using (4.173) it is not difficult to see that (5.102) reduces to (5.89a).

PROBLEM SET 5

Answers to relevant questions are given at the end of the text. More difficult questions are marked with an asterisk.

5.1 An exact relativistic quantum expression for Cerenkov emission is available. (Exact expressions are also available for Compton scattering and for gyromagnetic emission.) It is, cf. (13.175),

$$w^{\sigma}(\mathbf{p},\mathbf{k})=\frac{8\pi^2q^2c^4R_E^{\sigma}(\mathbf{k})}{|\varepsilon\varepsilon'\omega^{\sigma}(\mathbf{k})|}\left[|(\mathbf{p}-\tfrac{1}{2}\hbar\mathbf{k})\cdot\mathbf{e}^{\sigma}(\mathbf{k})|^2-\left\{\frac{\hbar\omega^{\sigma}(\mathbf{k})}{2c}\right\}^2+|\tfrac{1}{2}\hbar\mathbf{k}\times\mathbf{e}^{\sigma}(\mathbf{k})|^2\right]$$
$$\times\delta(\varepsilon'-\varepsilon+\hbar\omega^{\sigma}(\mathbf{k})),\quad(\text{P5-1})$$

with $\varepsilon' := [m^2c^4 + (\mathbf{p} - \hbar\mathbf{k})^2c^2]^{1/2}$. Show that this exact expression reproduces the classical result (5.28) in the classical limit.

5.2 The crossed process corresponding to Cerenkov emission is pair (electron plus positron) creation by a photon. The probability for pair creation is obtained from the probability for Cerenkov emission by reversing the signs of $\mathbf{p}$ and $\mathbf{k}$ in (P5-1), and assuming that the signs of ε and $\omega^\sigma(\mathbf{k})$ reverse when $\mathbf{p}$ and $\mathbf{k}$ are reversed respectively. Show that pair creation, like Cerenkov emission, is possible only for $\omega^\sigma(\mathbf{k}) < kc$, i.e., for refractive index greater than unity.

5.3 (a) Show that the isotropic diffusion equation, cf. (5.59),

$$\frac{\mathrm{d}f(p)}{\mathrm{d}t} = \frac{1}{p^2}\frac{\partial}{\partial p}\left[p^2 D_{pp}(p)\frac{\partial f(p)}{\partial p}\right] \tag{P5-2}$$

may be rewritten in the form

$$\frac{\mathrm{d}f(p)}{\mathrm{d}t} = \frac{1}{\varepsilon^2 v}\frac{\partial}{\partial \varepsilon}\left[\varepsilon^2 v D_{\varepsilon\varepsilon}(\varepsilon)\frac{\partial f(p)}{\partial \varepsilon}\right], \tag{P5-3}$$

with $D_{\varepsilon\varepsilon}$ defined by (5.60) with $(\Delta p)^2$ replaced by $(\Delta\varepsilon)^2 = \{\hbar\omega^\sigma(\mathbf{k})\}^2$, or by (5.65) with $\Delta p_A \Delta p_B$ replaced by $\{\hbar\omega^\sigma(\mathbf{k})\}^2$.

(b) Further show that in terms of the energy spectrum

$$N(\varepsilon) \propto p^2 \frac{\mathrm{d}p}{\mathrm{d}\varepsilon} f(p), \tag{P5-4}$$

the diffusion equation reduces to the form of a Fokker-Planck equation

$$\frac{\mathrm{d}N(\varepsilon)}{\mathrm{d}t} = -\frac{\partial}{\partial \varepsilon}[A(\varepsilon)N(\varepsilon)] + \frac{1}{2}\frac{\partial^2}{\partial \varepsilon^2}[D(\varepsilon)N(\varepsilon)], \tag{P5-5}$$

with

$$D(\varepsilon) = 2D_{\varepsilon\varepsilon}(\varepsilon), \quad A(\varepsilon) = \frac{1}{\varepsilon^2 v}\frac{\partial}{\partial \varepsilon}\{\varepsilon^2 v D_{\varepsilon\varepsilon}(\varepsilon)\}. \tag{P5-6}$$

5.4 By considering three states ($\mathbf{p} - \hbar\mathbf{k}$, $\mathbf{p}$ and $\mathbf{p} + \hbar\mathbf{k}$) generalize (5.36) and (5.37) to include the first quantum corrections to spontaneous emission.

5.5* An anisotropic distribution of fast electrons $f(p, \alpha)$ interacts with an isotropic distribution of Langmuir waves. Show that $f(p, \alpha)$ evolves according to

$$\frac{\mathrm{d}f(p,\alpha)}{\partial t} = \frac{1}{\sin\alpha}\frac{\partial}{\partial\alpha}\left[\sin\alpha D_{\alpha\alpha}\frac{\partial f(p,\alpha)}{\partial\alpha}\right] + \frac{1}{p^2}\frac{\partial}{\partial p}\left[p^2 D_{pp}\frac{\partial f(p,\alpha)}{\partial p}\right], \tag{P5-7}$$

with

$$\begin{bmatrix} D_{\alpha\alpha} \\ D_{pp} \end{bmatrix} = \frac{e^2}{v^3}\int_{\omega_p/v} \frac{\mathrm{d}k}{k}\,\hbar\omega_p N(k) \begin{bmatrix} \frac{1}{2}\left(\frac{kv}{p\sin\alpha}\right)^2\left(1 - \frac{\omega_p^2}{k^2v^2}\right) \\ \omega_p^2 \end{bmatrix}, \tag{P5-8}$$

where the frequency of the Langmuir waves has been assumed equal to ω_p.

5.6 Show that when spontaneous emission is included, there are additional terms

$$\frac{\partial}{\partial p_{\parallel}}[A_{\parallel}f(\mathbf{p})]+\frac{1}{p_{\perp}}\frac{\partial}{\partial p_{\perp}}[p_{\perp}A_{\perp}f(\mathbf{p})] \tag{P5-9}$$

on the right side of (5.64), and analogous terms

$$\frac{1}{\sin\alpha}\frac{\partial}{\partial\alpha}[\sin\alpha A_{\alpha}f(p,\alpha)]+\frac{1}{p^2}\frac{\partial}{\partial p}[p^2A_pf(p,\alpha)] \tag{P5-10}$$

on the right side of (5.67), and that one has

$$A_x=\sum_{s=-\infty}^{\infty}\int\frac{d^3\mathbf{k}}{(2\pi)^3}\Delta x w^{\sigma}(s,\mathbf{p},\mathbf{k}), \tag{P5-11}$$

with Δx given by (5.66) and (5.69) respectively.

5.7 Show explicitly that $\gamma_c^t(\omega)$, as given by (5.87) with (5.82), cannot be negative. (Firstly partially integrate and then show that $(\partial/\partial\varepsilon)\{p^2(dp/d\varepsilon)w^t(\mathbf{p},\mathbf{k})\}$ is non-negative.)

5.8 Show that for particles with speed v interacting with waves with phase speed $\omega/k \ll v$, the relative change in α is larger than the relative change in p by a factor kv/ω for $\alpha\sim 1$.

Remark: The wave-particle interaction is then said to lead to pitch-angle scattering, cf., Sec. 7.2.

5.9 Find an explicit solution to the equation

$$\frac{\partial\Phi(\alpha,t)}{\partial t}=\frac{v_0}{\sin\alpha}\frac{\partial}{\partial\alpha}\left[\sin\alpha\frac{\partial\Phi(\alpha,t)}{\partial\alpha}\right] \tag{P5-12}$$

which describes pitch-angle scattering in the case $D_{\alpha\alpha}$ is independent of α, by expanding in Legendre polynomials

$$\Phi(\alpha,t)=\sum_{n=0}^{\infty}P_n(\cos\alpha)\Phi_n(t), \tag{P5-13a}$$

$$\Phi_n(t)=\frac{2n+1}{2}\int_{-1}^{+1}d\cos\alpha\,\Phi(\alpha,t)P_n(\cos\alpha). \tag{P5-13b}$$

CHAPTER 6

The Absorption and Transfer of Radiation

The quasi-linear equations allow one to relate collisionless absorption to spontaneous emission. Both collisionless and collisional damping of waves occur in plasmas, and these two processes are discussed in the first part of Section 6.1. In order to use the theory developed in Section 5.3 (and below) for the damping or absorption of waves, it is necessary to have a transfer equation, which usually presupposes the use of geometric optics. The application of geometric optics to anisotropic media, and the derivation of a relevant transfer equation for a plasma are presented in the latter half of Section 6.1. At high frequencies, the two natural modes in a plasma have nearly the same refractive index and they are nearly transverse. The theory of the transfer of radiation needs to be generalized to treat this weak-anisotropy limit because the relative phase between the components in the two modes is an observable quantity. The relevant transfer equation is derived and discussed in some detail in Section 6.2. The theory of Section 6.2 is required to treat synchrotron absorption in Section 6.3. The counterpart of absorption for a scattering process is induced scattering, and induced Compton scattering is discussed in Section 6.4 for both relativistic and non-relativistic electrons.

6.1 DAMPING AND TRANSFER PROCESSES IN THERMAL PLASMAS

Two types of damping processes are important for waves in thermal plasmas: these are collisionless damping and collisional damping. Collisionless damping may be described in terms of the kinetic equations for waves derived in Section 5.3, e.g. (5.36). In the earlier part of this section, after discussing collisionless damping briefly, collisional damping is treated using several complementary approaches.

The kinetic equations derived in Section 5.3 give the total time-rate-of-change of the occupation number $N^{\sigma}(\mathbf{k})$. More generally one is concerned with the transfer of waves, by which is meant the change in both space and time of some quantity, e.g. the specific intensity, which describes the intensity of the waves. In the latter part of this section the transfer of waves in a weakly inhomogeneous anisotropic medium is discussed in some detail. By "weakly inhomogeneous" is meant that the conditions for geometric optics to apply are satisfied.

Collisionless Damping

Collisionless damping may be treated in two seemingly independent ways, as pointed out in Section 3.3. One approach involves the following steps:– (1) take the hermitian part of the dielectric tensor and calculate the wave properties, (2) determine the anti-hermitian part by imposing the causal condition, (3) include the anti-hermitian part and determine the imaginary part of the frequency. The alternative approach follows step (1), with steps (2) and (3) replaced by: (2′) calculate the power radiated spontaneously by a given particle, (3′) relate absorption to emission using thermodynamic arguments. The essential difference between the two approaches is that one appeals to causality while the other appeals to the second law of thermodynamics. Evidently the implications of these two powerful theoretical concepts are equivalent in the present context. The equivalence of these two approaches is established in Section 3.3, and more generally it reflects an equivalence between collective-medium and single-particle approaches in plasma physics. (It is not generally recognized that the single-particle approach may be used to derive the dielectric tensor and nonlinear response tensors, see Appendix B for an outline of the method.)

The physical significance of collisionless damping remains obscure in the collective-medium approach. One can say that the damping is associated with the existence of intrinsic time-irreversible processes in the plasma, but this offers little physical insight. On the other hand, the single-particle approach offers a simple physical interpretation of collisionless damping: it results from absorption of the waves by the particles due to the inverse of Cerenkov emission or gyromagnetic emission. The inverse of Cerenkov emission is called *Landau damping* and the inverse of gyromagnetic emission is called *gyromagnetic absorption.*

Landau damping by thermal electrons is particularly important for

Langmuir waves and for ion sound waves. Specific formulae describing Landau damping for these two modes are (2.110) and (2.114b) respectively. Landau damping can be negative under relatively mild conditions. The two stream instability (Section 10.1) (or more accurately, the quasi-linear form of it) is due to negative Landau damping of Langmuir waves by suprathermal electrons.

Gyromagnetic absorption in a thermal plasma is important for transverse waves in the solar corona (Section 12.2). Synchrotron reabsorption (Section 6.3) is gyromagnetic absorption by relativistic electrons, and it is important in synchrotron sources. Gyromagnetic absorption can be negative for transverse waves (Section 12.2) and such negative absorption is one possible mechanism for Jupiter's decametric radio emission (Section 13.2). Negative gyromagnetic absorption of low-frequency waves by anisotropic fast particles is one facet of the theory of resonant scattering (Section 7.2), which is important in a variety of astrophysical and geophysical contexts.

Collisional Damping

Collisional damping is the absorption process corresponding to bremsstrahlung emission. Collisional damping of magnetoionic, or of any waves in which the motion of the ions is unimportant, is due to the effects of electron-ion collisions. In astrophysical applications such collisional damping is sometimes referred to as free-free absorption particularly in connection with optical radiation. This name arises from the fact that bound-bound, bound-free, free-bound and free-free electronic transitions all contribute to the opacity of a stellar atmosphere. Collisional damping, in an otherwise collisionless plasma, is important when collisionless damping is anomalously weak or absent. In the absence of an ambient magnetic field, Landau damping is strictly zero for transverse waves (in an isotropic plasma) and is very weak for Langmuir waves for $\omega_p/k \gg V_e$, and collisional damping is then important. In the presence of an ambient magnetic field, both collisional damping of transverse waves and gyromagnetic absorption need to be considered: collisional damping is the relatively more important at higher frequencies. (Collisional damping of MHD waves in a collision-dominated plasma, cf. Section 12.3, must be treated in a quite different manner by firstly deriving transport coefficients and then attributing the damping to these coefficients being finite or non-zero, as the case may be, e.g. Braginskii (1965).)

Three different methods of treating collisional damping will be discussed here. In the first Kirchhoff's law is used to relate collisional damping to bremsstrahlung. Alternatively, the Einstein coefficients could have been used, as in the derivation of (5.86). The second method is to regard bremsstrahlung emission and collisional damping as occurring in the wings of collisionally broadened lines. The third method involves including a frictional term in the equations used to derive the cold plasma dielectric tensor; this term leads to an anti-hermitian part of the tensor which may be used to treat the damping. The most general approach is that based on the use of the Einstein coefficients or, for a thermal plasma, of Kirchhoff's law.

Kirchhoff's law

An expression $P(\omega)$ for the average power radiated in bremsstrahlung per thermal electron was derived in (3.81). It follows that $n_e P(\omega)/4\pi$ is the volume emissivity, i.e. the power radiated per unit volume, per unit solid angle and per unit frequency range. Kirchhoff's law (5.19) with $T = m_e V_e^2$ for thermal electrons, gives

$$2\gamma_c(\omega) = \frac{n_e P(\omega)}{4\pi m_e V_e^2} \left\{ \frac{n(\omega)\omega^2}{(2\pi c)^3} \right\}^{-1}, \tag{6.1}$$

where $k^2\, dk/d\omega = n(\omega)\omega^2/c^3$ applies provided one has $n(\omega) = (1 - \omega_p^2/\omega^2)^{1/2}$. The factor of 2 is introduced in (6.1) to take into account the two independent states of polarization for transverse waves. The resulting expression for the absorption coefficient may be written in the form

$$\gamma_c(\omega) = \frac{\omega_p^2}{\omega^2}\, \nu(\omega), \tag{6.2}$$

with

$$\nu(\omega) = \sum_i \frac{1}{3} \left(\frac{2}{\pi} \right)^{1/2} \frac{4\pi Z_i^2 n_i e^4}{m_e^2 V_e^3} \frac{\pi}{\sqrt{3}}\, G(T_e, \omega). \tag{6.3}$$

An appropriate expression for the Gaunt factor is given by (3.83). Numerically, the absorption coefficient per unit length in centimeters corresponding to (6.3) is

$$\mu(f) = 9.786 \times 10^{-3} \frac{n_e}{f^2 T_e^{3/2}} \sum_i Z_i^2 n_i \times \begin{cases} 18.2 + \ln T_e^{3/2} - \ln f & T_e < 1.6 \times 10^5\,\mathrm{K} \\ 24.2 + \ln T_e - \ln f & T_e > 1.6 \times 10^5\,\mathrm{K} \end{cases} \tag{6.3$'$}$$

where $f := \omega/2\pi$ is in hertz and T_e is in kelvins.

The result (6.2) applies only to transverse waves in an isotropic plasma. The collisional absorption coefficient for waves in an arbitrary mode in a thermal electron plasma may be derived by the following heuristic argument. Collisional damping may be attributed to the loss to the waves of that fraction of the energy associated with the forced oscillations of each electron which experiences a collision. Now the total energy in waves may be divided into the fraction R_E^σ in the electric field, the fraction R_M^σ in the magnetic fluctuations, with

$$R_M^\sigma = n_\sigma^2 R_E^\sigma (1 - |\boldsymbol{\kappa}\cdot\mathbf{e}^\sigma|^2), \tag{6.4}$$

and the remaining fraction R_p^σ in forced oscillations of the electrons (provided the motion of the ions is unimportant). For transverse waves in an isotropic plasma, one has $R_E = \frac{1}{2}$, $R_M = \frac{1}{2}(1 - \omega_p^2/\omega^2)$, and hence $R_p = \frac{1}{2}\omega_p^2/\omega^2$. Let us write (6.2) in the form

$$\gamma_c^\sigma(\mathbf{k}) = 2R_p^\sigma(\mathbf{k})\nu(\omega^\sigma(\mathbf{k})), \tag{6.5}$$

with

$$R_p^\sigma(\mathbf{k}) := 1 - R_E^\sigma(\mathbf{k}) - R_M^\sigma(\mathbf{k}). \tag{6.6}$$

At this stage (6.5) should be regarded as a speculative form for a general expression for the absorption coefficient for collisional damping. Further arguments for its validity are given below. The result (6.5) applies to both Langmuir waves and to magnetoionic waves. For Langmuir waves $R_E^l \approx \frac{1}{2}$ and $R_M^l = 0$ imply $R_P^l \approx \frac{1}{2}$, and (6.5) gives

$$\gamma_c^l(k) = \nu(\omega_p). \tag{6.7}$$

An explicit approximate expression for R_p^σ for the magnetoionic waves is given in equation (12.41) below.

Now that we have the collisional absorption coefficient for waves in any mode in a thermal electron plasma, Kirchhoff's law may be used to derive expressions describing the bremsstrahlung emission of these waves. For example, the average power radiated per thermal electron in the range $d^3\mathbf{k}/(2\pi)^3$ at $\mathbf{k}$ due to bremsstrahlung emission of Langmuir waves is

$$P^l(k) = \frac{m_e V_e^2}{n_e}\nu(\omega_p). \tag{6.8}$$

In the same way one obtains the emissivity per thermal electron in the magnetoionic waves due to bremsstrahlung,

$$\eta^\sigma(\omega, \theta) = \frac{\partial\{\omega n_\sigma(\omega, \theta)\}}{\partial\omega}\frac{2n_\sigma^2(\omega, \theta)\omega^2}{(2\pi c)^3} m_e V_e^2 R_p^\sigma(\omega, \theta)\nu(\omega). \tag{6.9}$$

Bremsstrahlung emission in a magnetized plasma is polarized because of the preferential emission of one of the magnetoionic waves, see Section 12.2.

Lorentzian Line Profiles

A significant effect of collisions on other radiation processes is that of line broadening. For example the probability (5.28) for Cerenkov emission involves the factor $\delta(\omega^\sigma(\mathbf{k})-\mathbf{k}\cdot\mathbf{v})$ which may be interpreted as implying a δ-function, or infinitely narrow, line profile. In the presence of collisions with a frequencing $\nu(v)$ the δ-function profile is replaced by a Lorentzian line profile (Lorentz, 1905) according to

$$\delta(\omega^\sigma(\mathbf{k})-\mathbf{k}\cdot\mathbf{v})\to\frac{\nu(v)}{\pi[(\omega^\sigma(\mathbf{k})-\mathbf{k}\cdot\mathbf{v})^2+\nu^2(v)]}. \tag{6.10}$$

(It may be noted that the RH side of (6.10) reduces to the LH side in the limit $\nu(v)\to 0$.)

The following physical model behind the replacement (6.10) is due to Lorentz and was applied to a plasma by Oster (1960). Two assumptions are involved. One is that the probability that a particle experience a collision in the interval $d(\Delta t)$ after an initially specified time $\Delta t=0$ is

$$dp(\Delta t)=\nu(v)\,e^{-\nu(v)\Delta t}\,d(\Delta t). \tag{6.11}$$

The other assumption is that the particle ceases radiating momentarily whenever a collision occurs, and then recommences with a random change in phase.

That this model does imply the replacement (6.10) can be seen as follows. Consider the Fourier transform of the current density due to a charge in constant rectilinear motion starting at time $t=0$, i.e.

$$\begin{aligned}[\mathbf{j}(\mathbf{k},\omega)]_{t>0}:=\int_0^\infty dt\, q\mathbf{v}\exp[i(\omega-\mathbf{k}\cdot\mathbf{v})t]&=\frac{iq\mathbf{v}}{\omega-\mathbf{k}\cdot\mathbf{v}+i0}\\ &=q\mathbf{v}\left\{\mathscr{P}\frac{i}{\omega-\mathbf{k}\cdot\mathbf{v}}+\pi\delta(\omega-\mathbf{k}\cdot\mathbf{v})\right\},\end{aligned} \tag{6.12}$$

where the Plemelj formula (3.12) has been used. Now consider the corresponding quantity for a particle which experiences collisions with these idealized properties, i.e.

$$\begin{aligned}[\mathbf{j}(\mathbf{k},\omega)]_{t>0}:=&\int_0^\infty d(\Delta t)\nu(v)\exp[-\nu(v)\Delta t]\int_0^{\Delta t} dt\, q\mathbf{v}\exp[i(\omega-\mathbf{k}\cdot\mathbf{v})t]\\ &=\frac{iq\mathbf{v}}{\omega-\mathbf{k}\cdot\mathbf{v}+i\nu(v)}\\ &=q\mathbf{v}\left\{\frac{i(\omega-\mathbf{k}\cdot\mathbf{v})}{(\omega-\mathbf{k}\cdot\mathbf{v})^2+\nu^2(v)}+\frac{\nu(v)}{(\omega-\mathbf{k}\cdot\mathbf{v})^2+\nu^2(v)}\right\}\end{aligned} \tag{6.13}$$

Comparison of the real parts of (6.12) and (6.13) leads to the replacement (6.10) as the appropriate one for collisionally broadened lines.

The problem of what value to take for the collision frequency $\nu(v)$ arises. One method by which the appropriate value may be identified is as follows. Emission in the wings of the broadened lines is due entirely to the effects of collisions and so may be identified as bremsstrahlung. Consider the probability (5.28) for Cerenkov emission, viz.

$$w^{\sigma}(\mathbf{p},\mathbf{k})=\frac{8\pi^2 q^2 R_E^{\sigma}(\mathbf{k})}{\hbar\omega^{\sigma}(\mathbf{k})}\quad|\mathbf{e}^{\sigma *}(\mathbf{k})\cdot\mathbf{v}|^2\delta(\omega^{\sigma}(\mathbf{k})-\mathbf{k}\cdot\mathbf{v}). \tag{6.14}$$

Suppose we make the dipole approximation ($k=0$), then average over the direction of $\mathbf{v}$ and finally make the replacement (6.10) with $k=0$ and $\omega^{\sigma}(\mathbf{k})\gg\nu(v)$. There results, for emission by electrons,

$$\overline{w^{\sigma}(\mathbf{p},\mathbf{k})}=\frac{8\pi e^2 v^2 R_E^{\sigma}(\mathbf{k})}{3\hbar\{\omega^{\sigma}(\mathbf{k})\}^3}\,\nu(v). \tag{6.15}$$

This may be equated to the average probability (5.80) to determine $\nu(v)$:

$$\nu(v)=\sum_i\frac{4\pi Z_i^2 n_i e^4\ln(b_2/b_1)}{m_e^2\gamma^2 v^3}. \tag{6.16}$$

It is interesting to note that the "collision frequency" (6.16) is related more closely to the deflection frequency than to the slowing-down frequency, cf. Section 7.1. That this should be so may be inferred from (5.79). In (5.79), the term involving $\mathbf{e}^{\sigma}(\mathbf{k})\cdot\mathbf{b}$ arises from the component of the force along $\mathbf{b}$ while the other term arises from the component along $\mathbf{v}$. It is clear that for $|\mathbf{k}\cdot\mathbf{v}|\ll\omega^{\sigma}(\mathbf{k})$, bremsstrahlung is due primarily to the component along $\mathbf{b}$. The condition $|\mathbf{k}\cdot\mathbf{v}|\ll\omega^{\sigma}(\mathbf{k})$ is satisfied for emission by non-relativistic particles. (The component of force along $\mathbf{v}$ does contribute significantly to bremsstrahlung from relativistic particles.) Thus bremsstrahlung is associated with the deflection of particles, and not with their slowing down. In other words "bremsstrahlung" is a misnomer.

Frictional Damping

The third method of treating collisional damping of waves in an electronic plasma is the simplest and perhaps the most familiar. The idea is to include a frictional term in the equation of fluid motion for the electrons in the derivation of the cold plasma approximation. Thus in place of (2.57) one has

$$-im_e(\omega\delta_{ij}+i\Omega_e\varepsilon_{ijk}b_k)u_j(\omega)=-eE_i(\omega)-\nu(\omega)m_e u_i(\omega). \tag{6.17}$$

It is straightforward to derive the generalization to the dielectric tensor for a cold electronic gas, but for present purposes it suffices to note that the first order correction in $\nu(\omega)$ leads to an anti-hermitian part given by

$$\varepsilon_{ij}^{a}(\omega)=\frac{i\nu(\omega)}{\omega}\left[\frac{\mathrm{d}}{\mathrm{d}\omega}\{\omega\varepsilon_{ij}^{h}(\omega)\}-\delta_{ij}\right], \tag{6.18}$$

where $\varepsilon_{ij}^{h}(\omega)$ is given by inserting (2.131) in (2.116).

This third approach complements the earlier approaches in the following sense. On explicit evaluation of the absorption coefficient using (2.86) with (6.18), the factor

$$e_i^{\sigma *}(\mathbf{k})e_j^{\sigma}(\mathbf{k})\left[\frac{\mathrm{d}}{\mathrm{d}\omega}\{\omega\varepsilon_{ij}^{h}(\omega)\}-\delta_{ij}\right]_{\omega=\omega^{\sigma}(k)}=R_P^{\sigma}(\mathbf{k}) \tag{6.19}$$

appears. Now in writing down (6.5) we speculated that γ_c^{σ} should be proportional to R_P^{σ}, and we already knew it was proportional to $\nu(\omega)$ as given by (6.3). In view of (6.19) the third approach shows that γ_c^{σ} is proportional to R_P^{σ}, but the value of $\nu(\omega)$ in (6.18) must be determined independently. The two methods combined establish the validity of (6.5).

Geometric Optics

Now let us return to consider the equation describing the effect of emission and absorption on the distribution of waves, e.g. (5.36) or (5.38), which may be written in the form

$$\frac{\mathrm{d}N^{\sigma}(\mathbf{k})}{\mathrm{d}t}=\frac{1}{\hbar\omega^{\sigma}(\mathbf{k})}\frac{\mathrm{d}P^{\sigma}(\mathbf{k})}{\mathrm{d}V}-\gamma^{\sigma}(\mathbf{k})N^{\sigma}(\mathbf{k}), \tag{6.20}$$

where $\mathrm{d}P^{\sigma}(\mathbf{k})/\mathrm{d}V$ is the power radiated per unit volume in the elemental range $\mathrm{d}^3\mathbf{k}/(2\pi)^3$ at $\mathbf{k}$ due to spontaneous emission. In (6.20), each of $\mathrm{d}P^{\sigma}(\mathbf{k})/\mathrm{d}V$ and $\gamma^{\sigma}(\mathbf{k})$ is the sum of contributions from all relevant emission and absorption processes. Now let us relax the assumption that the medium is homogeneous and time-independent and consider the propagation of waves through a slowly varying medium. From this viewpoint, (6.20) should be regarded as an equation of transfer for the waves.

The simplest theory of the propagation of radiation through a slowly varying medium is geometric optics. The basic idea in geometric optics is to introduce the concept of a ray. Although the theory applies only under steady state conditions (transient effects are excluded) one imagines the radiant energy to be transported along ray paths which are the paths of imaginary particles (photons) which propagate at the group velocity. The approach in which the radiation is regarded as a collection of photons is closely related to geometric optics from a conceptual viewpoint. Indeed geometric optics may be formulated by applying mechanical concepts to a system of photons. To be specific, if one regards the local dispersion relation $\omega=\omega^{\sigma}(\mathbf{k})$ as a function of position and time, e.g. as $\omega=\omega^{\sigma}(\mathbf{k};\,\mathbf{r},\,t)$, then the equations of geometric optics (in the Hamiltonian formalism) may be obtained by regarding $\hbar\omega^{\sigma}(\mathbf{k};\mathbf{r},t)$ as the Hamiltonian for a system of photons with energy $\hbar\omega$ and momentum $\hbar\mathbf{k}$. That is, the Hamiltonian equations of mechanics become the equations of geometric optics in the Hamiltonian form. In fact, from an historical viewpoint, Hamilton, see collected works (1931), first developed the formalism now associated with his name in the context of geometric optics (in 1837), and then later applied this formalism to mechanics. Because the Hamiltonian formalism is now more familiar in its mechanical context, it is appropriate to reverse the historical sequence and to write down the equations of geometric optics by analogy with their mechanical counterparts.

The Basic Equations

The procedure outlined above leads to the following equations (analogous to the equations of particle motion in the Hamiltonian form):

$$\frac{d\mathbf{r}}{dt}=\frac{\partial\omega^{\sigma}(\mathbf{k};\mathbf{r},t)}{\partial\mathbf{k}},\quad \frac{d\mathbf{k}}{dt}=-\frac{\partial\omega^{\sigma}(\mathbf{k};\mathbf{r},t)}{\partial\mathbf{r}},\tag{6.21a, b}$$

$$\frac{d\omega}{dt}=\frac{\partial\omega^{\sigma}(\mathbf{k};\mathbf{r},t)}{\partial t}.\tag{6.21c}$$

These are the equations for a ray in the Hamiltonian form. (The Hamiltonian equations were introduced into plasma physics by Haselgrove (1955); she was concerned with finding a simplified method for treating ray tracing for magnetoionic waves in the ionosphere.) The derivation of (6.21a, b, c) parallels the derivation of Hamilton's equations in mechanics, see below.

In a slowly varying medium one also regards $N^{\sigma}(\mathbf{k};\,\mathbf{r},\,t)$ in (6.20) as a slowly varying function of position and time. In addition, the LH member of (6.20) is to be interpreted according to

$$\frac{d}{dt}\to\frac{\partial}{\partial t}+\frac{d\mathbf{r}}{dt}\cdot\frac{\partial}{\partial\mathbf{r}}+\frac{d\mathbf{k}}{dt}\cdot\frac{\partial}{\partial\mathbf{k}}.\tag{6.22}$$

The Ray Tracing Problem

The equations (6.21a, b, c) determine a ray path for given initial conditions. Some care is required in interpreting these equations, particularly in connection with the parameter t. Suppose the properties of the medium do not depend on time so that (6.21c) reduces to $d\omega/dt=0$, i.e. ω is a constant. There is then no time variation of any quantity. In this case the parameter t in $d\mathbf{r}/dt$ and $d\mathbf{k}/dt$ should be regarded merely as a parameter labelling points along a ray path. To see this from a slightly different viewpoint, suppose a parameter s is the distance along the ray path from a fixed initial point. The ray path may then be described by equations

$$\mathbf{r}=\mathbf{r}(s), \quad \mathbf{k}=\mathbf{k}(s), \tag{6.23a, b}$$

which determine the path (through coordinate space) and the variation of $\mathbf{k}$ along the path respectively. The parameter t is some function of s. We may interpret (6.21a) as implying (i) that the tangent to the ray path is along the direction of the group velocity, i.e. $d\mathbf{r}(s)/ds$ is parallel to $\partial\omega^{\sigma}(\mathbf{k}, \mathbf{r}(s), t)/\partial\mathbf{k}$, and (ii) that the parameter t is that function of s which ensures that $d\mathbf{r}/dt$ is equal to the group velocity. Thus t may be regarded as the parameter defined as a function of s by (for a given time t_0 say)

$$\frac{ds}{dt}=v_g^{\sigma}(\mathbf{k}; \mathbf{r}(s), t_0), \tag{6.24}$$

where

$$\mathbf{v}_g^{\sigma}(\mathbf{k}; \mathbf{r}, t):=\frac{\partial\omega^{\sigma}(\mathbf{k}; \mathbf{r}, t)}{\partial\mathbf{k}}$$

is the group velocity. To interpret t as real time leads to conceptual difficulties.

With this interpretation of t, equation (6.21b) gives the change in $\mathbf{k}$ along the ray path. Snell's law is included in (6.21b). In effect, Snell's law states that in a time independent medium those components of $\mathbf{k}$ which are orthogonal to the gradient in the dispersion relation (e.g. in the refractive index) are constant along the ray path. The rate of change of the component of $\mathbf{k}$ along the gradient is also given by (6.21b).

Equation (6.21c) gives the rate of change (with time) of the frequency of a wave due to any intrinsic time-dependence in the properties of the medium. One is rarely concerned with time-dependent media, but an example of possible relevance to radio emission in the solar corona is the generation of Langmuir waves in a collisionless shock front. Across a shock front the plasma density changes, and a Langmuir wave generated as the front arrives would find itself in a plasma with plasma frequency increasing with time (Krall and Smith, 1974).

Derivation of Hamilton's Equations

It is instructive to consider the derivation of the equations (6.21a, b, c). The equations result from comparison of two different ways of considering variations along a ray.

The first approach is based on regarding the dielectric tensor, the dispersion equation and the dispersion relations as slowly varying functions of space and time. "Slowly varying" is not a well-defined concept. It is usually defined as meaning that the change over a wavelength or a wave period is negligible (but in comparison to what?). In practice any medium is slowly varying whenever one can justify applying geometric optics to it. Suppose the dispersion equation (2.65) is a function of position and time:

$$\Lambda(\mathbf{k}, \omega; \mathbf{r}, t)=0. \tag{6.25}$$

Along a ray path (6.25) must remain satisfied, and so the condition

$$\frac{\partial\Lambda}{\partial\mathbf{k}}\cdot \mathrm{d}\mathbf{k}+\frac{\partial\Lambda}{\partial\omega}\,\mathrm{d}\omega+\frac{\partial\Lambda}{\partial\mathbf{r}}\cdot \mathrm{d}\mathbf{r}+\frac{\partial\Lambda}{\partial t}\,\mathrm{d}t=0 \tag{6.26}$$

must be satisfied. (The arguments are omitted for simplicity in writing.) Along a ray path it is possible that a point (called a coupling point) may be encountered where two modes, which are otherwise distinct, have the same values of $\mathbf{k}$ and ω. (Mode coupling is mentioned in Section 6.2 and discussed in Section 12.3 and Section 12.4.) Apart from such points (6.25) and 6.26) are equivalent to the set of equations

$$\omega=\omega^\sigma(\mathbf{k}; \mathbf{r}, t) \tag{6.27}$$

and

$$\mathrm{d}\omega^\sigma=\frac{\partial\omega^\sigma}{\partial\mathbf{k}}\cdot \mathrm{d}\mathbf{k}+\frac{\partial\omega^\sigma}{\partial\mathbf{r}}\cdot \mathrm{d}\mathbf{r}+\frac{\partial\omega^\sigma}{\partial t}\,\mathrm{d}t, \tag{6.28}$$

respectively, for each distinct mode σ. It is clear that although (6.21a, b, c) ensure that (6.28) is satisfied, (6.28) itself does not suffice for the derivation of (6.21a, b, c).

The other approach is based on a variational principle. A heuristic derivation is as follows. The phase of a plane wave varies, by definition, according to $\exp\,[i(\mathbf{k}\cdot\mathbf{r}-\omega^\sigma t)]$. In an inhomogeneous time-dependent medium $\mathbf{k}$ and ω are functions of position and time. It is reasonable to suppose that the relative phase of the wave at time t_2 compared to the phase at t_1 is $\exp\,[i\Psi^\sigma(t_1, t_2)]$ with

$$\Psi^\sigma(t_1, t_2):=\int_{t_1}^{t_2}\mathrm{d}t(\mathbf{k}\cdot\dot{\mathbf{r}}-\omega^\sigma), \tag{6.29}$$

and $\dot{\mathbf{r}}:=\mathrm{d}\mathbf{r}/\mathrm{d}t$. The quantity Ψ^σ is called the *eikonal*. The variational principle is that the eikonal is an extremum along the ray path.

Consider paths between fixed end points at a particular time ($\partial\omega^\sigma/\partial t$ is then zero). As already explained the parameter t should be regarded as a parameter labelling points along a path between the two fixed end points which are labelled t_1 and t_2. Along neighboring paths the values of **k** and **r**, which are regarded as functions of t, are slightly different. Let the differences be $\delta\mathbf{k}$ and $\delta\mathbf{r}$ respectively. The variation principle, i.e. that the actual ray path corresponds to zero variation in the eikonal, requires

$$0=\delta\Psi^{\sigma}(t_1, t_2)=\int_{t_1}^{t_2} \mathrm{d}t\left(\delta\mathbf{k}\cdot\dot{\mathbf{r}}+\mathbf{k}\cdot\delta\dot{\mathbf{r}}-\frac{\partial\omega^{\sigma}}{\partial\mathbf{k}}\cdot\delta\mathbf{k}-\frac{\partial\omega^{\sigma}}{\partial\mathbf{r}}\cdot\delta\mathbf{r}\right)$$

$$=\int_{t_1}^{t_2} \mathrm{d}t\left\{\left(\dot{\mathbf{r}}-\frac{\partial\omega^{\sigma}}{\partial\mathbf{k}}\right)\cdot\delta\mathbf{k}-\left(\dot{\mathbf{k}}+\frac{\partial\omega^{\sigma}}{\partial\mathbf{r}}\right)\cdot\delta\mathbf{r}\right\}, \tag{6.30}$$

where a partial integration over the term $\mathbf{k}\cdot\delta\dot{\mathbf{r}}$ has been performed. (The integrated part is zero because $\delta\mathbf{r}$ vanishes at the end points.) Variations in **k** and **r** can be made independently along neighboring paths. Consequently (6.30) implies

$$\dot{\mathbf{r}}=\frac{\partial\omega^{\sigma}}{\partial\mathbf{k}}, \quad \dot{\mathbf{k}}=-\frac{\partial\omega^{\sigma}}{\partial\mathbf{r}},$$

which are just (6.21a, b) respectively. Finally (6.21c) follows from (6.28).

Fermat's Principle

The variation principle used above is Fermat's principle of least time. This is the counterpart of the principle of least action in mechanics. Just as in mechanics, where a Lagrangian (which is a function of the position vector **r**, the velocity $\dot{\mathbf{r}}$ and of time t) is introduced, Fermat's principle may be written in the form

$$\delta\int_{t_1}^{t_2} \mathrm{d}t\, F^{\sigma}(\mathbf{r}, \dot{\mathbf{r}}, t)=0, \tag{6.31}$$

where δ denotes a variation with fixed end points. The quantity $F^{\sigma}(\mathbf{r}, \dot{\mathbf{r}}, t)$ may be constructed by reversing the usual construction of a Hamiltonian from a Lagrangian, e.g. Goldstein (1959). This gives

$$F^{\sigma}(\mathbf{r}, \dot{\mathbf{r}}, t)=\mathbf{k}\cdot\dot{\mathbf{r}}-\omega^{\sigma}(\mathbf{k}; \mathbf{r}, t). \tag{6.32}$$

As one would expect, (6.31) and (6.32) are consistent with (6.29) and (6.30). Fermat's principle in the form (6.31) with (6.32) is very cumbersome because it is to be understood that **k** on the RH side of (6.32) is to be reexpressed as a function of **r**, $\dot{\mathbf{r}}$ and t by inverting (6.21a). (It is not possible to solve (6.21a) explicitly for **k** in general.)

There is a more familiar and useful form of Fermat's principle which applies to an anistropic medium which is time-dependent. For simplicity let us now drop the label σ denoting the wave mode. The alternative form of Fermat's principle is

$$\delta \int_{s_1}^{s_2} \mathrm{d}s \, \mathscr{M} = 0, \tag{6.33}$$

where s denotes distance along the ray path and where $\mathscr{M}$ is the so-called *ray refractive index*. Let us derive (6.33) from (6.31) and (6.32). Firstly, for a time independent medium ω is constant, and the final term in (6.32) does not contribute in (6.31). Then using (6.24), viz. $\mathrm{d}s/\mathrm{d}t = |\dot{\mathbf{r}}|$, (6.31) reduces to

$$\delta \int_{s_1}^{s_2} \mathrm{d}s \, \frac{\mathbf{k} \cdot \dot{\mathbf{r}}}{|\dot{\mathbf{r}}|} = 0. \tag{6.34}$$

This is of the form (6.33). The integrand in (6.34) may be made dimensionless by multiplying by c/ω. Thus one identifies

$$\mathscr{M} = n(\omega, \theta) \cos(\theta - \theta_r), \tag{6.35}$$

where n is the usual refractive index and where $\theta - \theta_r$ is the angle between the wave normal and the ray, i.e. between $\mathbf{k}$ and $\mathbf{v}_g$ (θ is the angle between $\mathbf{k}$ and $\mathbf{B}$, and θ_r is the angle between $\mathbf{v}_g$ and $\mathbf{B}$, where $\mathbf{B}$ is the ambient magnetic field).

The Transfer Equation

The transfer equation (6.20) with $\mathrm{d}/\mathrm{d}t$ interpreted according to (6.22) includes the effects of slow variations in the properties of the medium. Suppose the only variation in $N^\sigma(\mathbf{k})$ is a spatial one. Then (6.20) may be rewritten in terms of the rate of change of $N^\sigma(\mathbf{k})$ with distance s along the ray path:

$$v_g^\sigma(\mathbf{k}) \frac{\mathrm{d}N^\sigma(\mathbf{k})}{\mathrm{d}s} = \frac{1}{\hbar\omega^\sigma(\mathbf{k})} \frac{\mathrm{d}P^\sigma(\mathbf{k})}{\mathrm{d}V} - \gamma^\sigma(\mathbf{k}) N^\sigma(\mathbf{k}), \tag{6.36}$$

where (6.24) has been used and where the dependence of $\mathbf{r}$ and t is left understood. A standard general form of the transfer equation may be obtained from (6.36) simply by rewriting it in a more appropriate notation.

The specific intensity, $I^\sigma(\omega, \theta_r)$ say, is defined as the energy, per unit frequency range and per unit solid angle about the ray direction, which crosses unit surface area normal to the ray direction in unit time. We can relate $I^\sigma(\omega, \theta_r)$ to $N^\sigma(\mathbf{k})$ as follows. The Jacobian of the transformation from an element of solid angle $\mathrm{d}^2\Omega$ about the $\mathbf{k}$ to an element $\mathrm{d}^2\Omega_r$ of solid angle about the ray direction, i.e. about $\mathbf{v}_g^\sigma(\mathbf{k})$, is

$$\frac{d^2\Omega_r}{d^2\Omega}=\left|\frac{\partial \cos\theta_r}{\partial \cos\theta}\right|, \tag{6.37}$$

where θ is the angle between $\mathbf{k}$ and $\mathbf{B}$ and θ_r is the angle between $\mathbf{v}_g$ and $\mathbf{B}$. Now the energy density per unit frequency range and per unit solid angle about $\mathbf{k}$ is

$$\frac{k^2\hbar\omega^\sigma(\mathbf{k})N^\sigma(\mathbf{k})}{(2\pi)^3\partial\omega^\sigma(\mathbf{k})/\partial k}.$$

The energy density per unit solid angle about the ray direction is obtained by multiplying this quantity by the Jacobian (6.37). Then the energy crossing unit surface area in unit time may be obtained by multiplying the resulting expression by the group speed:

$$I^\sigma(\omega,\theta_r)=\frac{k^2\hbar\omega^\sigma(\mathbf{k})N^\sigma(\mathbf{k})v_g^\sigma(\mathbf{k})}{(2\pi)^3\partial\omega^\sigma(\mathbf{k})/\partial k}\left|\frac{\partial \cos\theta_r}{\partial \cos\theta}\right|, \tag{6.38}$$

where it is understood that the RH side is to be reexpressed as a function of ω and θ. Note that although one has $\theta_r=\theta$ and $v_g^\sigma(\mathbf{k})=\partial\omega^\sigma(\mathbf{k})/\partial k$ for an isotropic medium, neither equality obtains in general.

The transfer equation (6.36) may be rewritten in the form

$$n_{\sigma t}^2(\omega,\theta_r)\frac{d}{ds}\left(\frac{I^\sigma(\omega,\theta_r)}{n_{\sigma t}^2(\omega,\theta_r)}\right)=\alpha^\sigma(\omega,\theta_r)-\mu^\sigma(\omega,\theta_r)I^\sigma(\omega,\theta_r). \tag{6.39}$$

The various quantities introduced in (6.39) are

$$\alpha^\sigma:=\frac{\omega^2 n_\sigma^2}{(2\pi c)^3}\frac{dP^\sigma}{dV}\frac{\partial}{\partial\omega}(\omega n_\sigma)\left|\frac{\partial \cos\theta_r}{\partial \cos\theta}\right|, \tag{6.40a}$$

$$\mu^\sigma:=\frac{\gamma^\sigma}{v_g^\sigma}, \tag{6.40b}$$

and

$$n_{\sigma t}:=\left|\frac{v_g^\sigma}{c}n_\sigma^2\frac{\partial(\omega n_\sigma)}{\partial\omega}\frac{\partial\cos\theta}{\partial\cos\theta_r}\right|^{1/2}, \tag{6.41}$$

where the arguments are omitted, but where it is understood that the refractive index n_σ, the wave angle θ the group speed v_g^σ and also dP^σ/dV and γ^σ are to be expressed as functions of ω and θ_r. The quantity α^σ is the energy radiated due to spontaneous emission per unit frequency range and per unit solid angle about the ray direction, and μ^σ is the absorption coefficient per unit length along the ray path.

The quantity $n_{\sigma t}$ might be called the *transfer refractive index*. This quantity was first introduced by Bekefi (1966) who called it the "ray refractive index". However, this latter name is the standard name for the quantity (6.35). It might also be mentioned that in the literature, especially in connection with radio waves in the ionosphere, an additional quantity called the group refractive index is also introduced, see e.g. Budden (1961). In an isotropic medium one has $\theta_r = \theta$ and $v_g^\sigma = c/\partial(\omega n_\sigma)/\partial\omega$, (6.41) reduces to $n_{\sigma t} = n_\sigma$ and (6.39) reduces to the transfer equation of Wooley (1947). For an anisotropic medium, $n_{\sigma t}$ is not equal to n_σ.

The appropriate form of the transfer equation for waves in a plasma has been a subject of some controversy in the literature, see Oster (1963), Cronyn (1966), Zheleznyakov (1967, 1969, 1971), Énomé (1969) and Walsh (1970). The form (6.39) was derived by Bekefi (1966), Énomé (1969) and by Zhelzynakov (1971).

In most applications (6.39) is unnecessarily cumbersome, and one can usually justify either ignoring the ambient magnetic field (assuming the waves to be transverse or longitudinal), or using the transfer equation in the weak anisotropy limit.

6.2 THE TRANSFER EQUATION IN THE WEAK ANISOTROPY LIMIT

An essential feature of the transfer equation (6.39) is that each mode is assumed to propagate independently of other modes. This assumption is inadequate for the important case of high-frequency waves ($\omega \gg \omega_p$, Ω_e), and even radioastronomical observations involve waves which have $\omega \gg \omega_p$, Ω_e over most or all of their paths of propagation from their source to the Earth. At high frequencies the two magnetoionic modes have nearly equal refractive indices, $n(\omega) \approx 1 - \omega_p^2/2\omega^2$, and their polarizations are very nearly transverse. One can separate any electromagnetic disturbance into components in the two magnetoionic modes. However, the relative phase between the two components varies slowly and is an important observable quantity. Consequently, two transfer equations (one for each mode) of the form (6.39) are inadequate because they contain no information on the relative phase of the components. The change in the relative phase leads to "Faraday rotation", which is used here to mean the generic effect as well as the specific case of rotation of the plane of polarization (when the natural modes are circularly polarized).

In this section a transfer equation in the weak anisotropy limit is introduced and discussed in some detail. The weak anisotropy limit is that in which the two natural modes are regarded (a) as transverse, and (b) as having identical ray paths. The transfer equation in this limit may be written in two quite different notations. One is based on the use of polarization tensors and is called the Jones calculus. It is the more convenient for formal purposes. The other is based on the use of the Stokes parameters and is called the Mueller calculus. It is the more convenient for solving the transfer equation.

Three types of process should be included in a general transfer equation. These are absorption, Faraday rotation and polarization changes induced by the inhomogeneity of the medium. Two complementary approaches allow one to include all three effects. In one approach the medium is assumed to be homogeneous and both absorption and Faraday rotation are treated. In the other approach absorption is neglected and, by firstly separating into components in the two undamped modes, both Faraday rotation and changes due to spatial gradients in the plasma parameters may be taken into account.

As an aside it might be remarked that polarization changes, in a restricted sense, occur even in an isotropic inhomogeneous medium. As refraction occurs the polarization changes (a) such that the polarization ellipse is always orthogonal to the ray path, and (b) such that the component of the electric vector orthogonal to both the ray path and to the gradient in the refractive index remains constant. The shape of the polarization ellipse is unaffected. (A set of differential equations describing these changes is written down in (12.140) below.)

Motivation

The algebra involved below in the treatment of the transfer equation is detailed and lengthy, and it is appropriate to summarize the motivation for the development of such a detailed theory. The primary motivation is the understanding of the circularly polarized component in the emission from synchrotron sources. This topic is discussed in detail in Section 9.1.

Synchrotron emission is partially linearly polarized with a small degree of circular polarization (Section 4.2). On general grounds one expects synchrotron absorption to be polarization-dependent, favoring one linear polarization over another and with a slight preference for one circular polarization over the other. The question arises as to how one is to treat such polarization-dependent absorption. In earlier treatments of synchrotron absorption, e.g. Ginzburg and Syrovatski (1965), this problem was avoided simply by ignoring the circularly polarized component in absorption. From a formal viewpoint, the theory developed in Section 2.3 and in Section 4.1 allows one to write down an absorption coefficient $\gamma^{\alpha\beta}$ as a polarization tensor, and this obviously includes the circularly polarized component in the absorption. One question to be answered in constructing a transfer equation is how such an expression for $\gamma^{\alpha\beta}$ is to be used to treat absorption. Historically, the relevant transfer equation was first written down by Sazonov and Tsytovich (1968), see Unno (1956) however.

A related problem concerns the conversion of linear polarization into circular polarization as a propagation effect. Faraday rotation is a well-known process: in a medium with circularly polarized natural modes, the plane of polarization of any linearly polarized component rotates due to one circularly polarized component getting increasingly out of phase with the other. Another well-known effect, used in a quarter-wave plate, is the conversion of linear into circular polarization in a medium in which the natural modes are linearly polarized. The question arises as to how one treates "generalized" Faraday rotation, i.e. the polarization changes for arbitrarily polarized radiation in a medium in which the natural modes are arbitrarily elliptically polarized. This question is also answered by the theory presented below.

A final motivation concerns a dilemma in the treatment of polarization-dependent absorption. The following suggestion (Zheleznyakov, 1968) seems plausible: one should treat absorption by firstly separating into the two natural modes and regarding them as damping independently of each other. However, the results obtained in this way are inconsistent, in general, with the results obtained by using the general transfer equation of Sazonov and Tsytovich (1968). In fact, this suggested approach implies that absorption may be treated using only two parameters (the absorption coefficients for each mode) whereas three parameters are required in the general theory (e.g. μ_I, μ_Q and μ_V, with $\mu_U=0$ due to the choice of axes). This dilemma was overcome by Zheleznyakov, Sovorov and Shaposhnikov (1974) by arguing that the "natural modes" in the presence of absorption are not mutually orthogonal, as pointed out by Sazonov (1969). An alternative explanation is given below: it is legitimate to separate into the two natural modes (defined by ignoring the absorption), but the components in these two modes cannot be regarded as damping independently of each other. Put another way, an absorption-induced mode coupling exists.

The reader who is not interested in the details of how the foregoing problems are treated should take note of the introduction of the Jones calculus and the Mueller calculus, and proceed to Section 6.3.

Polarization Tensors: The Jones Calculus

A general way of describing the polarization of transverse waves is in terms of polarization tensors, as introduced in Section 4.1. The polarization tensor

$I^{\alpha\beta}$ for the specific intensity is proportional to $e_i^{\alpha} e_j^{\beta} \langle E_i E_j^* \rangle$ where $\langle E_i E_j^* \rangle$ denotes the (Fourier transform of) the auto-correlation function of the electric field. Suppose the two natural modes have polarization vectors $\mathbf{e}^+$ and $\mathbf{e}^-$. Then the specific intensities I^{++} and I^{--} in the two modes are given by

$$I^{\pm\pm} := (\mathbf{e}^{\pm} \cdot \mathbf{e}^{\alpha})^* (\mathbf{e}^{\pm} \cdot \mathbf{e}^{\beta}) I^{\alpha\beta}, \tag{6.47}$$

where the sum from 1 to 2 over the repeated indices α and β is implied. However, $I^{\alpha\beta}$ also contains information about the relative phase of the components in the two modes through the cross-correlation functions

$$I^{+-} := (\mathbf{e}^{+} \cdot \mathbf{e}^{\alpha})^* (\mathbf{e}^{-} \cdot \mathbf{e}^{\beta}) I^{\alpha\beta} = (I^{-+})^*, \tag{6.42$'$}$$

where the equality follows from the fact that $I^{\alpha\beta} = (I^{\beta\alpha})^*$ is hermitian. Thus the polarization tensor contains information which is not included in I^{++} and I^{--}, and a transfer equation for $I^{\alpha\beta}$ would retain all relevant information on the polarization.

The generic form for the transfer equation for $I^{\alpha\beta}$ is

$$n^2 \frac{\partial}{\partial s}\left(\frac{I^{\alpha\beta}}{n^2}\right) = \alpha^{\alpha\beta} + T^{\alpha\beta\gamma\delta} I^{\gamma\delta}, \tag{6.43}$$

which is the transfer equation in the Jones calculus. The second rank polarization tensor $\alpha^{\alpha\beta}$ describes the effect of spontaneous emission, and the fourth rank polarization tensor $T^{\alpha\beta\gamma\delta}$ describes absorption, Faraday rotation and any other process which causes the polarization to change. It is straightforward to write down $\alpha^{\alpha\beta}$ for any specific process, e.g. by writing the emissivity as a polarization tensor and integrating over the distribution of radiating particles (and dividing by the group speed). The construction of $T^{\alpha\beta\gamma\delta}$ is discussed in detail later in this section. Firstly, it is appropriate to discuss the Jones calculus, the Mueller calculus and the transfer equation from a general viewpoint.

The algebra of polarization tensors is that of hermitian 2×2 matrices. Let $I^{\alpha\beta}$ be the polarization tensor for the intensity. The total intensity I is the trace $I^{\alpha\alpha} = I^{11} + I^{22}$ of $I^{\alpha\beta}$. If the radiation is completely polarized, the determinant of $I^{\alpha\beta}$ is zero, and if the radiation is unpolarized $I^{\alpha\beta}$ is equal to $\frac{1}{2} I \delta^{\alpha\beta}$, where the $\frac{1}{2}$ is due to the fact that the trace of the unit 2×2 tensor $\delta^{\alpha\beta}$ is equal to two. More generally one may make a unique separation of $I^{\alpha\beta}$ into a part, $I_1^{\alpha\beta}$ say, whose determinant is zero, and a part proportional to $\delta^{\alpha\beta}$. The degree of r of polarization may be identified as the trace of $I_1^{\alpha\beta}$ divided by the trace of $I^{\alpha\beta}$.

From a more formal viewpoint, the only two independent invariants (i.e. quantities independent of the choice of $\mathbf{e}^1$ and $\mathbf{e}^2$) one can construct from

$I^{\alpha\beta}$ are the trace and the determinant. Any other invariant, e.g. $I^{\alpha\beta}I^{\beta\alpha}$ which is the trace of the square of the matrix $I^{\alpha\beta}$, may be expressed in terms of these two invariants. Specifically, one has

$$\det\,[I^{\alpha\beta}]=\tfrac{1}{2}\{(I^{\sigma\sigma})^2-I^{\sigma\tau}I^{\tau\sigma}\}. \tag{6.44}$$

The degree of polarization is given in terms of these invariants by

$$\tfrac{1}{4}(1-r)=\frac{\det\,[I^{\alpha\beta}]}{(I^{\sigma\sigma})^2},\quad r=\frac{(I^{\sigma\sigma})^2-4\det\,[I^{\alpha\beta}]}{(I^{\sigma\sigma})^2}. \tag{6.45}$$

In order to identify the degrees of linear and circular polarization one must presuppose a specific choice of $\mathbf{e}^1$ and $\mathbf{e}^2$. It is appropriate to leave the definitions of these degrees of polarization until after the Mueller calculus has been introduced.

Now let us return to (6.43) and discuss the tensor $T^{\alpha\beta\gamma\delta}$. The polarization tensor $I^{\alpha\beta}$ is proportional to an average (over phases) of $E^\alpha E^{*\beta}$, with $E^\alpha:=e_i^\alpha E_i$. Suppose we ignore spontaneous emission and assume that E^α evolves according to

$$\frac{\mathrm{d}E^\alpha}{\mathrm{d}s}=\tfrac{1}{2}T^{\alpha\beta}E^\beta, \tag{6.46}$$

where the factor $\frac{1}{2}$ is included for convenience. It then follows that $I^{\alpha\beta}$ evolves according to (6.43), with

$$T^{\alpha\beta\gamma\delta}=\tfrac{1}{2}\{T^{\alpha\gamma}\delta^{\beta\delta}+(T^{\beta\delta})^*\delta^{\alpha\gamma}\}. \tag{6.47}$$

In (6.47), $T^{\alpha\beta}$ is actually the average (over phases) of the $T^{\alpha\beta}$ introduced in (6.46), but this average is usually trivial. Thus (6.47) may be regarded as a general form of $T^{\alpha\beta\gamma\delta}$. It is convenient to separate $T^{\alpha\beta}$ into its hermitian part $-\mu^{\alpha\beta}$ and its anti-hermitian part $r^{\alpha\beta}$:

$$T^{\alpha\beta}=:-\mu^{\alpha\beta}+r^{\alpha\beta},\quad (\mu^{\alpha\beta})^*=\mu^{\beta\alpha},\quad (r^{\alpha\beta})^*=-r^{\beta\alpha}. \tag{6.48}$$

This separation is into dissipative ($\mu^{\alpha\beta}$) and non-dissipative ($r^{\alpha\beta}$) parts. We have

$$T^{\alpha\beta\gamma\delta}=-\mu^{\alpha\beta\gamma\delta}+r^{\alpha\beta\gamma\delta}, \tag{6.49a}$$

$$\mu^{\alpha\beta\gamma\delta}=\tfrac{1}{2}\{\mu^{\alpha\gamma}\delta^{\beta\delta}+\mu^{\delta\beta}\delta^{\alpha\gamma}\},\quad r^{\alpha\beta\gamma\delta}=\tfrac{1}{2}\{\gamma^{\alpha\gamma}\delta^{\beta\delta}-r^{\delta\beta}\delta^{\alpha\gamma}\} \tag{6.49b, c}$$

In general $\mu^{\alpha\beta}$ and $r^{\alpha\beta}$ each involves four separate functions (the diagonal components are real for $\mu^{\alpha\beta}$ and imaginary for $r^{\alpha\beta}$, and the off-diagonal components are complex conjugates of each other). However, the trace of $r^{\alpha\beta}$ can be assumed zero without loss of generality. This is due to the fact that any diagonal component of $r^{\alpha\beta}$ is assumed to be included in the zeroth order approximation in which the anisotropy is neglected. From another viewpoint, if one assumes $r^{\alpha\beta} \propto \delta^{\alpha\beta}$ and inserts this in (6.49c), then one finds that the resulting contribution to $r^{\alpha\beta\gamma\delta}$ is identically zero. Thus, in general, four parameters are required to describe the absorption and three parameters are required to describe the polarization-changing properties of the medium. (In a homogeneous magnetized plasma these can be reduced to three and two, respectively, by a suitable choice of $\mathbf{e}^1$ and $\mathbf{e}^2$.)

The Stokes Parameters and the Mueller Calculus

The Stokes parameters I, Q, U, V are more convenient than the polarization tensor $I^{\alpha\beta}$ for practical calculations. A definition of the Stokes parameters is as follows: choose $\mathbf{e}^1$ and $\mathbf{e}^2$ to be real and write

$$I^{\alpha\beta} =: \frac{1}{2}\begin{pmatrix} I+Q & U-iV \\ U+iV & I-Q \end{pmatrix}. \tag{6.50}$$

Note the sign of V. In optics, in contrast with radiophysics and plasma physics, V is often defined to have the other sign. Here V/I is the degree of circular polarization and it is positive for RH polarized waves, where RH means a RH screw sense relative to the direction $\boldsymbol{\kappa}$ of propagation. The opposite sign for V in (6.49) occurs often in the literature due to a variety of different conventions. For example, V can be positive for LH polarized radiation (an odd convention which arises from the historical definition, still used in optics, of handedness as a screw sense relative to the line of sight), or the opposite sign can appear because of the choice of sign in $e^{\pm i\omega t}$ when Fourier transforming (changing the sign in $e^{\pm i\omega t}$ is equivalent to complex conjugating in (6.50)).

The properties of the polarization are contained in the Stokes parameters. The degrees of linear and circular polarization are

$$r_l = \frac{(Q^2+U^2)^{1/2}}{I}, \quad r_c = \frac{V}{I}, \tag{6.51a, b}$$

with

$$r = (r_l^2 + r_c^2)^{1/2} = \frac{(Q^2+U^2+V^2)^{1/2}}{I}, \tag{6.51c}$$

and the angle χ between the major axis of the polarization ellipse and the direction $\mathbf{e}^1$ is

$$\chi = \tfrac{1}{2}\,\text{arc tan}\left(\frac{U}{Q}\right). \tag{6.51d}$$

The transfer equation in the Mueller calculus may be written in terms of the "Stokes vector" S_A, with $A=1, 2, 3, 4$ and with

$$S_1=I, \quad S_2:=Q, \quad S_3:=U, \quad S_4:=V. \tag{6.52}$$

The transfer equation (6.43), with $T^{\alpha\beta\gamma\delta}$ written in the form (6.49a), translates into

$$n^2 \frac{\partial}{\partial s}\left(\frac{S_A}{n^2}\right)=\alpha_A-\mu_{AB}S_B+r_{AB}S_B, \tag{6.53}$$

where the sum from 1 to 4 or the repeated index B is implied. In practice, to translate explicitly from one notation to the other it is convenient to write

$$\begin{aligned}&'S_1:=I^{11}, \quad 'S_2:=I^{12}, \quad 'S_3:=I^{21}, \quad 'S_4:=I^{22},\\ &'T_{11}:=T^{1111}, \quad 'T_{12}:=T^{1112}, \quad 'T_{13}:=T^{1121}, \quad 'T_{14}:=T^{1122},\\ &'T_{21}:=T^{1211}, \quad 'T_{22}:=T^{1212}, \text{ etc.,}\end{aligned} \tag{6.54}$$

i.e. to write the polarization tensors in a fourth-order matrix notation. One then constructs

$$S_A=L_{AB}'S_A, \quad T_{AB}=L_{AC}'T_{CD}L_{DB}^{-1}, \tag{6.55a, b}$$

with

$$L_{AB}=\begin{pmatrix}1 & 0 & 0 & 1\\ 1 & 0 & 0 & -1\\ 0 & 1 & 1 & 0\\ 0 & i & -i & 0\end{pmatrix}, \quad L_{AB}^{-1}=\frac{1}{2}\begin{pmatrix}1 & 1 & 0 & 0\\ 0 & 0 & 1 & -i\\ 0 & 0 & 1 & i\\ 1 & -1 & 0 & 0\end{pmatrix}. \tag{6.56}$$

Quite generally, the fact that μ_{AB} and r_{AB} are constructed from the terms describing time-irreversible and time-reversible processes is reflected in the symmetry properties

$$\mu_{AB}=\mu_{BA}, \quad r_{AB}=-r_{BA}. \tag{6.57a, b}$$

When the translation from (6.43) to (6.53) is made, one finds that the quantities in (6.53) may be identified by writing

$$\alpha^{\alpha\beta}=:\frac{1}{2}\begin{pmatrix}\alpha_I+\alpha_Q & \alpha_U-i\alpha_V\\ \alpha_U+i\alpha_V & \alpha_I-\alpha_Q\end{pmatrix}, \quad \mu^{\alpha\beta}=:\begin{pmatrix}\mu_I+\mu_Q & \mu_U-i\mu_V\\ \mu_U+i\mu_V & \mu_I-\mu_Q\end{pmatrix},$$

$$r^{\alpha\beta}=:-i\begin{pmatrix}r_Q & r_U-ir_V\\ r_U+ir_V & -r_Q\end{pmatrix}, \tag{6.58a, b, c}$$

with

$$\mu_{AB}=\begin{pmatrix}\mu_I & \mu_Q & \mu_U & \mu_V\\ \mu_Q & \mu_I & 0 & 0\\ \mu_U & 0 & \mu_I & 0\\ \mu_V & 0 & 0 & \mu_I\end{pmatrix},\quad r_{AB}=\begin{pmatrix}0 & 0 & 0 & 0\\ 0 & 0 & -r_V & r_U\\ 0 & r_V & 0 & -r_Q\\ 0 & -r_U & r_Q & 0\end{pmatrix} \qquad (6.59a, b)$$

Factors of $\frac{1}{2}$ are absent from (6.58b, c) because they have already been included in (6.46) or (6.49b, c). In a homogeneous magnetized plasma α_U, μ_U and r_U can be chosen to be zero by a suitable choice of $\mathbf{e}^1$ and $\mathbf{e}^2$. (The sign in (6.59b) is such that for $r_Q=0$, $r_U=0$ Faraday rotation is described by $d\chi/ds=\frac{1}{2}r_V$.)

Transfer in a Homogeneous Absorbing Medium

The theory of polarization transfer in a weakly anisotropic, homogeneous medium was developed by Sazonov and Tsytovich (1968). The basic idea is to solve the homogeneous wave equation under the assumption that the anisotropic part of the dielectric tensor may be treated as a perturbation. The homogeneous wave equation, cf. (2.64) with (2.44), may be written in the form

$$\left\{-\frac{k^2c^2}{\omega^2}\,\delta^{\alpha\beta}+\varepsilon^{\alpha\beta}\right\}E^\beta=0 \qquad (6.60)$$

for waves which are assumed to be transverse. In (6.60) we have

$$E^\beta:=e_i^\beta E_i,\quad \varepsilon^{\alpha\beta}:=e_i^\alpha e_j^\beta\varepsilon_{ij}, \qquad (6.61)$$

and so on, and all arguments are omitted. Now $\varepsilon^{\alpha\beta}$ is separated into an isotropic part and an anisotropic part:

$$\varepsilon^{\alpha\beta}=\varepsilon_0\delta^{\alpha\beta}+\Delta\varepsilon^{\alpha\beta}. \qquad (6.62)$$

The first step in the solution of (6.60) is to solve for the wave properties in the isotropic limit, i.e. for $\Delta\varepsilon^{\alpha\beta}=0$. One has $kc/\omega=n:=\varepsilon_0^{1/2}$, with $\mathbf{e}\cdot\boldsymbol{\kappa}=0$ and $R_E=\frac{1}{2}n\partial(\omega n)/\partial\omega$. Next assume that the anisotropic correction $\Delta\varepsilon^{\alpha\beta}$ leads to a small rate of temporal change in the amplitude. The small rate of change may be introduced by allowing ω to have an imaginary part, $\omega\to\omega+i\omega_I$, and solving to first order in ω_I and $\Delta\varepsilon^{\alpha\beta}$:

$$\omega_I E^\alpha=i\omega R_E\Delta\varepsilon^{\alpha\beta}E^\beta. \qquad (6.63)$$

Then one may reinterpret $\omega_I\mathrm{E}^\alpha$ as the secular rate of change of E^α, and replace it by $v_g \partial E^\alpha/\partial s$, with the group speed given by $v_g = c/\partial(\omega n)/\partial\omega$. Thus one finds

$$\frac{\mathrm{d}E^\alpha}{\mathrm{d}s} = \frac{i\omega}{2nc}\Delta\varepsilon^{\alpha\beta}E^\beta, \tag{6.64}$$

which is of the form (6.46). The next step is the same as from (6.46) to (6.47), and one identifies

$$T^{\alpha\beta\gamma\delta} = \frac{i\omega}{2nc}\{\Delta\varepsilon^{\alpha\gamma}\delta^{\beta\delta} - (\Delta\varepsilon^{\beta\delta})^*\delta^{\alpha\gamma}\}. \tag{6.65}$$

Sazonov and Tsytovich (1968) also used this method to derive an expression for $\alpha^{\alpha\beta}$ in (6.43), but it is straightforward to write down $\alpha^{\alpha\beta}$ directly in practice.

Let us neglect absorption for the moment. This means neglecting the antihermitian part of the dielectric tensor, in which case we may solve for the wave properties in terms of the hermitian part of $\Delta\varepsilon^{\alpha\beta}$. It is convenient to do so, and to reexpress the hermitian part of $\Delta\varepsilon^{\alpha\beta}$ in terms of these wave properties. In general, the hermitian part of $\Delta\varepsilon^{\alpha\beta}$ involves three independent functions, namely $\Delta\varepsilon^{11} = -\Delta\varepsilon^{22}$ and the real and imaginary parts of $\Delta\varepsilon^{12} = (\Delta\varepsilon^{21})^*$. However, for a magnetized plasma, if we choose, as in (4.31),

$$\mathbf{e}^1 = (\cos\theta, 0, -\sin\theta), \quad \mathbf{e}^2 = (0, 1, 0), \tag{6.66}$$

then $\Delta\varepsilon^{12}$ is imaginary and only two functions $\Delta\varepsilon^{11} = -\Delta\varepsilon^{22}$ and $\Delta\varepsilon^{12} = -\Delta\varepsilon^{21}$ are involved. It is an elementary exercise to find the difference Δn between the refractive indices of the two modes and the axial ratio T of their polarization ellipses in terms of these two functions. Inverting the solutions one finds

$$\Delta\varepsilon^{11} = -\Delta\varepsilon^{22} = (\Delta n)n\frac{T^2-1}{T^2+1}, \quad \Delta\varepsilon^{12} = -\Delta\varepsilon^{21} = -i(\Delta n)n\frac{2T}{T^2+1}. \tag{6.67}$$

One then obtains explicit expressions for the three parameters in (6.59b):

$$r_V = -\frac{2T}{T^2+1}\Delta k, \quad r_U = 0, \quad r_Q = -\frac{(T^2+1)}{T^2+1}\Delta k, \tag{6.68}$$

with $\Delta k := \Delta n\omega/c$. ($r_U$ vanishes because we have chosen $\mathbf{e}^1$ and $\mathbf{e}^2$ to lie along the axes of the polarization ellipse.) The sign of r_V is such that $r_V > 0$ implies rotation of a plane of polarization in the positive sense (RH screw sense relative to $\boldsymbol{\kappa}$) and rotation in the opposite sense occurs if the RH polarized mode ($T > 0$) has the larger refractive index ($\Delta k > 0$).

Absorption may be treated simply by writing the general formula (2.86) for the absorption coefficient as a polarization tensor,

$$\gamma^{\alpha\beta} = -2i\omega R_E[\Delta\varepsilon^{\alpha\beta}]^a, \tag{6.69}$$

and dividing it by the group speed, as in the derivation of (6.64),

$$\mu^{\alpha\beta} := \frac{\gamma^{\alpha\beta}}{v_g} = -\frac{i\omega}{nc}\,[\Delta\varepsilon^{\alpha\beta}]^a. \tag{6.70}$$

Then comparison of (6.47), (6.65) and (6.70) leads to the identification (6.49c), viz.

$$\mu^{\alpha\beta\gamma\delta} = \tfrac{1}{2}(\mu^{\alpha\gamma}\delta^{\beta\delta} + \mu^{\delta\beta}\delta^{\alpha\gamma}). \tag{6.71}$$

The parameters in (6.59a) may be identified by writing $\mu^{\alpha\beta}$ in the form (6.58b). In a magnetized plasma with $\mathbf{e}^1$ and $\mathbf{e}^2$ chosen as in (6.66), μ_U vanishes.

It is informative to discuss the vanishing of μ_U and r_U for the choice (6.66) from a slightly different viewpoint. Let us choose some other pair of vectors $\mathbf{e}^{1'}$ and $\mathbf{e}^{2'}$ related to $\mathbf{e}^1$ and $\mathbf{e}^2$ by say

$$\begin{aligned} \mathbf{e}^{1'} &= \cos\phi\;\mathbf{e}' + \sin\phi\;\mathbf{e}^2, \\ \mathbf{e}^{2'} &= -\sin\phi\;\mathbf{e}' + \cos\phi\;\mathbf{e}^2. \end{aligned} \tag{6.72}$$

By constructing the matrices (they are not polarization tensors)

$$R^{\alpha\alpha'} := \mathbf{e}^\alpha \cdot \mathbf{e}^{\alpha'}, \quad R^{\alpha'\alpha} := \mathbf{e}^{\alpha'} \cdot \mathbf{e}^\alpha, \tag{6.73}$$

i.e.

$$R^{\alpha\alpha'} = \begin{pmatrix} \cos\phi & -\sin\phi \\ \sin\phi & \cos\phi \end{pmatrix}, \quad R^{\alpha'\alpha} = \begin{pmatrix} \cos\phi & \sin\phi \\ -\sin\phi & \cos\phi \end{pmatrix}, \tag{6.73$'$}$$

for the particular case (6.72), any second rank polarization tensors $f^{\alpha\beta}$ transform according to

$$f^{\alpha'\beta'} = R^{\alpha'\alpha}R^{\beta'\beta}f^{\alpha\beta}, \tag{6.74}$$

and a fourth rank tensor transforms like the product of two such tensors. (In a matrix formulation, one pre-multiplies by $R^{\alpha'\alpha}$ and post-multiplies by $R^{\beta\beta'}$; no simple matrix formulation is possible for a fourth rank tensor.) By inspection it is apparent that one can always choose basic polarization vectors $\mathbf{e}^1$ and $\mathbf{e}^2$ to ensure that the U-component, in (6.71) say, vanishes. In a system with only one unique direction, as for a homogeneous magnetized plasma, the U-components of all intrinsic polarization tensors must vanish simultaneously when the appropriate choice (6.66) of basic vectors is made.

The transformation (6.72) must also lead to a transformation of the Stokes vectors. Let f_A, with $A=1$ to 4, be the Stokes vector corresponding to $f^{\alpha\beta}$ and let $f_{A'}$ be that corresponding to $f^{\alpha'\beta'}$. The transformation from f_A to $f_{A'}$ is

$$f_{A'} = R_{A'A}f_A \tag{6.75}$$

with

$$R_{A'A}=\begin{pmatrix}1 & 0 & 0 & 0\\ 0 & \cos 2\phi & \sin 2\phi & 0\\ 0 & -\sin 2\phi & \cos 2\phi & 0\\ 0 & 0 & 0 & 1\end{pmatrix}, \tag{6.76}$$

and $R_{AA'}$ is the transpose of $R_{A'A}$. The transformation of the matrices μ_{AB} and r_{AB} in (6.54) is then given by

$$\mu_{A'B'}=R_{A'A}R_{B'B}\mu_{AB}, \quad r_{A'B'}=R_{A'A}R_{B'B}r_{AB}. \tag{6.77}$$

Specifically, for $U=0$ at $\phi=0$, one has $U'=-Q\sin^2\phi$ and $Q'=Q\cos^2\phi$ at $\phi\neq 0$, according to (6.75) with (6.76).

Transfer in an Inhomogeneous Non-Absorbing Medium

The complementary approach in which one neglects absorption but takes the inhomogeneity into account has been developed by Cheng and Fung (1977). The method is based on an approach developed by Kai (1965), Zheleznyakov (1968), and Fung (1969). The essential idea is to write $I^{\alpha\beta}$ as a sum of the components I^{++}, I^{+-}, I^{-+}, I^{--} in the natural modes, and to assume that $I^{+-}=(I^{-+})^*$ evolves with a phase factor given by the eikonal (6.79). The spatial derivative of $I^{\alpha\beta}$ then involves the derivative of the eikonal, which leads to Faraday rotation, and derivatives of the components I^{++}, etc., themselves. The latter derivatives are attributed to "mode coupling" (Section 12.3).

[Cheng and Fung (1977) also included the effects of derivatives of the polarization vectors, but this additional effect is non-physical, as explained in detail below.]

Let the two natural modes be labelled + and −, let T be the axial ratio of the + mode and let $\Delta n(=:\Delta kc/\omega)$ be the difference between the refractive indices of the + and − modes. The polarization vectors may be written

$$\mathbf{e}^{+}=\frac{T\mathbf{t}+i\mathbf{a}}{(T^2+1)^{1/2}}, \quad \mathbf{e}^{-}=\frac{T}{|T|}\frac{(-\mathbf{t}+iT\mathbf{a})}{(T^2+1)^{1/2}}. \tag{6.78a, b}$$

Without loss of generality we may choose $T>0$. Let $\mathbf{e}^1$ and $\mathbf{e}^2$ be arbitrary mutually orthonormal (to $\mathbf{k}$) vectors. Let $\mathbf{b}$ be at polar angles θ and ϕ to $\mathbf{k}$, where $\phi=0$ corresponds to the projection of $\mathbf{b}$ (on the plane orthogonal to $\mathbf{k}$) being along $\mathbf{e}^1$. Then one has

$$\mathbf{t}=-\cos\phi\,\mathbf{e}^1-\sin\phi\,\mathbf{e}^2, \quad \mathbf{a}=\sin\phi\,\mathbf{e}^1-\cos\phi\,\mathbf{e}^2. \tag{6.79a, b}$$

Along the polarization vectors may change due to changes in T or to changes in ϕ. In the weak anisotropy limit refraction is unimportant and changes in the angles between **k** and **b** are due primarily to changes in the direction of the magnetic field. T may change due to changes in the plasma parameters (Ω_e and ω_p here) or to changes in θ, and the latter are usually the more important.

Let us write

$$a_\sigma^\alpha := \mathbf{e}^\sigma \cdot \mathbf{e}^\alpha \tag{6.80}$$

with $\sigma = +$ or $-$. Then at any point along the ray path we may separate $I^{\alpha\beta}$ into components in the two modes

$$I^{\alpha\beta} = a_\sigma^\alpha a_{\sigma'}^{*\beta},\ I^{\sigma\sigma'} \tag{6.81}$$

with

$$I^{\sigma\sigma'} = a_\sigma^{*\,\alpha} a_{\sigma'}^\beta I^{\alpha\beta}. \tag{6.82}$$

(The sum over σ and σ' in (6.81) is implied.) Suppose we make the separation (6.81) at some point, $s = s_0$ say. Then $a_\sigma^\alpha(s_0)$ and $a_{\sigma'}^\beta(s_0)$ are constants. However $I^{\alpha\beta}$ varies along the ray path due to the fact that the two modes get out of phase. This may be described by

$$I^{+-}(s) = I^{+-}(s_0)\exp(i\Delta\Psi), \quad \Delta\Psi = \int_{s0}^{s} ds'\,\Delta k(s'), \tag{6.83}$$

with $I^{-+}(s)$ the complex conjugate of $I^{+-}(s)$ and with $I^{++}(s)$ and $I^{--}(s)$ remaining constant. This particular variation leads to Faraday rotation. On substitution (6.83) in (6.81) and differentiating the resulting equation, one obtains

$$\frac{d}{ds} I^{\alpha\beta} = i\Delta k\{a_+^\alpha\, a_-^{*\beta} I^{+-} - a_-^\alpha a_+^{*\beta} I^{-+}\}. \tag{6.84}$$

Now using (6.82), (6.84) may be written in the form (6.49a, c) with

$$r_{\rm F}^{\alpha\beta} = i\Delta k(2a_+^\alpha a_+^{*\beta} - \delta^{\alpha\beta}), \tag{6.85}$$

where "FR" denotes Faraday rotation. In deriving (6.85) the orthogonality of the two modes has been used in the form

$$a_+^\alpha a_+^{*\beta} + a_-^\alpha a_-^{*\beta} = \delta^{\alpha\beta}. \tag{6.86}$$

An explicit expression for $\gamma_{\rm FR}^{\alpha\beta}$ may be obtained by constructing explicit expressions for a_+^α using (6.78a), (6.79a, b), and (6.80). One finds

$$r_{\rm FR}^{\alpha\beta} = \frac{i\Delta k}{T^2+1}\begin{pmatrix} (T^2-1)\cos 2\phi & (T^2-1)\sin 2\phi - 2iT \\ (T^2-1)\sin 2\phi + 2iT & -(T^2-1)\cos 2\phi \end{pmatrix} \tag{6.87}$$

The corresponding components of r_{AB} in Mueller calculus are given by (6.58c) with

$$r_V = -\frac{2T}{T^2+1}\Delta k, \quad r_U = -\frac{T^2-1}{T^2+1}\Delta k \sin 2\phi, \quad r_Q = -\frac{T^2-1}{T^2+1}\Delta k \cos 2\phi. \tag{6.88}$$

The result (6.88) reduces to (6.68) for $\phi=0$ or π, and (6.88) may be obtained from (6.68) by a rotation of axes, cf. (6.76) and (6.77).

The additional term which appears in the present approach is that due to mode coupling. As pointed out above, Cheng and Fung (1977) were in error in including a further term associated with spatial gradients of the polarization vectors. To see this, suppose that one were to apply the present method to waves *in vacuo*. Then one would have $\Delta k=0$ and the term (6.87) which describes Faraday rotation is zero, as it must be. Following Cheng and Fung (1977), there are two additional terms which arise from (6.81). These are

$$\frac{dI^{\alpha\beta}}{ds} = I^{\sigma\sigma'}\frac{d}{ds}(a^{\alpha}_{\sigma}a^{*\beta}_{\sigma'}) + a^{\alpha}_{\sigma}a^{*\beta}_{\sigma'}\frac{d}{ds}I^{\sigma\sigma'},$$

which describe the polarization changes and mode coupling respectively. these terms are not obviously zero *in vacuo* because the a^{α}_{σ} are arbitrary and their spatial variation is arbitrary. However, the transfer equations *in vacuo* is $dI^{\alpha\beta}/ds=0$, and hence the polarization changes and mode-coupling terms must cancel. This cancellation is independent of the actual form of $a^{\alpha}_{\sigma'}$ and hence it must always apply. Thus, rather than being independent terms, as assumed by Cheng and Fung (1977), one of these two terms must be incorrect. Put another way, these two terms must describe the same physical effect from complementary viewpoints, and only one of them is to be retained.

As pointed out above, it is the polarization-changes term which is non-physical. However, Cheng and Fung's (1977) approach, together with the above-deduced cancellation provides a convenient way of treating mode coupling in the weak anisotropy limit: calculate $r^{\alpha\beta}$ for polarization changes, then identify $r^{\alpha\beta}_{MC}$ for mode coupling (MC) as minus the result calculated. This gives

$$r^{\alpha\beta}_{MC} = -2(a'^{\alpha}_{+}a^{*\beta}_{+} + a'^{\alpha}_{-}a^{*\beta}_{-}), \tag{6.89}$$

where the prime denotes a derivative with respect to s. Alternatively, if one differentiates (6.82) keeping $I^{\alpha\beta}$ constant, then one reobtains (6.89). That is, one assumes that (ignoring the phase factors which lead to Faraday rotation) $I^{\alpha\beta}$ evolves according to

$$I^{\alpha\beta}(s) = a^{\alpha}_{\sigma}(s_0)a^{*\beta}_{\sigma'}(s_0)a^{*\gamma}_{\sigma}(s)a^{\delta}_{\sigma'}(s)I^{\gamma\delta}(s_0),$$

which is to be differentiated with respect to s before setting $s_0=s$.

Carrying out the differentiations explicitly, one finds

$$a'^{\alpha}_{+}=-\frac{T'}{T^2+1}a^{\alpha}_{-}-\frac{i\phi'}{T^2+1}\{2Ta^{\alpha}_{+}+(T^2-1)a^{\alpha}_{-}\}, \tag{6.90a}$$

$$a'^{\alpha}_{-}=\frac{T'}{T^2+1}a^{\alpha}_{+}-\frac{i\phi'}{T^2+1}\{-2Ta^{\alpha}_{-}+(T^2-1)a^{\alpha}_{+}\}, \tag{6.90b}$$

and hence

$$r^{\alpha\beta}_{MC}=-\frac{2iT'}{T^2+1}\begin{pmatrix}\sin 2\phi & -\cos 2\phi\\ -\cos 2\phi & -\sin 2\phi\end{pmatrix}-2i\phi'\begin{pmatrix}0 & i\\ -i & 0\end{pmatrix} \tag{6.91}$$

In the Mueller calculus (6.91) corresponds to

$$r^{(MC)}_{V}=-2\phi',\quad r^{(MC)}_{U}=-\frac{2T'}{T^2+1}\cos 2\phi,\quad r^{(MC)}_{Q}=\frac{2T'}{T^2+1}\sin 2\phi. \tag{6.92}$$

It might be remarked that the derivatives of ϕ and T should be regarded as independent in the sense that $r^{\alpha\beta}_{MC}$, or the equivalent quantity $r^{(MC)}_{AB}$ in the Mueller calculus, are total derivatives of s. That is, mode coupling should be regarded as resulting from two independent contributions, one for $\phi'\neq 0$ with $T'=0$, and other for $T'\neq 0$ and $\phi'=0$.

It is interesting that the foregoing approach is perhaps the simplest approach to mode-coupling theory. The coupling ratio Q, defined in Section 12.3 below, can be calculated from

$$Q=\left\{\frac{|r^{(MC)}_{Q}|^2+|r^{(MC)}_{U}|^2+|r^{(MC)}_{V}|^2}{|r_Q|^2+|r_U|^2+|r_V|^2}\right\}^{1/2}=\frac{\{(4\phi')^2+(2T')^2/(T^2+1)^2\}^{1/2}}{|\Delta k|}. \tag{6.93}$$

Apart from geometric factors, (6.93) follows from (12.123), (12.126), (12.127), and (12.128). (The geometric factors are different because "vertical incidence" is implicit in the present approach.) In the limit of weak mode-coupling, i.e., $Q\ll 1$, the inhomogeneity is unimportant. In the opposite limit of strong mode coupling ($Q\gg 1$), the separation into magnetoionic components is inappropriate to a first approximation, and the radiation propagates as though it were *in vacuo* (i.e., with $\mathrm{d}I^{\alpha\beta}/\mathrm{d}s=0$). Just how mode coupling leads to this result has not been explained clearly in the literature.

Solutions of the Transfer Equation

The transfer equation (6.53) in the Mueller calculus is a matrix equation. One may solve it by firstly finding the eigenvalues and eigenfunctions of the matrix $-\mu_{AB}+r_{AB}$. That is, one finds four new variables $S_{1'}$ to $S_{4'}$ such that (6.53) separates into four uncoupled equations. Each of these four equations may then be integrated separately. However, it is usually not convenient to solve the transfer equation in this way. In practice, either absorption dominates or the polarization-changing processes dominate. In the former case it is convenient to solve by setting $r_{AB}=0$ and finding the eigenvalues and eigenfunctions of μ_{AB}. Alternatively, if the polarization-changing processes, e.g. Faraday rotation, dominate then it is convenient to set $\mu_{AB}=0$ and find the eigenvalues and eigenfunctions of r_{AB}. One may then include absorption to find how these eigenfunctions damp.

Firstly, consider the case in which the eigenvalues and eigenfunctions of $-\mu_{AB}+r_{AB}$ are found. The eigenvalues are found from

$$\det\left[-\mu_{AB}+r_{AB}-\lambda\delta_{AB}\right]=0, \tag{6.95}$$

where δ_{AB} is the unit 4×4 matrix. One finds

$$\begin{aligned}(\lambda+\mu_I)^4&+(\lambda+\mu_I)^2(r_Q^2+r_U^2+r_V^2-\mu_Q^2-\mu_U^2-\mu_V^2)\\&-\mu_Q^2r_Q^2-\mu_U^2r_U^2-\mu_V^2r_V^2\\&-2\mu_Qr_Q\mu_Ur_U-2\mu_Qr_Q\mu_Vr_V-2\mu_Ur_U\mu_Vr_V=0\end{aligned} \tag{6.96}$$

The four solutions of (6.96) are the square roots of

$$\begin{aligned}(\lambda+\mu_I)^2=&\tfrac{1}{2}(\mu_Q^2+\mu_U^2+\mu_V^2)-\tfrac{1}{2}(r_Q^2+r_U^2+r_V^2)\\&\pm\tfrac{1}{2}[(\mu_Q^2+\mu_U^2+\mu_V^2-r_Q^2-r_U^2-r_V^2)^2\\&+4(\mu_Qr_Q+\mu_Ur_U+\mu_Vr_V)^2]^{1/2}\end{aligned} \tag{6.97}$$

The eigenfunctions may be found by inserting the relevant solution for $\lambda':=\lambda+\mu_I$ in (from the first row of the matrix which is the inverse of $-\mu_{AB}+r_{AB}-\lambda\delta_{AB}$)

$$\begin{aligned}I(\lambda'):=&-\lambda'(\lambda'^2+r_Q^2+r_U^2+r_V^2)I\\&+\{\mu_Q(\lambda'^2+r_Q^2)-\lambda'(\mu_Ur_V-r_U\mu_V)+r_Q(\mu_Ur_U+\mu_Vr_V)\}Q\\&+\{\mu_U(\lambda'^2+r_U^2)-\lambda'(\mu_Vr_Q-r_V\mu_Q)+r_U(\mu_Vr_V+\mu_Qr_Q)\}U\\&+\{\mu_V(\lambda'^2+r_V^2)-\lambda'(\mu_Qr_U-r_Q\mu_U)+r_V(\mu_Qr_Q+\mu_Ur_U)\}V.\end{aligned} \tag{6.98}$$

Zheleznyakov, Sovorov and Shaposhnikov (1974) have discussed the solution of the transfer equation in terms of these eigenfunctions for $r_U=0$, $\mu_U=0$.

The derivation of (6.98) may be presented in a form closely analogous to the solution of the wave equation in Section 2.3. Specifically, write

$$\Lambda_{AB}:=-\mu_{AB}+r_{AB}-\lambda\delta_{AB}$$

and then (6.96) corresponds to $\Lambda=0$ with

$$\Lambda:=\det \Lambda_{AB}.$$

Next construct λ_{AB} such that it satisfies

$$\Lambda_{AB}\lambda_{BC}=\Lambda\delta_{AC},$$

where δ_{AB} is the unit 4×4 matrix. Contracting with S_C, it follows that $\lambda_{BC}S_C$ is an eigenfunction for $\Lambda=0$. The result (6.98) is obtained from $\lambda_{1C}S_C$. The eigenfunctions obtained from the fourth row, i.e. from $\lambda_{4C}S_C$, are

$$\begin{aligned} I(\lambda') = &\{\mu_V(\lambda'^2+r_V^2)-\lambda'(\mu_U r_Q - r_U\mu_Q)+r_V(\mu_Q r_Q+\mu_U r_U)\}I \\ &+\{-r_U(\lambda'^2-\mu_U^2)-\lambda'(r_Q r_V+\mu_Q\mu_V)+\mu_U(\mu_Q r_Q+\mu_V r_V)\}Q \\ &+\{r_Q(\lambda'^2-\mu_Q^2)-\lambda'(r_V r_U+\mu_V\mu_U)-\mu_Q(\mu_U r_U+\mu_V r_V)\}U \\ &-\lambda'(\lambda'^2+r_V^2-\mu_Q^2-\mu_V^2)V. \end{aligned} \tag{6.98'}$$

For $\mu_{AB}=0$ (6.98) cannot be used to derive the eigenfunctions and (6.98′) or the analogous results obtained from rows 2 or 3, may be used.

When absorption is the dominant process one can usually neglect Faraday rotation entirely. For $r_Q=r_U=r_V=0$, the eigenvalues are

$$\lambda=-\mu_I,\ -\mu_I,\ -\mu_I\pm(\mu_Q^2+\mu_U^2+\mu_V^2)^{1/2}. \tag{6.99}$$

The two eigenfunctions corresponding to the double eigenvalue are degenerate, and both must be orthogonal to the other two eigenfunctions, which are proportional to

$$I_\pm^{\text{abs}} := I\mp\frac{\mu_Q Q+\mu_U U+\mu_V V}{(\mu_Q^2+\mu_U^2+\mu_V^2)^{1/2}}. \tag{6.100}$$

The two eigenfunctions (6.100) damp with absorption coefficients $\mu_I\mp(\mu_Q^2+\mu_U^2+\mu_V^2)^{1/2}$, and the two degenerate eigenfunctions damp with absorption coefficient μ_I.

Perhaps the most informative method of solution of the transfer equation is in terms of the eigenfunction of r_{AB}. This method is a direct analog of the method developed in treating weakly damped waves in Chapter 2. Briefly one neglects absorption initially, finds the eigenvalues and eigenfunctions, and then includes absorption to find how each eigenfunction damps. Before discussing the significance of this method further, let us proceed to develop the method.

For $\mu_I=\mu_Q=\mu_U=\mu_V=0$, the eigenvalues (6.97) include two which are zero and another two with

$$\lambda=\pm i(r_Q^2+r_U^2+r_V^2)^{1/2}=\pm i|\Delta k|. \tag{6.101}$$

In this case (6.98) gives identically zero, and one could use say (6.98′) to find the eigenfunctions. However, on physical grounds one knows that the four eigenfunctions may be chosen as I^{++} and I^{--}, which have eigenvalue zero, and I^{+-} and I^{-+}, which have eigenvalues $\pm i\Delta k$. These four eigenfunctions are defined by (6.42) and (6.42′), or by (6.82) with $\Delta\Psi=0$. In fact, (6.82) is obtained by integrating the resulting transfer equations for the four eigenfunctions. The four eigenfunctions are, explicitly,

$$\begin{pmatrix} I^{++} \\ I^{--} \end{pmatrix} = \tfrac{1}{2}I \pm \frac{1}{2(T^2+1)}\{(T^2-1)(Q\cos 2\phi + U\sin 2\phi) + 2TV\} \tag{6.102}$$

$$I^{+-} = (I^{-+})^* = \frac{i}{2}\left[(U\cos 2\phi + Q\sin 2\phi) + \frac{i}{T^2+1} \times\{2T(Q\cos 2\phi + U\sin 2\phi) - (T^2-1)V\}\right].$$

Using (6.88), one finds that I^{+-} corresponds to the eigenvalue $i\Delta k$ in accord with (6.82).

Now let us include spontaneous emission and absorption. It is convenient to define four absorption coefficients which correspond to the four eigenfunctions:

$$\begin{pmatrix} \mu^{++} \\ \mu^{--} \end{pmatrix} = \mu_I \pm \frac{1}{T^2+1}\{(T^2-1)(\mu_Q\cos 2\phi + \mu_U\sin 2\phi) + 2T\mu_V\} \tag{6.103}$$

$$\mu^{+-} = (\mu^{-+})^* = i\left[\mu_U\cos 2\phi - \mu_Q\sin 2\phi + \frac{i}{T^2+1} \times\{2T/\mu_Q\cos 2\phi + \mu_U\sin 2\phi) - (T^2-1)\mu_V\}\right].$$

Similarly, one defines four emission coefficients, but with extra factors of $\frac{1}{2}$, as in (6.102). The transfer equation then becomes

$$n^2\frac{d}{ds}\left(\frac{I^{++}}{n^2}\right) = \alpha^{++} - \mu^{++}I^{++} - \tfrac{1}{2}(\mu^{+-}I^{-+} + \mu^{-+}I^{+-}), \tag{6.104a}$$

$$n^2\frac{d}{ds}\left(\frac{I^{+-}}{n^2}\right) = \alpha^{+-} - i\Delta k I^{+-} - \tfrac{1}{2}(\mu^{++} + \mu^{--})I^{+-} - \tfrac{1}{2}\mu^{+-}(I^{++} + I^{--}), \tag{6.104b}$$

$$n^2\frac{d}{ds}\left(\frac{I^{-+}}{n^2}\right) = \alpha^{-+} + i\Delta k I^{-+} - \tfrac{1}{2}(\mu^{++} + \mu^{--})I^{-+} - \tfrac{1}{2}\mu^{-+}(I^{++} + I^{--}), \tag{6.104c}$$

$$n^2\frac{d}{ds}\left(\frac{I^{--}}{n^2}\right) = \alpha^{--} - \mu^{--}I^{--} - \tfrac{1}{2}(\mu^{-+}I^{+-} + \mu^{+-}I^{-+}) \tag{6.104d}$$

An important implication of the form (6.104) of the transfer equation is that it implies an absorption induced coupling between the components I^{++} and I^{--} in the two natural modes. Consider, for example, the transfer equation (6.39), which is derived under the assumption that the two modes are absorbed independently of each other. The components I^{++} and I^{--} in (6.104) are the counterparts of I^o and I^x, for the o-mode and x-mode say, in (6.39), but there is no counterpart in (6.39) of I^{+-} and I^{-+}. It is clear that a subtle assumption is implicit in (6.39). Specifically, it is implicit that I^{+-} and I^{-+} do not contribute in (6.104a) and (6.104d). In effect, what is assumed is that Faraday rotation occurs much more rapidly than does absorption, i.e. $|\Delta k| \gg \mu_Q$ is implicit. (The relevant absorption coefficients are μ^{+-} or μ^{-+}, which are assumed of order μ_Q for the sake of discussion.) This limiting case has been called the *strong rotativity limit* by Jones and O'Dell (1977). In this limit I^{+-} and I^{-+} oscillate rapidly (over distances $1/\Delta k$) compared with an absorption length (a distance μ_Q^{-1} here).

It is apparent from (6.104) that in the opposite limit ($\Delta k \ll \mu_Q$) the two modes do not damp independently. One explanation for this is that absorption may be regarded as implying that the wavenumber has an imaginary part of order μ_Q. In other words, wavenumbers cannot be defined meaningfully to within less than an uncertainty of order μ_Q in the presence of absorption. Consequently, if the intrinsic difference Δk in the wavenumbers of the two modes (defined by neglecting absorption) is less than about μ_Q, then the absorption must wash out the distinction between the two modes, i.e. it can be said to induce a coupling between the modes.

From a slightly different viewpoint, it was pointed out in connection with (6.26) that the two modes propagate independently except in the neighborhood of a coupling point. Now one can regard the two dispersion curves, $\omega^+(\mathbf{k})$ and $\omega^-(\mathbf{k})$ say, as curves in $\mathbf{k}$-space and regard Δk as the distance between them. The "neighborhood" of a coupling point (where the two curves meet) may be defined as any region in which the separation Δk is less than the intrinsic uncertainty in wavenumber. Here the uncertainty is of order μ_Q and is due to absorption. The inhomogeneity of the medium also leads to an uncertainty in k, and mode coupling is "strong" when this uncertainty exceeds Δk.

Generalized Faraday Rotation

Generalized Faraday rotation is implicit in (6.82) with (6.102). For example, suppose one is given I, Q, U and V at an initial point in the medium where the axial ratio of the modes is T and the azimuthal angle of the magnetic field is ϕ. Then one may construct I^{+-} and I^{-+} using (6.102). Now by performing the integral (6.80) along the ray path from that initial point to some final point, one can determine $\Delta\Psi$ given Δk as a function of position along the ray path. At the final point I^{+-} and I^{-+} are given by (6.82). Now one may use (6.102) again to find I, Q, U and V at the final point, provided one knows T and ϕ there. Note that one needs to know T and ϕ only at the initial and final points, and that one needs to know Δk at every point along the ray path. Before treating this general case, let us consider the simplest and most familiar example.

Faraday rotation, in the specific sense rather than the generic sense, applied to media in which the modes are circularly polarized. For a magnetized plasma, the properties of the magnetoionic waves (2.132) in the quasi-circular limit give

$$T_\sigma = -\frac{\sigma \cos\theta}{|\cos\theta|}, \quad \Delta n = \frac{\sigma\omega_p^2\Omega_e|\cos\theta|}{n\omega^3}, \tag{6.105}$$

with $\sigma = 1$ for the o-mode and $\sigma = -1$ for the x-mode. With $\Delta k := \Delta n\,\omega/c$, and with the polarization vectors $\mathbf{e}^1$ and $\mathbf{e}^2$ chosen as in (6.66), i.e. $\phi = 0$ in (6.79a, b), the quantities (6.88) which describe Faraday rotation reduce to

$$r_V = \frac{\omega_p^2\Omega_e \cos\theta}{\omega_p^2 c(1-\omega_p^2/\omega^2)^{1/2}}, \quad r_U = 0, \quad r_Q = 0. \tag{6.106}$$

In this case the transfer equation leaves I and V unaffected, with U and Q changing only such that the angle $\chi := \text{arc tan}\,(U/Q)$, cf. (6.51d), changes according

$$\frac{d\chi}{ds} = \frac{\omega_p^2\Omega_e \cos\theta}{2\omega^2 c(1-\omega_p^2/\omega^2)^{1/2}} \left(= \frac{2.36\times 10^4 n_e B\cos\theta}{(\omega/2\pi)^2(1-\omega_p^2/\omega^2)^{1/2}}\ \text{cm}^{-1}\right). \tag{6.107}$$

The rate $d\chi/ds$ is the number of radians per unit length through which the plane of polarization rotates in a RH screw sense relative to $\boldsymbol{\kappa}$.

Now let us treat the general case where the natural modes are elliptically polarized. Given I_0, Q_0, U_0 and V_0 at a point s_0 where the axial ratio is T_0 and the azimuthal angle of the magnetic field is ϕ_0, the value of I, Q, U and V at any other point s where the axial ratio is T and the azimuthal angle is ϕ may be determined by using the fact that I^{++} and I^{--} are constants, and that I^{+-} evolves with a phase factor $\exp(i\Delta\Psi)$. Thus, using (6.102), one has

$$I = I_0, \tag{6.108a}$$

$$\frac{1}{T^2+1}\{(T^2-1)(Q\cos 2\phi + U\sin 2\phi) + 2TV\} = \frac{1}{T_0^2+1}\{(T_0^2-1)(Q_0\cos 2\phi_0 + U_0\sin 2\phi_0) + 2T_0V_0\}, \tag{6.108b}$$

and

$$U \cos 2\phi - Q \sin 2\phi + \frac{i}{T^2+1} \{2T(Q \cos 2\phi + U \sin 2\phi) - (T^2-1)V\}$$

$$= \exp(i\Delta\Psi)\left[U_0 \cos 2\phi_0 - Q_0 \sin 2\phi_0 + \frac{i}{T_0^2+1} \times \{2T_0(Q_0 \cos 2\phi_0 + U_0 \sin 2\phi_0) - (T_0^2-1)V_0\}\right] \quad (6.108c)$$

with

$$\Delta\Psi := \int_{s0}^{s} ds' \, \Delta k(s'). \quad (6.108d)$$

The particular case of a partial conversion of linear into circular polarization due to slightly elliptical modes ($T^2 \neq 1$) has been called "circular repolarization" by Pacholczyk (1973). This effect is included in (6.108) and is discussed specifically in Section 9.1 in connection with the circular polarization of synchrotron sources. In the special case of circularly polarized modes, the result (6.108) for $T=0$, $\phi=0$ may be obtained by integrating (6.107) and noting that I and V are constant in this case.

Another special case is that in which the natural modes are plane polarized. ("Plane" and "linear" have the same meaning in connection with polarization.) Assuming $T_0=\infty$, $\phi_0=0$ and $U_0=0$ in (6.108) one finds that V oscillates as $\cos(\Delta ks)$ along the ray path. This has been called Faraday pulsation by Pacholczyk and Swihart (1970). On the other hand, again with $T_0=\infty$, $\phi_0=0$ and $U_0=0$ inside the source, and $T=1$ outside the source, one finds $V=Q_0$. That is, complete conversion of linear into circular polarization is possible if the natural modes are planar within the source (they are circular outside sources in the interplanetary interstellar or intergalactic media). This point is discussed further in Section 9.1 below.

6.3 SYNCHROTRON ABSORPTION

Let us refer to the absorption process corresponding to synchrotron emission as "synchrotron absorption". From a formal viewpoint, synchrotron absorption is almost unique (in astrophysics) in that it is about the only example of an absorption process (for transverse waves) which is strongly dependent on the polarization of the radiation. In practice, synchrotron absorption is usually considered only in the context of synchrotron re-absorption, when both emission and absorption are due to the synchrotron process in a given source.

It is interesting to contemplate the effect of synchrotron absorption alone, e.g. to consider the effect of an optically thick source as an absorbing region between the observer and a bright background source. This situation is of little interest in practice because optically thick sources are typically much smaller than any bright background source, and they would be ineffective in obscuring any background source. Because synchrotron emission is partially linearly polarized with a small circularly polarized component, synchrotron absorption must have analogous properties. For example, if the incident radiation is decomposed into its linearly polarized components along $\boldsymbol{\kappa} \times \mathbf{B}$ and $\boldsymbol{\kappa} \times (\boldsymbol{\kappa} \times \mathbf{B})$, where $\mathbf{B}$ refers to the magnetic field in the absorption region, then because the latter component would be the more strongly emitted, it should also be the more strongly absorbed. Likewise, if the incident radiation is decomposed into oppositely circularly polarized components then one component (the more strongly emitted one) would be more strongly absorbed than the other. Consequently, if the initial radiation were completely linearly polarized along a direction other than $\boldsymbol{\kappa} \times \mathbf{B}$ or $\boldsymbol{\kappa} \times (\boldsymbol{\kappa} \times \mathbf{B})$, then the plane of polarization would rotate toward $\boldsymbol{\kappa} \times \mathbf{B}$ as a result of the component along $\boldsymbol{\kappa} \times (\boldsymbol{\kappa} \times \mathbf{B})$ being the more rapidly absorbed. Also a small degree of circular polarization would build up due to one circularly polarized component being preferentially absorbed. These and related effects would be significant in inhomogeneous self-absorbed sources.

Historically, Twiss (1958) discussed synchrotron absorption and evaluated the absorption coefficient using the Einstein relations. le Roux derived the absorption coefficient using a different approach which essentially reproduced Twiss' result. Later, Kawabata (1964) developed an ingeneous technique for taking the polarization into account properly: by noting that the emission at given s, ω and θ is completely polarized, one may apply the Einstein coefficients to emission with the inferred polarization vectors, cf. (4.28), retained inside the sums and integrals. Kawabata's (1964) procedure was criticized by Zheleznyakov (1968) but the criticism seems to be without foundation (Melrose 1971c). The complete expression for the synchrotron absorption coefficient including the circularly polarized component was derived by Melrose (1971c).

The Synchrotron Absorption Coefficient

The methods available for treating such polarization-dependent absorption are summarized in Section 6.2. It follows from (6.43) and (6.49) that all the relevant information on the absorption can be included by writing the absorption coefficient as a polarization tensor $\mu^{\alpha\beta}$, which is just $\gamma^{\alpha\beta}$ divided by the group speed. The absorption coefficient $\gamma^{\alpha\beta}(\omega, \theta)$ may be written down as follows. Firstly define the probability $w^{\alpha\beta}$ as a polarization tensor in (5.46) and regard (5.46) as a definition of $\gamma^{\alpha\beta}$. The probability $w^{\alpha\beta}$ may be related to the emissivity $\eta^{\alpha\beta}$ using (4.39) with

$$P^{\alpha\beta}=\int \frac{d^3\mathbf{k}}{(2\pi)^3}\,\hbar\omega w^{\alpha\beta}(\mathbf{p},\mathbf{k}).$$

In this way one obtains

$$\gamma^{\alpha\beta}(\omega,\theta)=-\sum_{s=1}^{\infty} 2\pi \int_0^{\infty} dp\, p^2 \int_{-1}^{+1} d\cos\alpha \left(\frac{2\pi c}{\omega}\right)^3 \eta^{\alpha\beta}(s,\omega,\theta)$$
$$\times\frac{\omega}{n(\omega)\beta c}\left\{\frac{\partial}{\partial p}+\frac{\cos\alpha-n(\omega)\beta\cos\theta}{p\sin\alpha}\frac{\partial}{\partial\alpha}\right\}f(p,\alpha), \quad (6.109)$$

where the emissivity $\eta^{\alpha\beta}(s,\omega,\theta)$ is given by (4.33). The absorption coefficient (6.109) may be evaluated for synchrotron radiation by using the method developed in Section 4.2. Suppose that $f(p,\alpha)$ is separable into $f_0(p)\phi(\alpha)$ as in (4.62). The (differential) energy spectrum $N(\varepsilon)$ is the number of particles in the range $d\varepsilon$ at ε, and it is related to the distribution $f_0(p)$ by

$$f_0(p)=\frac{c^3}{4\pi\varepsilon^2}N(\varepsilon), \quad (6.110)$$

which is correct to $O(\gamma^{-2})$. The absorption coefficient (6.109) separates into the part which involves only the derivative with respect to energy, and the part which involves only the differentiation with respect to pitch angle:

$$\gamma^{\alpha\beta}(\omega,\theta)=\gamma_\varepsilon^{\alpha\beta}(\omega,\theta)+\gamma_A^{\alpha\beta}(\omega,\theta). \quad (6.111)$$

The calculation of the former term is analogous to the calculation of the emissivity in synchrotron radiation:

$$\gamma_\varepsilon^{\alpha\beta}(\omega,\theta)=-\frac{(2\pi c)^3}{\omega^2}\phi(\theta)\int_0^{\infty} d\varepsilon\,\varepsilon^2\bar{\eta}^{\alpha\beta}(\omega,\theta)\frac{d}{d\varepsilon}\left\{\frac{N(\varepsilon)}{\varepsilon^2}\right\}, \quad (6.112)$$

with $\bar{\eta}^{\alpha\beta}(\omega,\theta)$ given by (4.81), or by (4.87) when the Airy integral approximation is made. The other term in (6.111) may be reduced to the form

$$\gamma_A^{11}(\omega,\theta)=0=\gamma_A^{22}(\omega,\theta),$$

$$\gamma_A^{12}(\omega,\theta)=-\gamma_A^{21}(\omega,\theta)=-\frac{(2\pi c)^3\phi(\theta)}{\omega^2}\int_0^{\infty}\frac{d\varepsilon}{\varepsilon}\frac{e^2\Omega_e s_0^2}{2\pi c\gamma^2}$$
$$\times N(\varepsilon)i\epsilon\cos\theta g(\theta)\frac{1}{2s_0}\int_0^{\beta'} dy\,J_{2s}(2sy), \quad (6.113)$$

with β' given by (4.71) and s_0 by (4.80), or to the form

$$\gamma_A^{12}(\omega,\theta) = -\gamma_A^{21}(\omega,\theta) = -\frac{(2\pi c)^3}{\omega^2}\phi(\theta)\int_0^\infty \frac{d\varepsilon}{\varepsilon}\frac{\sqrt{3}}{16\pi^2 c}\frac{e^2\Omega_e \xi \sin\theta}{\gamma^2} \times N(\varepsilon)\left\{\frac{2i\epsilon\cot\theta}{3\xi}g(\theta)\int_{\omega/\omega_c}^\infty dt\, K_{1/3}(t)\right\}, \quad (6.112')$$

where the Airy integral approximation has been made.

The Possibility of Amplified Synchrotron Emission

It was first suggested by Twiss (1958) that synchrotron absorption could be negative. However, his analysis contained an error, and it was later shown that synchrotron absorption *in vacuo* cannot be negative. Negative absorption is possible in principle under two rather extreme conditions. The first is an extreme pitch-angle anisotropy, when $d\phi(\alpha)/d\alpha$ is nearly singular. Specifically, if $d\phi(\alpha)/d\alpha$ has a value of order γ or larger, then the conventional treatment of synchrotron emission, in which one expands in powers of γ^{-1}, breaks down and negative absorption is possible (Sazonov, 1972; Zheleznyakov and Suvorov, 1972). However, the resonant scattering discussed in Chapter 7 below makes it implausible that a strong pitch-angle anisotropy could persist, as discussed in detail in Section 9.4.

The second case where amplified synchrotron emission can occur was pointed out by McCray (1966) and Zheleznyakov (1966), see also Yukon (1968) and Zheleznyakov and Suvorov (1972). In the presence of a plasma, amplified emission can occur near the Razin-Tsytovich frequency (4.131) provided that $N(\varepsilon)/\varepsilon^2$ is an increasing function of ε in the neighborhood of the energy of electrons radiating at this frequency. Two conditions need to be satisfied in order for (6.112) to predict negative absorption. The first is that $d\{N(\varepsilon)/\varepsilon^2\}/d\varepsilon$ be positive over some finite range. The second is obtained by partially integrating (6.112). Thus the two conditions required for negative absorption are

$$\frac{d}{d\varepsilon}\left\{\frac{N(\varepsilon)}{\varepsilon^2}\right\} > 0, \quad \frac{d}{d\varepsilon}\{\varepsilon^2\bar{\eta}^{\alpha\beta}(\omega,\theta)\} < 0. \qquad (6.113a, b)$$

Actually, in (6.113b) one should really insert the emissivity in a specific polarization. That is, for negative absorption of a polarization described by polarization **e**, the quantity which must be negative is $e^{*\alpha}e^{\beta}$ times the LH side of (6.113b).

The proof that synchrotron absorption is strictly positive *in vacuo*, and that negative absorption is possible in a plasma, are based on the requirement

(6.113b). With

$$t := \gamma/\xi, \quad t = \left(1 + \frac{\gamma^2\omega_p^2}{\omega^2}\right)^{1/2}, \tag{6.114}$$

one has

$$\frac{dt}{d\varepsilon} = \frac{1}{\varepsilon}\frac{(t^2-1)}{t}, \quad \frac{d}{d\varepsilon}\left(\frac{\omega}{\omega_c}\right) = \frac{1}{\varepsilon}\frac{\omega}{\omega_c}\left(2 - \frac{3(t^2-1)}{t^2}\right), \tag{6.115}$$

and with $z := \omega/\omega_c$, the LH side of (6.113b) for the two linear polarizations is proportional to

$$\frac{\partial}{\partial\varepsilon}\left[\frac{\varepsilon^2}{t}\left\{z\int_z^\infty dt\, K_{5/3}(t) \pm K_{2/3}(z)\right\}\right]$$
$$= \frac{2\varepsilon}{t}\left[\frac{t^2-1}{t^2}\left\{z\int_z^\infty dt\, K_{5/3}(t) - \frac{3z^2}{2}\begin{pmatrix} K_{5/3}(z) + K_{1/3}(z) \\ (4/3z)K_{2/3}(z)\end{pmatrix}\right\}\right.$$
$$\left. + z^2\begin{pmatrix} \frac{3}{2}K_{5/3}(z) + \frac{1}{2}K_{1/3}(z) \\ (2/3z)K_{2/3}(z)\end{pmatrix}\right], \tag{6.116}$$

where the following identities have been used:

$$K_{2/3}(z) - \frac{3z}{2}K'_{2/3}(z) = \frac{3z}{2}K_{5/3}(z),$$

$$K_{2/3}(z) + \frac{3z}{2}K'_{2/3}(z) = -\frac{3z}{2}K_{1/3}(z).$$

The functions $K_\nu(z)$ are positive for all $z > 0$ and all real ν. It follows from (6.116) that the LH side of (6.114b) is strictly positive for $t = 1$, i.e. for absorption *in vacuo*. In essence, this proof was first given by Wild, Smerd and Weiss (1963), although their actual proof is based on emissivity with an incorrect energy dependence. For $t > 1$, both the $\alpha = \beta = 1$ and $\alpha = \beta = 2$ terms become negative at low frequencies. The functions on the RH side of (6.116) were plotted by Yukon (1968) and Zheleznyakov and Suvorov (1972), who found that negative absorption is possible for $\omega \lesssim \frac{1}{2}\gamma\omega_p$ (see Figure 6.1).

It should be pointed out that the conditions under which significant amplification of synchrotron radiation could occur appear to be very restrictive. Three conditions must be satisfied. First, the Razin effect must be important (the amplification could occur only for $\omega < \gamma\omega_p$) whereas there is no source, with the possible exception of the solar corona, for which this is known to be the case. Second, the condition (6.113a) must be satisfied, and it would appear that some special mechanism is required to create an energy spectrum which increases with ε faster than ε^2. Third, the source must be optically thick for $\omega \lesssim \frac{1}{2}\gamma\omega_p$. It seems unlikely that such a combination of preconditions would be satisfied simultaneously in any given source.

Power-Law Energy Spectra

The absorption coefficient $\gamma^{\alpha\beta}$ may be calculated for synchrotron absorption *in vacuo* by electrons with a power-law energy distribution using the same methods as in the calculation of the volume emissivity (4.102). One finds

$$\gamma^{\alpha\beta}(\omega, \theta) = \frac{(2\pi c)^3}{\omega^2} \frac{K(m_e c^2)^{-a}}{16\pi^2 c} \sqrt{3}\, e^2 \Omega_e \sin\theta \phi(\theta) L^{\alpha\beta}(a) \left(\frac{2\omega}{3\Omega_e \sin\theta}\right)^{-a/2}. \tag{6.117}$$

One has $L^{\alpha\beta}(a) = (a+2)J^{\alpha\beta}(a+1)$ for an isotropic distribution, where $J^{\alpha\beta}(a)$ is given by (4.103). For an anisotropic distribution, i.e. $g(\theta) \neq 0$, one has

$$L^{11}(a) = (a+2)J^{11}(a+1),\ L^{22}(a) = (a+2)J^{22}(a+1),$$

with

$$L^{12}(a) = -L^{21}(a) = (a+2+g(\theta))\left(\frac{a+3}{a+3+g(\theta)}\right) J^{12}(a+1). \tag{6.118}$$

The quantities which appear in the transfer equation (6.53) for synchrotron emission and absorption by power-law distributions *in vacuo* are

$$\alpha_I = \frac{1}{c}\left(\frac{d\eta^{11}}{dV} + \frac{d\eta^{22}}{dV}\right),\quad \alpha_Q = \frac{1}{c}\left(\frac{d\eta^{11}}{dV} - \frac{d\eta^{22}}{dV}\right),\ \alpha_U = 0,\ \alpha_V = \frac{2i}{c}\frac{d\eta^{12}}{dV}, \tag{6.119}$$

with $d\eta^{\alpha\beta}/dV$ given by (4.102), and

$$\mu_I = \frac{1}{2c}(\gamma^{11} + \gamma^{22}),\ \mu_Q = \frac{1}{2c}(\gamma^{11} - \gamma^{22}),\ \mu_U = 0,\ \mu_V = \frac{i}{c}\gamma^{12}, \tag{6.120}$$

with $\gamma^{\alpha\beta}$ given by (6.117).

Emission from Self-Absorbed Sources

In the absence of any effects other than absorption, the transfer equation (6.53) simplifies in two ways. Firstly, one has $n^2 = 1$ and $r_{AB} = 0$. Secondly, one may choose the basic polarization vectors $\mathbf{e}^1$ and $\mathbf{e}^2$ such that α_U and γ_U are zero, in which case U is zero provided the magnetic field in the source is assumed homogeneous. Thus the transfer equation for a homogeneous source reduces to

$$\frac{d}{ds}\begin{pmatrix} I \\ Q \\ V \end{pmatrix} = \begin{pmatrix} \alpha_I \\ \alpha_Q \\ \alpha_V \end{pmatrix} - \begin{pmatrix} \mu_I & \mu_Q & \mu_V \\ \mu_Q & \mu_I & 0 \\ \mu_V & 0 & \mu_I \end{pmatrix} \begin{pmatrix} I \\ Q \\ V \end{pmatrix}. \tag{6.121}$$

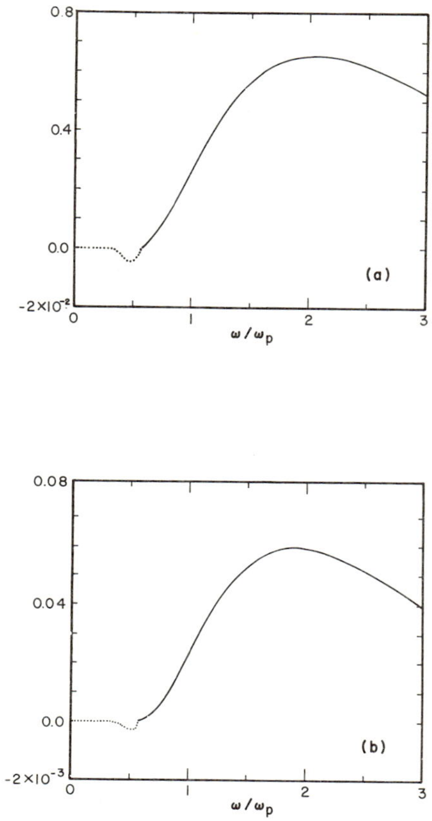

FIGURE 6.1 The functions

$$\frac{1}{2\gamma}\frac{\partial}{\partial\gamma}\left[\frac{\gamma^2}{t}z\left\{\int_z^\infty \mathrm{d}x\,K_{5/3}(x)\pm K_{2/3}(z)\right\}\right]$$

cf. (6.116), are plotted as functions of $\omega/\gamma\omega_p$ for $\frac{3}{2}\Omega_e\gamma^2 \sin\theta/\gamma\omega_p=1$.

These functions may be interpreted as, apart from a constant factor, the absorption coefficients for a monoenergetic distribution of electrons for the polarizations corresponding to the electric vector (a) perpendicular, and (b) parallel to the ambient magnetic field. Note the scales: the scale in figure (b) is expanded by a factor of 10 over that in figure (a), and in both figures the scale below the axis, i.e., where the absorption coefficient is negative, is expanded by a factor of 10 over the scale above the axis.

Let L be the thickness of the source, i.e. the depth along the line of sight. For $\mu_I L \ll 1$ the source is optically thin, and absorption may be neglected. For a homogeneous optically thin source one has $I = \alpha_I L$ and the degrees of linear and circular polarization are given by (4.105) and (4.106) respectively.

In the opposite limit $\mu_I L \gg 1$ the source is optically thick to its own synchrotron emission, and is said to be *self-absorbed*. For a homogeneous source of thickness L the solution of (6.121) may be found by diagonalizing the 3×3 matrix. For this purpose it is useful to note that corrections of order γ^{-2} are ignored in synchrotron theory. Hence, because α_V/α_I, μ_V/μ_I and V/I are all of order γ^{-1}, products of these terms can be neglected. One finds, either by using (6.99) and (6.100) or directly,

$$\frac{\mathrm{d}}{\mathrm{d}s}\begin{pmatrix} \frac{1}{2}(I+Q) \\ \frac{1}{2}(I-Q) \\ \mu_Q V - \mu_V Q \end{pmatrix} = \begin{pmatrix} \frac{1}{2}(\alpha_I + \alpha_Q) \\ \frac{1}{2}(\alpha_I - \alpha_Q) \\ \mu_Q \alpha_V - \mu_V \alpha_Q \end{pmatrix} - \begin{pmatrix} \mu_I + \mu_Q & 0 & 0 \\ 0 & \mu_I - \mu_Q & 0 \\ 0 & 0 & \mu_I \end{pmatrix} \begin{pmatrix} \frac{1}{2}(I+Q) \\ \frac{1}{2}(I-Q) \\ \mu_Q V - \mu_V Q \end{pmatrix}, \tag{6.122}$$

the solution of which is

$$\tfrac{1}{2}(\mathrm{I} \pm Q) = \frac{\frac{1}{2}(\alpha_I \pm \alpha_Q)}{\mu_I \pm \mu_Q}\{1 - \exp[-i(\mu_I \pm \mu_Q)L]\}, \tag{6.123a}$$

$$\mu_Q V - \mu_V Q = \frac{\mu_Q \alpha_V - \mu_V \alpha_Q}{\mu_I}\{1 - \exp[-i\mu_I L]\}. \tag{6.123b}$$

In the limit $\mu_I L \gg 1$, (6.123a, b) give

$$I(\omega, \theta) = \left[\frac{m_e}{(2\pi)^3}\left(\frac{2}{3\Omega_e \sin\theta}\right)^{1/2} \frac{1}{a+2}\left\{\frac{J^{11}(a)}{J^{11}(a+1)} + \frac{J^{22}(a)}{J^{22}(a+1)}\right\}\right]\omega^{5/2}, \tag{6.124}$$

where $J^{\alpha\beta}(a)$ is given by (4.103). For a homogeneous self-absorbed source, the degrees of polarization are

$$r_l = \frac{Q}{I} = \frac{3}{6a+13}, \tag{6.125}$$

and

$$r_c = -\, r_{co} F(a), \tag{6.126}$$

where r_{co} is the degree of circular polarization for an optically thin source.

In (6.126), one has

$$F(a) := \frac{3a+7}{3a+10} \times \left[\frac{a(a+3)(3a-1)}{12(a+1)^2} \frac{\left\{\Gamma\left(\frac{3a-1}{12}\right)\Gamma\left(\frac{3a+7}{12}\right)\right\}^2}{\Gamma\left(\frac{3a+10}{12}\right)\Gamma\left(\frac{3a+2}{12}\right)\Gamma\left(\frac{3a+8}{12}\right)\Gamma\left(\frac{3a+4}{12}\right)} - \frac{2(3a+8)}{6a+13} \right]. \tag{6.127}$$

Some values of $F(a)$ are summarized in Table 6.1.

TABLE 6.1

a	1.0	1.5	2.0	3	4
$F(a)$	0.65	0.43	0.33	0.24	0.19

The linear and circular polarizations change sign as the source becomes optically thick.

There is a simple explanation for the $\omega^{5/2}$-dependence of the intensity from an optically thick source. The intensity from a black body at temperature T_b is (in the Rayleigh-Jeans limit)

$$I(\omega) = \frac{2\omega^2 T_b}{(2\pi)^3 c^2}. \tag{6.128}$$

A source becomes optically thick when the brightness temperature is of the order of the energy of the radiating particles. (Recall that a temperature of 1 eV is 1.1×10^4 K and that the rest energy of an electron is 5.11×10^5 eV.) Thus one expects $T_b \approx \varepsilon$ in (6.128). However, electrons with an energy $\varepsilon = \gamma m_e c^2$ radiate at a frequency $\omega \approx \Omega_e \gamma^2 \sin\theta$, cf. (4.96a), and hence one has $\varepsilon \propto \omega^{1/2}$ and $T_b \propto \omega^{1/2}$. Thus one arrives at $I(\omega) \propto \omega^{5/2}$ for a self-absorbed synchrotron source.

The Effect of Faraday Rotation

In deriving the results (6.123a, b) for an optically thick source, Faraday rotation has been neglected. It is not difficult to solve the transfer equation including Faraday rotation. Assuming the natural modes to be circularly polarized ($r_V \neq 0$, $r_Q = r_U = 0$) and that μ_V and V are of order γ^{-1} (implying that μ_V^2 and $\mu_V V$ are to be neglected), the transfer equation for V is of order γ^{-1}, and may be neglected to a first approximation. By diagonalizing

$$\frac{d}{ds}\begin{pmatrix} I \\ Q \\ U \end{pmatrix}=\begin{pmatrix} \alpha_I \\ \alpha_Q \\ 0 \end{pmatrix}-\begin{pmatrix} \mu_I & \mu_Q & 0 \\ \mu_Q & \mu_I & r_V \\ 0 & -r_V & \mu_I \end{pmatrix}\begin{pmatrix} I \\ Q \\ U \end{pmatrix}$$

and integrating the resulting equations from the back of the source ($s=0$) to an arbitrary point s, one finds

$$r_V I(s)+\mu_Q U(s)=\frac{r_V\alpha_I}{\mu_I}(1-e^{-\mu_I s}), \tag{6.129a}$$

$$\mu_Q I(s)+r_V U(s)\pm\lambda Q=\frac{\mu_Q\alpha_I\pm\lambda\alpha_Q}{\mu_I\pm\lambda}(1-e^{-(\mu_I\pm\lambda)s}), \tag{6.129b}$$

with

$$\lambda:=\begin{cases}\pm(\mu_Q^2-r_V^2)^{1/2} & \text{for } \mu_Q^2>r_V^2, \\ \pm i(r_V^2-\mu_Q^2)^{1/2} & \text{for } r_V^2>\mu_Q^2.\end{cases} \tag{6.130}$$

The transfer equation for V, namely

$$\frac{dV}{ds}=\alpha_V-\mu_I V-\mu_V I,$$

may now be integrated to find

$$V(s)=\frac{\alpha_V}{\mu_I}(1-e^{-\mu_I s})-\mu_V\, e^{-\mu_I s}\int_0^s ds'\, e^{\mu_I s'} I(s'). \tag{6.131}$$

By setting $r_V=0$, one finds that (6.129b) reproduces (6.123a).

The effect of Faraday rotation for $r_V L\gg 1$ is to wash out the linear polarization. This is called Faraday depolarization. It is due to the emission from different points through the source having its plane of polarization rotated through different angles.

For an optically thick source, all the emission observed is effectively from a surface region of about one optical depth, i.e. of thickness μ_I^{-1}. For $r_V\gg\mu_I$ Faraday rotation should wash out the linear polarization even in this case. However, Faraday rotation has virtually no effect on the circular polarization. In the optically thin case this is obvious from (6.131). In the optically thick case, for $r_V\gg\mu_I$, (6.129a, b) imply

$$I\approx\frac{\alpha_I}{\mu_I}\gg Q,\, U, \tag{6.132}$$

and then (6.131) gives

$$V\approx\frac{\alpha_V}{\mu_I}-\frac{\mu_V I}{\mu_I}. \tag{6.133}$$

Thus the degree of circular polarization in the strong-rotativity, optically thick case is

$$r_c = \frac{\alpha_V}{\alpha_I} - \frac{\mu_V}{\mu_I} = -r_{cO}(1 - \tilde{F}(a)) \tag{6.134}$$

with

$$\tilde{F}(a) := F(a) - \frac{2(3a+7)(3a+8)}{(3a+10)(6a+13)}, \tag{6.135}$$

where $F(a)$ is given by (6.127). It follows from numerical values tabulated by Melrose (1971c) that the result (6.134) differs little from the optically thick case in (6.134) in the absence of Faraday rotation, i.e. from (6.126).

Whether or not Faraday rotation is important in any particular synchrotron source can be determined by measuring the position angle of the plane of polarization at several different frequencies. The observational evidence indicates little Faraday rotation in most sources, i.e. the position angle does not vary significantly with frequency. This implies that $d\chi/ds$ in (6.107) is less than about the inverse of the thickness of the source. Consequently, the case which is likely to be of most interest in practice corresponds to $|\lambda s| \ll 1$ in (6.129b). Nevertheless Faraday rotation is of interest especially when the possibility of elliptically polarized modes is taken into account. This general topic is discussed further in Section 9.1.

6.4 INDUCED COMPTON SCATTERING

Induced Compton scattering is related to spontaneous Compton scattering in a way analogous to the relation between absorption and spontaneous emission. However, a misconception with induced inverse-Compton-emission is the following: spontaneous scattering causes transfer of photons from an initial frequency ω to a final frequency of order $\gamma^2\omega$ and one might expect that the corresponding induced scattering should transfer photons from $\gamma^2\omega$ back to ω. Such is not the case in practice. A qualitative reason for this is given below after the formal treatment of induced scattering by relativistic electrons. Induced scattering by thermal electrons is also discussed below as it is also of interest for synchrotron sources. The discussion in this section follows that given by Melrose (1971d). Other discussions of induced scattering have been given by Oster (1968a, b), Goldreich *et al.* (1968), Kaplan and Tsytovich (1969, 1973), Sunyaev (1970, 1971) Oster and Altschuler (1971) and Blandford and Scharlemann (1975).

Induced Compton (or Thomson) scattering may be treated using the transfer equation (5.89a) with the relevant probability taken to be that for Thomson scattering of transverse waves *in vacuo*, i.e. (4.133) with (4.134). For isotropic distributions of waves and of particles, a lengthy calculation gives (Melrose, 1971d)

$$\frac{dN(k)}{dt}=\frac{2\pi^2 e^4}{c}\int_0^\infty dp\int_0^\infty \frac{dy}{y}\, g(y)\times\left[\{N(yk)-N(k)\}f(p)+N(yk)N(k)(1-y)\hbar k\,\frac{\varepsilon}{pc}\frac{\partial f(p)}{\partial p}\right], \tag{6.136}$$

with

$$g(y)=\begin{cases} g_1(y) & \text{for } (1-\beta)/(1+\beta)\leqslant y\leqslant 1,\\ y^3 g_1(y) & \text{for } 1\leqslant y\leqslant (1+\beta)/(1-\beta),\\ 0 & \text{otherwise,}\end{cases} \tag{6.137a}$$

$$g_1(y):=-y(1+y)\frac{(1-\beta^2)(3-\beta^2)}{\beta^4}\left(\ln\frac{1+\beta}{1-\beta}+\ln y\right)+\{(1+\beta)y-(1-\beta)\}\times\left[2y+\frac{(1-\beta^2)}{2\beta^4}\{(1+\beta)+10y+(1-\beta)y^2\}\right]. \tag{6.137b}$$

The equation describing the evolution of the particles, i.e. (5.89c), reduces to

$$\frac{df(p)}{dt}=\frac{\pi^2 e^4}{c}\frac{1}{p^2}\frac{\partial}{\partial p}\left[\left[\int_0^\infty \frac{dk k^2}{(2\pi)^3}\int_0^\infty\frac{dy}{y}\,g(y)\frac{\hbar k\varepsilon}{pc}\times\left[\{N(yk)-N(k)\}f(p)+N(yk)N(k)(1-y)\frac{\hbar k\varepsilon}{pc}\frac{\partial f(p)}{\partial p}\right]\right]\right]. \tag{6.138}$$

An extra factor of $\frac{1}{2}$ has been inserted on the RH side of (6.138) to take account of the fact that the scattered and unscattered waves are in the same mode, and hence that one would count each scattering twice using (5.89c) as it stands.

From the fact that the results derived in Section 4.4 for spontaneous scattering, e.g. (4.140) and (4.144) must be reproduced by (6.136), one infers the following values for integrals over $g(y)$:

$$\int_0^\infty \frac{dy}{y}\,g(y)=\frac{16}{3}\frac{\beta^2}{1-\beta^2},\quad \int_0^\infty dy\, g(y)=\frac{16}{9}\frac{\beta^2(3+\beta^2)}{(1-\beta^2)^2}, \tag{6.139a, b}$$

$$\int_0^\infty dy\, y g(y)=\frac{16}{45}\frac{\beta^2(15+20\beta^2+7\beta^4)}{(1-\beta^2)^3}. \tag{6.139c}$$

Needless to add, the explicit evaluation of these integrals using (6.136) with (6.137a, b) is tedious.

Three comments on these equations are appropriate. First, $g(y)$ is non-zero only for a finite range of y due to the fact that there is a finite range of scattered frequencies ω for a given unscattered frequency. Specifically, the δ-function in the scattering probability, e.g. in (4.133), implies

$$\frac{\omega'(1-\beta)}{1+\beta}\leqslant\omega\leqslant\omega'\frac{(1+\beta)}{1-\beta}, \tag{6.140}$$

which corresponds to the range in which $g(y)$ is non-zero. Second, for spontaneous scattering, which is described by the terms involving $N(yk)$ alone on the RH sides of (6.136) and (6.138), one may identify $k=\omega/c$ as the scattered wave and $yk=\omega'/c$ as the unscattered wave. However this identification of "scattered" and "unscattered" waves is not possible for induced scattering. Induced scattering is described by the terms involving $N(yk)\,N(k)$ on the RH sides of (6.136) and (6.138). Neither wave can be identified as the scattered wave or as the unscattered wave; there is simply a net transfer of energy (photons) from one frequency range to another. Third, the terms involving $N(k)$ alone on the RH sides of (6.136) and (6.138) describe the effects of rescattering of the already scattered waves, and are usually negligible, cf. (9.48) however.

Relativistic Electrons

When considering induced scattering by ultrarelativistic electrons, $g(y)$ may be approximated by

$$g(y)=\begin{cases}4y^2 & \text{for } (4\gamma^2)^{-1}\leqslant y\leqslant 1,\\ 4y & \text{for } 1\leqslant y\leqslant 4\gamma^2,\\ 0 & \text{otherwise.}\end{cases} \tag{6.141}$$

The term describing induced scattering in (6.136) then reduces to

$$\frac{1}{I(\omega)}\frac{\mathrm{d}I(\omega)}{\mathrm{d}t}=\frac{(2\pi)^5e^4c^3}{2\omega^2}\int_0^\infty \mathrm{d}\varepsilon\,\frac{\mathrm{d}}{\mathrm{d}\varepsilon}\left(\frac{N_e(\varepsilon)}{\varepsilon^2}\right)\int_{1/4\gamma^2}^{4\gamma^2}\frac{\mathrm{d}y}{y}\,g(y)\,\frac{1-y}{y^3}\quad I(y\omega) \tag{6.142}$$

where the occupation number $N(k)$ has been reexpressed in terms of the specific intensity using

$$I(\omega)=\frac{1}{4\pi}\frac{\hbar\omega^3}{(2\pi)^3c^2}\,N(\omega/c), \tag{6.143}$$

and where $f(p)$ has been reexpressed in terms of the energy spectrum using

$$N_e(\varepsilon)=\frac{4\pi\varepsilon^2}{c^3} f(\varepsilon/c). \tag{6.144}$$

After a partial integration (6.142) reduces to

$$\frac{1}{I(\omega)}\frac{\mathrm{d}I(\omega)}{\mathrm{d}t}=\frac{8(2\pi)^5e^4c}{m_e\omega^2}\int_0^\infty \mathrm{d}\varepsilon\,\frac{N_e(\varepsilon)}{\varepsilon^2}\left[\frac{1}{(2\gamma)^3}I(4\gamma^2\omega)-2\gamma I(\omega/4\gamma^2)\right]. \tag{6.145}$$

Consider the particular case of a self-absorbed synchrotron spectrum, approximated by

$$I(\omega)=I_0\begin{cases}(\omega/\omega_b)^{5/2} & \text{for } \omega<\omega_b,\\ (\omega/\omega_b)^{-\alpha} & \text{for } \omega>\omega_b.\end{cases} \tag{6.146}$$

Then (6.145) gives

$$\frac{1}{I(\omega)}\frac{\mathrm{d}I(\omega)}{\mathrm{d}t}=\frac{8I_0(2\pi)^5e^4c}{m_e\omega^2}\int_0^\infty \mathrm{d}\varepsilon\,\frac{N_e(\varepsilon)}{\varepsilon^2}\begin{cases}(2\gamma)^2(\omega/\omega_b)^{5/2} & \text{for } \omega<\omega_b,\\ -(2\gamma)^{2\alpha+1}(\omega/\omega_b)^{-\alpha} & \text{for } \omega>\omega_b.\end{cases} \tag{6.147}$$

It follows that there is a transfer of photons from above the maximum ($\omega>\omega_b$) to below the maximum ($\omega<\omega_b$).

It seems unlikely that induced scattering by relativistic particles could be important in practice. For example, the maximum rate at which induced scattering occurs is for $\omega\approx\omega_b$. At this maximum (6.147) gives

$$\left|\frac{1}{I(\omega)}\frac{\mathrm{d}I(\omega)}{\mathrm{d}t}\right|\approx 6\pi^2r_0^2c\gamma_b^2m_ec^2N_e(\varepsilon_b), \tag{6.148}$$

where ε_b is the energy of electrons radiating at $\omega=\omega_b$ and where (6.128) has been used to estimate $I_0\approx2\omega_b^2\varepsilon_b/(2\pi)^3c^2$. An estimate of the synchrotron absorption coefficient at $\omega=\omega_b$ shows it to be roughly $m_e^6c^{13}/r_0\Omega_e\varepsilon_b^6$ times the RH member of (6.148). This ratio may be written as $10^8\ (10^2m_ec^2/\varepsilon_b)^6 \times(B/10^{-4}G)^{-1}$, where $\varepsilon_b\approx10^2m_ec^2$ and $B\approx10^{-4}G$ are thought to be typical values, see Section 9.2. Even for extreme values $\varepsilon_b\approx10^3m_ec^2$ and $B\approx10^{-3}G$, induced scattering could be only marginally significant.

It is evident from (6.147), and from (6.145) more generally, that induced scattering does not have an effect opposite to that of spontaneous scattering in the following sense. Spontaneous scattering has, as its primary effect, transfer of photons from a frequency range ω to a range of order $\gamma^2\omega$. On the other hand, induced scattering has as its primary effect a net transfer from just above to just below the peak in the frequency spectrum. The two results may be understood from the following qualitative viewpoint.

Consider scattering by a specific relativistic electron. The scattering may be classified according to whether the initial and final photons are or are not in the forward cone (of half angle γ^{-1}) of the scattering particle. Spontaneous scattering can be understood in terms of two simple results. The first is that the rate of scattering is roughly independent of both the initial and final directions of the photons. The second result follows from the δ-function in the probability (4.133):

$$\omega'(1-\beta\cos\theta')=\omega(1-\beta\cos\theta).$$

It is that if neither the initial nor the final photons is in the forward cone ($\theta'>\gamma^{-1}, \theta>\gamma^{-1}$), then their frequencies are roughly equal ($\omega\approx\omega'$), but if one is in the forward cone and the other is not, then their frequencies differ by a factor γ^2. With these qualitative results we may draw up the following table in which ticks and crosses in the first two columns indicate whether the photons are or are not in the forward cone. The third column gives the relative rate of the processes and the fourth column gives the energy of the scattered photon. The fifth column is the product of the entries in columns 3 and 4, and gives the relative power radiated spontaneously. It follows from the entries in the fifth column that most of the power is radiated into the forward cone at frequencies of order $\omega'\gamma^2$.

1	2	3	4	5	6
$\theta'>\gamma^{-1}$?	$\theta>\gamma^{-1}$?	relative rate	$\dfrac{\omega}{\omega'}$	relative power	induced power
✓	✓	1	1	1	1
✓	X	γ^{-1}	γ^2	γ	1
X	✓	γ^{-1}	γ^{-2}	γ^{-3}	γ^{-4}
X	X	γ^{-2}	1	γ^{-2}	γ^{-3}

However, when induced scattering is considered, the rate is also proportional to the occupation number of the "scattered" photons. Only a fraction γ^{-1} of these are in the forward cone. Consequently, for induced scattering the relative rates in the third column would be 1, γ^{-2}, γ^{-1} and γ^{-3} respectively. Similarly the relative powers (column 5) are to be reduced by γ^{-1} when the final photon is in the forward cone. Thus for induced scattering, the relative power is given by the entries in the sixth column. Note that there is no preference for the power to be in the forward cone and hence no preference for a large frequency change. In fact, there is no intrinsic preference for a transfer of power from any two frequency ranges differing by a ratio less than γ^2. Hence the dominant induced scattering occurs where the density of photons is highest, i.e. near the peak in frequency spectrum.

Finally it is appropriate to mention a related effect, namely induced scattering by relativistic electrons of transverse waves into Langmuir waves in the presence of Langmuir turbulence. This has been discussed in connection with synchrotron sources by Tsytovich and Chikhachev (1969) and by Kaplan and Tsytovich (1973, Section 13).

Induced Scattering by Thermal Electrons

Now consider scattering by non-relativistic electrons. In this case (6.137a) implies that $g(y)$ is non-zero only for $|1-y| \leqslant 2\beta$. Because $1-y$ is necessarily small, it is appropriate to expand (6.136) in powers of $1-y$. It is convenient in this case to include the first quantum correction to spontaneous scattering. The first quantum correction may be obtained using only semi-classical arguments (Melrose, 1968). With the first quantum correction, (6.136) becomes

$$\frac{\mathrm{d}N(k)}{\mathrm{d}t} = \frac{2\pi^2 e^4}{c} \int_0^\infty \mathrm{d}p \int_0^\infty \frac{\mathrm{d}y}{y}\, g(y)(y-1)k \left\{1+\tfrac{1}{2}(y-1)k\frac{\mathrm{d}}{\mathrm{d}k}\right\} \times \left[\frac{\mathrm{d}N(k)}{\mathrm{d}k} f(p) - N(k)\{1+N(k)\}\frac{\hbar mc}{p}\frac{\partial f(p)}{\partial p}\right]. \tag{6.149}$$

to lowest non-trivial order in $1-y$. Use of (6.136) as it stands would lead to (6.149) with $N^2(k)$ in place of $N(k)\{1+N(k)\}$.

The y-integral in (6.149) may be performed using (6.139a, b, c):

$$\frac{\mathrm{d}N(k)}{\mathrm{d}t} = \frac{32\pi^2 e^4}{9m^4c^5} \int_0^\infty \mathrm{d}p\, p^4 k \left(4+k\frac{\mathrm{d}}{\mathrm{d}k}\right) \times \left[\frac{\mathrm{d}N(k)}{\mathrm{d}k} f(p) - N(k)\{1+N(k)\}\frac{\hbar mc}{p}\frac{\partial f(p)}{\partial p}\right]. \tag{6.150}$$

Here we may set $p = mc\beta$. For a thermal distribution of electrons at temperature T the integrals in (6.150) are trivial, and (6.150) reduces to (Kompaneets, 1956; Dreicer, 1964; Weymann, 1965)

$$\frac{\mathrm{d}N(k)}{\mathrm{d}t} = \frac{\sigma_T n_e T}{m_e c} k\left(4+k\frac{\mathrm{d}}{\mathrm{d}k}\right)\left[\frac{\mathrm{d}N(k)}{\mathrm{d}k} + N(k)\{1+N(k)\}\frac{\hbar c}{T}\right], \tag{6.151}$$

with $\sigma_T := (8\pi/3)r_0^2$, $r_0 := e^2/m_e c^2$.

In a stationary state (i.e. for $\mathrm{d}N(k)/\mathrm{d}t=0$) (6.151) has the solution

$$N(k)=[\exp\{(\hbar kc+\mu)/T\}-1]^{-1}, \tag{6.152}$$

where μ is a constant of integration. If the first quantum correction to spontaneous scattering is not included in (6.151), the corresponding result is

$$N(k)=\frac{T}{\hbar kc+\mu}. \tag{6.152$'$}$$

The case $\mu=0$ is a Planck distribution, and for $\mu\neq0$ it is a Bose distribution. Weymann (1965) first pointed out that a given initial distribution of photons approaches a distribution of the form (6.152) with $\mu\neq0$, rather than a Planck distribution, as a result of Compton scattering by thermal electrons. The reason that a strictly thermal distribution (the Planck distribution) is not approached is that the number of photons is conserved. A Planck distribution at temperature T would be approached only if there were a mechanism which allowed photons to be created or destroyed. In practice suitable such mechanisms exist, e.g. bremsstrahlung and double Compton scattering, but Compton scattering alone conserves photons.

The term in (6.151) which describes the effect of induced scattering is

$$\frac{\partial N(k,t)}{\partial t}=\frac{\sigma_T n_e\hbar}{m_e}\frac{1}{k^2}\frac{\partial}{\partial k}\{k^2N(k,t)\}^2, \tag{6.153}$$

where the total time-derivative has been replaced by a partial time-derivation under the assumption that the distribution of photons is homogeneous. A remarkable feature of (6.153) is that it does not involve the temperature of the electrons. The solution (6.153) was given by Zeldovich and Levich (1968):

$$k=F(k^2N(k,t))-\frac{2\sigma_T n_e\hbar}{m_e}k^2N(k,t)t, \tag{6.154}$$

where the function F is determined from the initial conditions.

Sunyaev (1970, 1971) and Levich and Sunyaev (1970) have discussed some of the effects of induced scattering by thermal electrons in synchrotron sources and in pulsars. One possible application, to low frequency turnovers in synchrotron spectra, is discussed briefly in Section 9.2.

PROBLEM SET 6

Answers to relevant questions are given at the end of the text. More difficult questions are marked with an asterisk.

6.1* Derive a probability for emission of bremsstrahlung in the magnetoionic waves by analogy with the derivation of (6.15). In other words, find the probability of emission by cold electrons in the wings of the Lorentzian line profiles for gyromagnetic emission.

6.2 (a) Show that the autocorrelation of the current density in a time-independent homogeneous plasma, i.e.,

$$\langle j_i j_j \rangle(\boldsymbol{\xi}, \tau) := \langle j_i(\mathbf{r}, t) j_j(\mathbf{r}+\boldsymbol{\xi}, t+\tau) \rangle, \tag{P6-1}$$

where the angular brackets on the right side imply an average over the distribution of particles, has Fourier transform given by

$$\langle j_i j_j^* \rangle(\mathbf{k}, \omega) = \begin{cases} 2\pi q^2 \int d^3p\, v_i v_j \delta(\omega - \mathbf{k}\cdot\mathbf{v}) f(\mathbf{p}), \\ \displaystyle\sum_{s=-\infty}^{\infty} 2\pi q^2 \int d^3\mathbf{p}\, V_i(s, \mathbf{p}, \mathbf{k}) V_j^*(s, \mathbf{p}, \mathbf{k}) \delta(\omega - s\Omega - k_{\parallel} v_{\parallel}) f(\mathbf{p}), \end{cases} \tag{P6-2}$$

in the absence and presence of an ambient magnetic field respectively.
(b) Hence show that the level of electric fluctuations in a stable plasma is given by

$$\langle E_i E_j^* \rangle(\mathbf{k}, \omega) = \frac{16\pi^2}{\omega^2} \frac{\lambda_{ir}(\mathbf{k}, \omega) \lambda_{js}^*(\mathbf{k}, \omega) \langle j_r j_s^* \rangle(\mathbf{k}, \omega)}{\Lambda^2(\mathbf{k}, \omega)}. \tag{P6-3}$$

6.3* The Fluctuation-Dissipation Theorem
(a) Show that when emission and absorption are in balance the energy density in waves in the mode σ is given by

$$W^\sigma(\mathbf{k}) = \frac{4\pi R_E^\sigma(\mathbf{k})}{\gamma^\sigma(\mathbf{k})} e_i^{\sigma *}(\mathbf{k}) e_j^\sigma(\mathbf{k}) \langle j_i j_j^* \rangle(\mathbf{k}, \omega^\sigma(\mathbf{k})). \tag{P6-4}$$

(b) Show that (P6-4) implies that in a thermal plasma at a temperature T one has

$$\langle j_i j_j^* \rangle(\mathbf{k}, \omega) = -\frac{i\omega}{2\pi} T \varepsilon_{ij}^{a}(\mathbf{k}, \omega). \tag{P6-5}$$

(c) Rederive (P6-5) directly by comparison of (2.62) or (2.63) with (P6-2).
(d) Rederive (P6-4) from (P6-3).

6.4 Derive (6.67) by solving the two-dimensional wave equation

$$\Lambda^{\alpha\beta}(\mathbf{k}, \omega) E^\beta(\mathbf{k}, \omega) = 0, \quad \Lambda^{\alpha\beta}(\mathbf{k}, \omega) = e_i^\alpha e_j^\beta \Lambda_{ij}(\mathbf{k}, \omega), \tag{P6-6}$$

with $\varepsilon^{\alpha\beta}$ given by (6.63). The polarization vectors are to be found by constructing $\lambda^{\alpha\beta}(\mathbf{k}, \omega)$.

6.5 Derive (6.104a to d) by writing out the transfer equation in the Jones calculus, cf. (6.43) with (6.49), in terms of basis vectors $\mathbf{e}^+$ and $\mathbf{e}^-$ for the two natural modes.

Hint: The initial rewriting is trivial, and involves replacing $\alpha\beta\gamma\delta$ by $\sigma\sigma'\sigma''\sigma''$ in (6.43), and the problem may be reduced to showing $r^{\sigma\sigma'} = 0$ for $\sigma = \sigma'$ and $r^{+-} = (r^{-+})^* = i\Delta k$.

6.6 In the transfer equation in the Mueller calculus, i.e.,

$$n^2 \frac{\partial}{\partial s}\left(\frac{S_A}{n^2}\right)=\alpha_A+r_{AB}S_B-\mu_{AB}S_B, \tag{P6-7}$$

one may write

$$r_{AB}=n^2 \frac{\partial}{\partial s}\left(\frac{f_{AB}}{n^2}\right) \tag{P6-8}$$

with

$$f_{AB}=\begin{pmatrix} 0 & 0 & 0 & 0 \\ 0 & 0 & a & b \\ 0 & -a & 0 & c \\ 0 & -b & -c & 0 \end{pmatrix}. \tag{P6-9}$$

Show that (P6-7) may be rewritten in the form

$$n^2 \frac{\partial}{\partial s}\left(\frac{F_{AB}S_B}{n^2}\right)=F_{AB}(\alpha_B-\mu_{BC}S_C), \tag{P6-10}$$

with

$$F_{AB}=\delta_{AB}-\frac{\sin\alpha}{\alpha}f_{AB}+\frac{(1-\cos\alpha)}{\alpha^2}f_{AC}f_{CB}, \tag{P6-11}$$

$$\alpha:=(a^2+b^2+c^2)^{1/2}. \tag{P6-12}$$

6.7 The Poincaré Sphere

Let $\mathbf{e}^1$, $\mathbf{e}^2$ and $\boldsymbol{\kappa}$ be mutually orthonormal real vectors. Consider elliptically polarized radiation with axial ratio T with the major axis (for $|T|>1$) along an axis at an angle $\psi/2$ to $\mathbf{e}^1$. Write

$$\beta:=2\arctan\left(\frac{1-T}{1+T}\right). \tag{P6-13}$$

(a) Show that one has

$$\frac{Q}{I}=\sin\beta\cos\psi,\quad \frac{U}{I}=\sin\beta\sin\psi,\quad \frac{V}{I}=\cos\beta, \tag{P6-14}$$

and hence that the polarization may be represented by a point on a unit sphere, called the Poincaré sphere.

(b) Assume that the natural modes in the medium have axial ratios T_0 and $T_x=-1/T_0$, and assume that the projection of the magnetic field on the plane orthogonal to $\boldsymbol{\kappa}$ is at an angle $\psi'/2$ to $\mathbf{e}^1$, implying $\mathbf{e}^1\cdot\mathbf{t}=\cos(\psi'/2)$. Show that the representative points for the two modes are at $\beta=\beta_0$, $\psi_0=\psi'$ and $\beta=\beta_x=\pi-\beta_0$, $\psi_x=\psi'\pm\pi$, where β_0 and β_x are defined by analogy with (P6-13).

(c)* Show that the relative intensities of the components of radiation described by (P6-14) in the two natural modes are in the ratio

$$|\mathbf{e}^*\cdot\mathbf{e}^o|^2 : |\mathbf{e}^*\cdot\mathbf{e}^x|^2=\tfrac{1}{2}(1+\cos\beta\cos\beta_0+\sin\beta\sin\beta_0\cos(\psi-\psi')) : \tfrac{1}{2} \times(1-\cos\beta\cos\beta_0-\sin\beta\sin\beta_0\cos(\psi-\psi')). \tag{P6-15}$$

6.8* Show that generalized Faraday rotation may be described as follows in terms of the Poincaré sphere (problem 6.7):

Draw a line between the two points which correspond to the two natural modes (which line passes through the center of the sphere). Construct a plane orthogonal to this line and through the representative point of the radiation of interest. Generalized Faraday rotation causes the representative point to move around the circle defined by the intersection of this plane with the Poincaré sphere.

6.9 A receiver has a bandwidth Δf and is used to observe radiation which has a degree of polarization r_0 and an axial ratio T_0 at the source. Faraday rotation (for circularly polarized natural modes) along the path from the source to the receiver leads to a total rotation through Ψ radians. Show that the detected degree of polarization r and axial ratio T are given by

$$r=\left[r_0^2+\frac{T_0^2-1}{T_0^2+1}\left\{\left(\frac{\sin\Delta\Psi}{\Delta\Psi}\right)^2-1\right\}\right]^{1/2}, \tag{P6-16}$$

$$T=\frac{T_0^2-1}{2T_0}\,\frac{\sin\Delta\Psi}{\Delta\Psi}+\frac{T_0}{|T_0|}\left\{\left[\frac{T_0^2-1}{2T_0}\,\frac{\sin\Delta\Psi}{\Delta\Psi}\right)^2+1\right\}^{1/2}, \tag{P6-17}$$

with $\Delta\Psi=\Psi\Delta f/f$.

6.10* (a) Generalize the transfer equation (4.81a) for scattering of waves to the case of polarized transverse waves. You may assume that the waves are described by $N^{\alpha\beta}(\mathbf{k})$ and $N^{\alpha'\beta'}(\mathbf{k}')$ and that the probability is of the form $w^{\alpha\beta\alpha'\beta'}(\mathbf{p}, \mathbf{k}, \mathbf{k}')$.

(b) Write down the corresponding form of the transfer equation in the Mueller calculus.

APPENDIX A

Bessel Functions

Note: In the following the order of Bessel functions is denoted by ν where the formula applies to all real ν, and by $s=0, \pm 1, \pm 2, \ldots$ in formulae which apply only for integral orders.

Bessel Functions

The two solutions of the differential equation

$$Z_\nu''(z)+\frac{1}{z}Z_\nu'(z)+\left(1-\frac{\nu^2}{z^2}\right)Z_\nu(z)=0, \tag{A.1}$$

where a prime denotes differentiation with respect to the argument, are usually chosen to be either (a) $J_\nu(z)$ and $N_\nu(z)$, called Bessel functions of the first and second kinds respectively, or (b) $H_\nu^{(1)}(z)$ and $H_\nu^{(2)}(z)$, called Bessel functions of the third kind or Hankel functions:

$$H_\nu^{(1)}(z):=J_\nu(z)+iN_\nu(z), \quad H_\nu^{(2)}(z)=J_\nu(z)-iN_\nu(z). \tag{A.2}$$

Solutions of the differential equation

$$U_\nu''(z)+\frac{1}{z}U_\nu'(z)-\left(1+\frac{\nu^2}{z^2}\right)U_\nu(z)=0 \tag{A.3}$$

are called modified Bessel functions. The modified Bessel functions of the first kind are

$$I_\nu(z)=\begin{cases} e^{-i\nu\pi/2}\,J_\nu(ze^{i\pi/2}) & \text{for } -\pi<\arg z\leqslant\pi/2 \\ e^{i\nu 3\pi/2}\,J_\nu(ze^{-i3\pi/2}) & \text{for } \pi/2<\arg z\leqslant\pi, \end{cases} \tag{A.4}$$

with

$$z=:(\text{Re}\, z)\exp\left[i(\arg z)\right].$$

in particular, for ν an integer (A.4) reduces to

$$I_s(z)=i^{-s}J_s(iz). \tag{A.5}$$

The second solution may be chosen to be the Macdonald function $K_\nu(z)$:

$$K_\nu(z):=\frac{i\pi}{2}e^{i\nu\pi/2}H_\nu^{(1)}(iz) \tag{A.6a}$$

$$=\frac{i\pi}{2}e^{-i\nu\pi/2}H_{-\nu}^{(1)}(iz) \tag{A.6b}$$

$$=\frac{\pi}{2}\frac{I_{-\nu}(z)-I_\nu(z)}{\sin\nu\pi}. \tag{A.6c}$$

(All three definitions are equivalent.)

Recurrence Relations

Ordinary Bessel functions (J_ν, N_ν, $H_\nu^{(1)}$ and $H_\nu^{(2)}$) all satisfy the recurrence relations

$$J_{\nu-1}(z)+J_{\nu+1}(z)=2\frac{\nu}{z}J_\nu(z), \tag{A.7a}$$

$$J_{\nu-1}(z)-J_{\nu+1}(z)=2J'_\nu(z). \tag{A.7b}$$

The recurrence relations for I_ν and K_ν are different from (A.7a, b) and different from each other:

$$I_{\nu-1}(z)+I_{\nu+1}(z)=2I'_\nu(z), \tag{A.8a}$$

$$I_{\nu-1}(z)-I_{\nu+1}(z)=2\frac{\nu}{z}I_\nu(z); \tag{A.8b}$$

$$K_{\nu-1}(z)+K_{\nu+1}(z)=-2K'_\nu(z), \tag{A.9a}$$

$$K_{\nu-1}(z)-K_{\nu+1}(z)=-2\frac{\nu}{z}K_\nu(z). \tag{A.9b}$$

Changes in the sign of the argument follow from

$$J_\nu(ze^{i\pi m})=e^{i\pi m\nu}J_\nu(z) \tag{A.10a}$$

$$I_\nu(ze^{i\pi m})=e^{i\pi m\nu}I_\nu(z), \tag{A.10b}$$

$$K_\nu(ze^{i\pi m})=e^{-i\pi m\nu}K_\nu(z)-i\pi\frac{\sin m\pi\nu}{\sin\pi\nu}I_\nu(z), \tag{A.10c}$$

where m is an integer. Change of the sign of the order leads to a simple relation only in the following cases:

$$J_{-s}(z)=(-)^s J_s(z), \quad I_{-s}(z)=I_s(z), \tag{A.11a, b}$$

$$K_{-\nu}(z)=K_\nu(z). \tag{A.11c}$$

Series Expansions

Series expansions of J_ν and I_ν are

$$J_\nu(z)=\sum_{n=0}^{\infty}\frac{(-)^n}{n!\Gamma(\nu+n+1)}\left(\frac{z}{2}\right)^{2n+\nu} \tag{A.12}$$

$$I_\nu(z)=\sum_{n=0}^{\infty}\frac{1}{n!\Gamma(\nu+n+1)}\left(\frac{z}{2}\right)^{2n+\nu}. \tag{A.13}$$

The series expansion of $K_\nu(z)$ follows from (A.6c) and (A.13) for non-integral ν, and for integral $\nu=s$ is

$$K_s(z)=\frac{1}{2}\sum_{n=0}^{s-1}\frac{(-)^s(s-n-1)!}{n!}\left(\frac{z}{2}\right)^{2n-s}+(-)^{s+1}\sum_{n=0}^{\infty}\frac{1}{n!(n+s)!}\left(\frac{z}{2}\right)^{2n+s} \times\left\{\ln\frac{z}{2}-\frac{1}{2}\psi(n+1)-\frac{1}{2}\psi(s+n+1)\right\}, \tag{A.14}$$

where

$$\psi(x):=\frac{\mathrm{d}}{\mathrm{d}x}\ln\Gamma(x) \tag{A.15}$$

has the following values

$$\psi(1)=-\gamma, \quad \psi(m+1)=\frac{1}{1}+\frac{1}{2}+\cdots+\frac{1}{m}-\gamma, \tag{A.16}$$

where $\gamma=0.5772\ldots$ is Euler's constant.

Generating Functions

The generating functions for J_s and I_s are

$$\exp\left(\frac{1}{2}\left[t-\frac{1}{t}\right)z\right]=\sum_{s=-\infty}^{\infty}.t^s J_s(z), \tag{A.17}$$

and

$$\exp\left[\frac{1}{2}\left(t+\frac{1}{t}\right)z\right]=\sum_{s=-\infty}^{\infty}t^s I_s(z). \tag{A.18}$$

respectively.

Infinite Sums

Infinite sums of Bessel functions include the following:

$$\sum_{s=-\infty}^{\infty} J_s^2(z)=1, \quad \sum_{s=-\infty}^{\infty} J_s'^2(z)=\tfrac{1}{2}, \quad \sum_{s=-\infty}^{\infty} s^2 J_s^2(z)=\tfrac{1}{2}z^2; \tag{A.19a, b, c}$$

$$\sum_{s=-\infty}^{\infty} I_s(z)=e^z, \quad \sum_{s=-\infty}^{\infty} I_s'(z)=e^z, \quad \sum_{s=-\infty}^{\infty} sI_s(z)=ze^z. \tag{A.20a, b, c}$$

Other types of infinite sum are Kapteyn series. Let us write, for $0<x<1$,

$$g_m(x):=\sum_{s=1}^{\infty} s^{2m} J_{2s}(2sx), \tag{A.21}$$

$$h_m(x):=\sum_{s=1}^{\infty} s^{2m} J_s^2(sx). \tag{A.22}$$

Differentiation of (A.21) and (A.22) gives

$$\sum_{s=1}^{\infty} s^{2m+1} J_{2s}'(2sx)=\tfrac{1}{2}g_m'(x) \tag{A.23}$$

and

$$\sum_{s=1}^{\infty} s^{2m+2} J_s'^2(sx)=\tfrac{1}{2}h_m''(x)+\frac{1}{2x}h_m'(x)+\left(1-\frac{1}{x^2}\right)h_{m+1}(x) \tag{A.24}$$

respectively. Particular Kapteyn series have the following sums:

$$g_{-1}(x)=\tfrac{1}{2}x^2, \quad g_0(x)=\frac{x^2}{2(1-x^2)}, \quad g_1(x)=\frac{x^2(1+x^2)}{2(1-x^2)^4}; \tag{A.25a, b, c}$$

$$h_{-1}(x)=\frac{1}{4}x^2, \quad h_0(x)=\frac{1}{2}\left(\frac{1}{\sqrt{1-x^2}}-1\right), \quad h_1(x)=\frac{x^2(4+x^2)}{16(1-x^2)^{7/2}}. \tag{A.26a, b, c}$$

Asymptotic Expansions

For large values of z the following asymptotic expansion may be used.

$$J_{\pm\nu}(z)=\left(\frac{2}{\pi z}\right)^{1/2}\left\{\left(1-\frac{(\nu^2-\frac{1}{4})(\nu^2-\frac{9}{4})}{2(2z)^2}+\cdots\right)\cos\left(z\pm\frac{\nu\pi}{2}-\frac{\pi}{4}\right)\right.$$
$$\left.-\left(\frac{(\nu^2-\frac{1}{4})}{2z}+\cdots\right)\sin\left(z\pm\frac{\nu\pi}{2}-\frac{\pi}{4}\right)\right\}, \tag{A.27}$$

$$I_\nu(z) = \frac{e^z}{(2\pi z)^{1/2}}\left(1 - \frac{(\nu^2 - \frac{1}{4})}{2z} + \cdots\right) \tag{A.28}$$

$$K_\nu(z) = \left(\frac{\pi}{2z}\right)^{1/2} e^{-z}\left(1 + \frac{(\nu^2 - \frac{1}{4})}{2z} + \cdots\right). \tag{A.29}$$

The Airy-integral approximation applies for $z < s$, $z - s \ll s$, $s \gg 1$:

$$J_s(z) \simeq \frac{1}{\pi}\left\{\frac{2(s-z)}{3z}\right\}^{1/2} K_{1/3}\left(\frac{2^{3/2}(s-z)^{3/2}}{3z^{1/2}}\right). \tag{A.30}$$

APPENDIX B

Specific Dielectric Tensors

In this Appendix the different methods used to calculate dielectric tensors are summarized and specific results are quoted.

Methods of Calculation

There are three general methods of calculating dielectric tensors. The most familiar is the Vlasov theory outlined in Section 1.2 and further in Section 10.4. The general results of calculations using this method are (2.62) and (2.63), in the absence and presence, respectively, of a magnetic field. Before expanding in Bessel functions, the latter result is

$$\varepsilon_{ij}(\mathbf{k}, \omega) = \delta_{ij} + \sum_{\alpha} 4\pi i q \int_0^{2\pi} d\phi \int_{-\infty}^{\infty} dp_{\parallel} \int_0^{\infty} dp_{\perp} p_{\perp} v_i$$

$$\times \frac{|q|}{\Omega\omega} \int_{\epsilon\infty}^{\phi} d\phi' e^{P(\phi') - P(\phi)} g_{sj}(\mathbf{k}, \omega; \mathbf{v}') \frac{\partial f(p_{\perp}, p_{\parallel})}{\partial p'_s}, \quad \text{(B.1)}$$

with

$$\mathbf{p}' := (p_{\perp} \cos \phi', \quad p_{\perp} \sin \phi', p_{\parallel}),$$

$$P(\phi) := \frac{i}{\varepsilon\Omega} \int^{\phi} d\phi' (\omega - \mathbf{k} \cdot \mathbf{v}')$$

$$g_{ij}(\mathbf{k}, \omega; \mathbf{v}) := (\omega - \mathbf{k} \cdot \mathbf{v}) \delta_{ij} + k_i v_j, \quad \text{(B.2)}$$

and where the lower limit $\epsilon\infty$ incorporates the causal condition.

An even more general approach is Tsytovich's (1961) method using relativistic quantum theory. This method is developed in Section 13.4. The results are written in the form

$$\varepsilon_{ij}(\mathbf{k}, \omega) = \delta_{ij} + \frac{4\pi c}{\omega^2} \alpha_{ij}(\mathbf{k}, \omega) \quad \text{(B.3)}$$

with α_{ij} given in units with $\hbar=c=1$ by (13.160) in the absence of an ambient magnetic field and by (13.162) in the presence of an ambient magnetic field.

The third general approach appears not to have been stated in the literature. The idea is to regard the response of the medium as the collective effect of forward scattering by all the particles, as in Tsytovich's (1961) method, with the forward-scattering amplitude calculated classically. The current associated with a scattering charge in the absence of a magnetic field is given by (3.96). Integrating the forward amplitude (i.e. for $\omega=\omega'$ and $\mathbf{k}=\mathbf{k}'$) over the distributions of particles in the various species, and writing the result in the form

$$j_i(\mathbf{k}, \omega)=\sigma_{ij}(\mathbf{k}, \omega)E_j(\mathbf{k}, \omega), \quad \varepsilon_{ij}(\mathbf{k}, \omega)=\delta_{ij}+\frac{4\pi i}{\omega}\sigma_{ij}(\mathbf{k}, \omega), \tag{B.4}$$

one finds

$$\varepsilon_{ij}(\mathbf{k}, \omega)=\delta_{ij}-\sum_{\alpha}\frac{8\pi^2q^2}{m\omega^2}\int d^3\mathbf{p}\, a_{ij}(\mathbf{k}, \omega; \mathbf{k}, \omega; \mathbf{v})f(\mathbf{p}), \tag{B.5}$$

with, from (3.99),

$$a_{ij}(\mathbf{k}, \omega; \mathbf{k}, \omega; \mathbf{v})=\frac{1}{\gamma}\left[\delta_{ij}+\frac{k_iv_j+k_jv_i}{(\omega-\mathbf{k}\cdot\mathbf{v})}+\frac{(|\mathbf{k}|^2-\omega^2/c^2)v_iv_j}{(\omega-\mathbf{k}\cdot\mathbf{v})^2}\right]. \tag{B.6}$$

The result (B.5) with (B.6) is equivalent to that obtained using Vlasov theory, e.g. (B.5) with (B.6) may be obtained by partially integrating in (2.62). (The causal condition must be imposed separately in both cases.) The generalization to the magnetized case follows by using (3.107) *et seq.* in place of (3.96) *et seq.*, and the result is of the form (B.5) with a_{ij} given by (3.109) for $\omega=\omega'$, $\mathbf{k}=\mathbf{k}'$ and summed over s and s'. The result is equivalent to one obtained by partially integrating in (2.63).

This final single-particle forward-scattering approach is relatively simple to generalize to treat the nonlinear responses, provided one performs the calculations of the current for a scattering particle at rest and then Lorentz transforms it to the frame where its momentum is $\mathbf{p}$.

The equivalence of the single-particle approach and the Vlasov approach is of formal significance in that it implies the complete equivalence of single-particle and collective-medium approaches in plasma theory.

Relativistic Maxwellian Distributions

Now let us summarize some results for specific distributions starting with a relativistic Maxwellian distribution:

$$f(p)=\frac{n\mu}{4\pi(mc)^3K_2(\mu)}\exp\left(-\frac{\mu\varepsilon}{mc^2}\right) \tag{B.7}$$

with

$$\varepsilon=(m^2c^4+p^2c^2)^{1/2},\quad \mu:=\frac{mc^2}{T}, \tag{B.8}$$

where T is the temperature and Boltzmann's constant is unity. The dielectric tensor for an unmagnetized plasma was derived for this case by Silin (1960, 1961) and by Prentice (1967) and it can also be found as a limiting case of Trubnikov's (1958) result for a magnetized plasma. Specific expressions are

$$\varepsilon^l(\alpha,\omega)=1-\sum_\alpha\frac{\omega_p^2}{\omega kc}\frac{1}{K_2(\mu)}\int_{-kc}^{kc}\frac{d\omega}{\omega-\omega'+i0}\frac{(\gamma_0^2-1)}{\gamma_0^2}(\tfrac{1}{2}\mu\gamma_0^2+\gamma_0+\mu^{-1})e^{-\mu\gamma_0}$$

$$=1-\sum_\alpha\frac{i\omega_p^2}{\omega^2}\frac{\sigma^2}{K_2(\sigma)}\int_0^\infty d\xi\left\{\frac{K_2(R^{1/2})}{R}-\left(\frac{kc\xi}{\omega}\right)^2\frac{K_3(R^{1/2})}{R^{3/2}}\right\}, \tag{B.9a}$$

$$\varepsilon^t(k,\omega)=1-\sum_\alpha\frac{\omega_p^2}{\omega kc}\frac{1}{2K_2(\mu)}\int_{-kc}^{kc}\frac{d\omega'}{\omega-\omega'+i0}\left\{\frac{1}{\gamma_0}+\frac{1}{\mu\gamma_0^2}e^{-\mu\gamma_0}\right\} \tag{B.9c}$$

$$=1-\sum_\alpha\frac{i\omega_p^2}{\omega^2}\frac{\mu^2}{K_2(\mu)}\int_0^\infty d\xi\,\frac{K_2(R^{1/2})}{R} \tag{B.9d}$$

with

$$\gamma_0:=\left\{1-\left(\frac{\omega'}{kc}\right)^2\right\}^{-1/2},\quad R:=(\mu+i\xi)+\frac{k^2c^2}{\omega^2}\xi^2. \tag{B.10a, b}$$

Trubnikov's (1958) result (see also Drummond and Rosenbluth (1960) and Shkarofsky (1966a)) for a magnetized plasma is

$$\varepsilon_{ij}(\mathbf{k},\omega)=\delta_{ij}-\sum_\alpha\frac{i\omega_p^2}{\omega\Omega_0}\frac{\mu^2}{K_2(\mu)}\int_0^\infty d\zeta\left[\frac{K_2(R^{1/2})}{R}T_{ij}^{(1)}(\zeta)-\frac{K_3(R^{1/2})}{R^{3/2}}T_{ij}^{(2)}(\zeta)\right] \tag{B.11}$$

with

$$R:=\left(\mu+i\frac{\omega}{\Omega_0}\zeta\right)^2+2\left(\frac{k_\perp c}{\Omega_0}\right)^2(1-\cos\zeta)+\left(\frac{k_\parallel c}{\Omega_0}\right)^2\zeta^2, \tag{B.12}$$

$$T_{ij}^{(1)}(\zeta)=\begin{pmatrix}\cos\zeta & \epsilon\sin\zeta & 0\\ -\epsilon\sin\zeta & \cos\zeta & 0\\ 0 & 0 & 1\end{pmatrix}, \tag{B.13a}$$

$$T_{ij}^{(2)}(\zeta)=\frac{c^2}{\Omega_0^2}\begin{pmatrix} k_\perp^2\sin^2\zeta & \epsilon k_\perp^2 \sin\zeta(1-\cos\zeta) & k_\perp k_\parallel \zeta\sin\zeta \\ -\epsilon k_\perp^2\sin\zeta(1-\cos\zeta) & k_\perp^2(1-\cos\zeta)^2 & -\epsilon k_\perp k_\parallel\zeta(1-\cos\zeta) \\ k_\perp k_\parallel\zeta\sin\zeta & \epsilon k_\perp k_\parallel \zeta(1-\cos\zeta) & k_\parallel^2\zeta^2 \end{pmatrix} \tag{B.13b}$$

and where Ω_0 is the non-relativistic gyrofrequency and ϵ is the sign of the charge.

The non-relativistic limit corresponds to $\mu(=c^2/V^2)\gg 1$ when the asymptotic expansion (A.29) of the Macdonald functions may be used. The result (12.28) to (12.30) for $U=0$ and $V_\perp=V_\parallel=V$ may be reproduced by using (A.29) and the following results:

$$R^{1/2}\simeq\mu+i\frac{\omega}{\Omega_0}\zeta+\left(\frac{k_\perp V}{\Omega_0}\right)^2(1-\cos\zeta)+\frac{1}{2}\left(\frac{k_\parallel V}{\Omega_0}\right)^2\zeta^2 \tag{B.14a}$$

$$\exp\left[-\gamma(1-\cos\zeta)\right]=\exp(-\gamma)\sum_{s=-\infty}^{\infty}\exp(is\zeta)\,I_s(\gamma), \tag{B.14b}$$

and

$$\int_0^\infty \mathrm{d}t e^{-izt-t2/4}=\frac{i}{z}\{\phi(z)-i\sqrt{\pi}\,ze^{-z^2}\}$$

where the function $\phi(z)$ is given by (2.105), specifically,

$$\phi(z)=2ze^{-z^2}\int_0^z \mathrm{d}te^{t^2}=\frac{1}{\sqrt{\pi}}P\int_{-\infty}^{\infty}\frac{\mathrm{d}t}{t-z}e^{-t^2}. \tag{13.16}$$

Bi-Maxwellian Streaming Distribution

A general distribution for which the dielectric tensor can be evaluated in terms of known functions in the non-relativistic case is the bi-Maxwellian streaming distribution (12.26). The general form of the resulting dielectric tensor is given by (12.28) to (12.30). An important limiting case of this general result is the small gyroradius limit.

In the small gyroradius limit, the power series expansion (A.13) of the modified Bessel functions converges rapidly. To lowest non-trivial order in an expansion in λ_α (12.28) reduces to

$$\varepsilon_{ij}(k,\omega)=\begin{pmatrix} \frac{1}{2}(R_++R_-) & -i(R_+-R_-)/2 & \frac{1}{2}(E_++E_-) \\ i(R_+-R_-)/2 & \frac{1}{2}(R_++R_-) & i(E_+\ aE_-)/2 \\ \frac{1}{2}(E_++E_-) & -i(E_+-E_-)/2 & P \end{pmatrix} \tag{B.17}$$

with

$$R_{\pm}=1-\sum_{\alpha}\frac{\omega_{p\alpha}^2}{\omega^2}\{A_\alpha+G_{\alpha\pm}\tilde{\phi}(y_{\alpha\pm})\}, \tag{B.18a}$$

$$E_{\pm}=\sum_{\alpha}\frac{\omega_{p\alpha}^2}{\omega^2}\frac{k_\perp}{k_\parallel}\left[A_\alpha\mp\frac{\omega}{\epsilon_\alpha\Omega_\alpha}(1-A_\alpha)\tilde{\phi}(y_{\alpha 0})\pm\frac{\omega\pm\epsilon_\alpha\Omega_\alpha}{\epsilon_\alpha\Omega_\alpha}G_{\alpha\pm}\tilde{\phi}(y_{\alpha\pm})\right] \tag{B.18b}$$

$$P=1-\frac{\omega_{p\alpha}^2}{\omega^2}\frac{k_\perp^2}{k_\parallel^2}\left[A_\alpha-\frac{\omega^2}{k_\parallel^2 V_{\alpha\parallel}^2}\{1-\tilde{\phi}(y_{\alpha 0})\}-\frac{\omega^2}{\Omega_\alpha^2}(1-A_\alpha)\tilde{\phi}(y_{\alpha 0})\right.$$
$$\left.+\frac{1}{2}\left\{\left(\frac{\omega-\epsilon_\alpha\Omega_\alpha}{\Omega_\alpha}\right)^2 G_{\alpha+}\tilde{\phi}(y_{\alpha+})+\left(\frac{\omega+\epsilon_\alpha\Omega_\alpha}{\Omega_\alpha}\right)^2 G_{\alpha-}\tilde{\phi}(y_{\alpha-})\right\}\right] \tag{B.18c}$$

and with

$$A_\alpha:=1-V_{\alpha\perp}^2/V_{\alpha\parallel}^2,$$
$$G_{\alpha\pm}:=\frac{\omega-k_\parallel U_\alpha}{\omega\pm\epsilon_\alpha\Omega_\alpha-k_\parallel U_\alpha}\quad -A_\alpha, \tag{B.19}$$

$$\tilde{\phi}(y):=\phi(y)-i\sqrt{\pi}\,ye^{-y^2}, \tag{B.20}$$

and $y_{\alpha 0}$, $y_{\alpha\pm}$ given by $s=0$, $\pm\epsilon_\alpha$ in (12.29), viz.

$$y_{\alpha s}:=\frac{\omega-s\Omega_\alpha-k_\parallel U_\alpha}{\sqrt{2}k_\parallel V_{\alpha\parallel}}.$$

For a thermal distribution ($V_{\alpha\perp}=V_{\alpha\parallel}=:V_\alpha$, $U_\alpha=0$, (B.17) with (B.18) reduces to (2.137) in the limit stated.

Cold Plasmas

The cold plasma theory is outlined in (2.57) *et seq.* and in (10.95) *et seq.* The generalization of the usual cold plasma dielectric tensor (2.116) to the case where the particles are streaming may be derived in a straightforward way using cold plasma theory. It may also be derived from (B.17) and (B.18). In the limit $y_{\alpha 0}^2\gg 1$, $y_{\alpha\pm}^2\gg 1$, the thermal motions are negligible and (B.18a, b, c) become

$$R_{\pm}=1-\sum_{\alpha}\frac{\omega_{p\alpha}^2}{\omega^2}\frac{\omega-k_\parallel U_\alpha}{\omega\pm\epsilon_\alpha\Omega_\alpha-k_\parallel U_\alpha}, \tag{B.21a}$$

$$E_{\pm}=-\sum_{\alpha}\frac{\omega_{p\alpha}^2}{\omega^2}\frac{k_\perp U_\alpha}{\omega\pm\epsilon_\alpha\Omega_\alpha-k_\parallel U_\alpha}, \tag{B.21b}$$

$$P=1-\sum_{\alpha}\frac{\omega_{p\alpha}^2}{\omega^2}\left[\frac{\omega^2}{(\omega-k_\parallel U_\alpha)^2}+\frac{k_\perp^2 U_\alpha^2}{(\omega-k_\parallel U_\alpha)^2-\Omega_\alpha^2}\right] \tag{B.21c}$$

This result for a cold plasma with streaming motions may also be obtained from the relativistic quantum approach of Section 13.4.

In the absence of a magnetic field, the dielectric tensor for a cold plasma in which the different components are streaming is

$$\varepsilon_{ij}(k,\omega)=\delta_{ij}-\sum_{\alpha}\frac{\omega_{p\alpha}^2}{\omega^2}\left[\delta_{ij}-\frac{1}{\omega-\mathbf{k}\cdot\mathbf{U}}(k_iU_j+k_jU_i)+\frac{(k^2-\omega^2/c^2)U_iU_j}{(\omega-\mathbf{k}\cdot\mathbf{U})^2}\right]. \tag{B.22}$$

One way of obtaining (B.22) is to note that in its rest frame each component contributes $-\omega_{p\alpha}^2\delta_{ij}/c$ to α_{ij}, cf. (B.3), and then to construct $\alpha^{\mu\nu}$ using (13.130) and to apply a Lorentz transformation to the frame in which the streaming speed is **U**.

Degenerate Electron Gas

The dielectric tensor for a degenerate electron gas was calculated using non-relativistic quantum theory by Lindhard (1954). It is interesting to derive Lindhard's result using the general expression (13.160) for the tensor $\alpha_{ij}(\mathbf{k},\omega)$ which includes both relativistic and quantum effects. Firstly, for an isotropic electron gas, (13.160) reduces to (Tsytovich, 1961)

$$\alpha_{ij}(\mathbf{k},\omega)=\alpha^l(\mathbf{k},\omega)\kappa_i\kappa_j+\alpha^t(\mathbf{k},\omega)(\delta_{ij}-\kappa_i\kappa_j),$$

$$a^{l,t}(\mathbf{k},\omega)=-e^2\int\frac{d^3\mathbf{p}}{(2\pi)^3}f(\mathbf{p})\,\frac{\varepsilon(\mathbf{p})-\varepsilon(\mathbf{p}-\mathbf{k})}{\{\varepsilon(\mathbf{p})-\varepsilon(\mathbf{p}-\mathbf{k})\}^2-\omega^2}\,\Lambda_+^{l,t}(\mathbf{p},\mathbf{k})$$
$$+\frac{\varepsilon(\mathbf{p})+\varepsilon(\mathbf{p}-\mathbf{k})}{\{\varepsilon(\mathbf{p})+\varepsilon(\mathbf{p}-\mathbf{k})\}^2-\omega^2}\,\Lambda_-^{l,t}(\mathbf{p},\mathbf{k}), \tag{B.23}$$

with

$$f(\mathbf{p}):=2\{n^+(\mathbf{p})+n^-(\mathbf{p})\},$$
$$\Lambda_\pm^l(\mathbf{p},\mathbf{k})=1\pm\frac{2(\mathbf{p}\cdot\mathbf{k})^2/|\mathbf{k}|^2-\mathbf{p}\cdot\mathbf{k}-\varepsilon^2(\mathbf{p})}{\varepsilon(\mathbf{p})\varepsilon(\mathbf{p}-\mathbf{k})}, \tag{B.24}$$
$$\Lambda_\pm^t(\mathbf{p},\mathbf{k})=1\pm\frac{\mathbf{p}\cdot\mathbf{k}-m^2-(\mathbf{p}\cdot\mathbf{k})^2/|\mathbf{k}|^2}{\varepsilon(\mathbf{p})\varepsilon(\mathbf{p}-\mathbf{k})}.$$

The completely degenerate non-relativistic limit corresponds to

$$n^+(\mathbf{p})=\begin{cases}1 \text{ for } p<p_F\\ 0 \text{ for } p>p_F\end{cases}, \tag{B.25}$$

where p_F is the Fermi momentum, with $n^-(\mathbf{p})=0$. The integrals may be performed directly in the non-relativistic case for the hermitian part, i.e. the real parts of α^l and α^t:

$$\alpha^l(\mathbf{k}, \omega) = \frac{3e^2 n_e u^2}{8mq^2}\left[1 + \frac{1}{2q}\left\{1 - \frac{1}{4}\left(q - \frac{u}{q}\right)^2\right\} \ln \frac{q(q+2)-u}{q(q-2)-u}\right.$$
$$\left. + \frac{1}{2q}\left\{1 - \frac{1}{4}\left(q + \frac{u}{q}\right)^2\right\} \ln \frac{q(q+2)+u}{q(q-2)-u}\right] \tag{B.26a}$$

$$\alpha^t(\mathbf{k}, \omega) = \frac{3e^2 n_e}{8m}\left[1 + \frac{q^2}{4} + \frac{3u^2}{4q^2} - \frac{1}{2q}\left\{1 - \frac{1}{4}\left(q - \frac{u}{q}\right)^2\right\} \ln \frac{q(q+2)-u}{q(q-2)-u} - \frac{1}{2q}\right.$$
$$\left. \times \left\{1 - \frac{1}{4}\left(q + \frac{u}{q}\right)^2\right\}^2 \ln \frac{q(q+2)+u}{q(q-2)+u}\right] \tag{B.26b}$$

with

$$u := \frac{|\omega|}{\varepsilon_F}, \quad q := \frac{|\mathbf{k}|}{p_F}, \quad \varepsilon_F := \frac{p_F^2}{2m}. \tag{B.27}$$

The anti-hermitian may be obtained using the Kramers-Kronig relations (Sitenko, 1967).

The dielectric tensor for a completely degenerate electron gas in a magnetic field has been written down by Svetozarova and Tsytovich (1962) and, in the non-relativistic limit, by Canuto and Ventura (1972). The non-degenerate case for a thermal magnetized quantum plasma has been discussed recently by Gurevich and Panov (1976).

Answers to Problems

1.1 1.2×10^{-11} sterad.

1.2 1.5×10^{11} K,

1.3 241 f.u.

1.5 $\mathbf{v}/\!/^l\ \delta\mathbf{B}/\!/^l\mathbf{k}\times\mathbf{B}_0,\quad \mathbf{E}/\!/^l(\mathbf{k}\times\mathbf{B}_0)\times\mathbf{B}_0.$

1.6 Defining $\boldsymbol{\kappa}$ and $\mathbf{a}$ as in (2.71), $\omega=kv_A|\cos\theta|$ in

$$\frac{\partial\omega}{\partial\mathbf{k}}=\frac{\partial\omega}{\partial k}\boldsymbol{\kappa}+\frac{1}{k}\frac{\partial\omega}{\partial\theta}\mathbf{t}$$

gives

$$\mathbf{v}_g=v_A\left(|\cos\theta|\boldsymbol{\kappa}+\frac{\cos\theta\sin\theta}{|\cos\theta|}\mathbf{t}\right)=v_A\frac{\cos\theta}{|\cos\theta|}\mathbf{b},$$

with $\mathbf{b}:=\mathbf{B}_0/|\mathbf{B}_0|$.

1.8 (i) $v_A=7.0\times10^4(\Omega_e/\omega_p)$ km s^{-1}
(ii) 4.37×10^4 km/arc min
(iii) 0.98 km s^{-1}/pc (10^6 yr)$^{-1}$.

2.2 Writing μ_{ij}^{-1} for the inverse of μ_{ij}, i.e. $H_i=\mu_{ij}^{-1}B_j$, one has

$$\varepsilon_{ij}(\mathbf{k},\omega)=\varepsilon_{ij}+\frac{k^2c^2}{\omega^2}\left[\mu_{ji}^{-1}+(\delta_{ij}-\kappa_i\kappa_j)(1-\mu_{ss}^{-1})+\delta_{ij}\kappa_r\kappa_s\mu_{rs}^{-1}-\kappa_i\kappa_s\mu_{js}^{-1}\right.$$
$$\left.-\kappa_r\kappa_j\mu_{ri}^{-1}\right].$$

2.7(d)

$$\mathbf{v}_g^\sigma(\omega,\theta,\phi)=\frac{c}{\partial\{\omega n_\sigma(\omega,\theta,\phi)\}/\partial\omega}$$
$$\times\left[\boldsymbol{\kappa}-\left\{\frac{\partial}{\partial\theta}\ln n_\sigma(\omega,\theta,\phi)\right\}\hat{\theta}-\frac{1}{\sin\theta}\left\{\frac{\partial}{\partial\phi}\ln n_\sigma(\omega,\theta,\phi)\right\}\hat{\phi}\right],$$

with

$$\boldsymbol{\kappa}=(\sin\theta\cos\phi, \sin\theta\sin\phi, \cos\theta),$$
$$\hat{\theta}=(\cos\theta\cos\phi, \cos\theta\sin\phi, -\sin\theta),$$
$$\hat{\phi}=(-\sin\phi, \cos\phi, 0).$$

2.9 The Cayley-Hamilton theorem implies

$$\Lambda\delta_{ij}-\Lambda_{ij}\lambda_{ss}+\Lambda_{ir}\Lambda_{rj}\Lambda_{ss}-\Lambda_{ir}\Lambda_{rs}\Lambda_{sj}=0.$$

The trace gives

$$3\Lambda-\Lambda_{rr}\lambda_{ss}+\Lambda_{rs}\Lambda_{sr}\Lambda_{tt}-\Lambda_{rs}\Lambda_{st}\Lambda_{tr}=0.$$

Contracting the original equation with λ_{ja}, relabelling dummy indices, and using (2.75), gives

$$\lambda_{ij}-\lambda_{ss}\delta_{ij}+\Lambda_{ij}\Lambda_{ss}-\Lambda_{ir}\Lambda_{rj}=0,$$

whose trace is

$$-2\lambda_{ss}+(\Lambda_{ss})^2-\Lambda_{rs}\Lambda_{sr}=0.$$

These equations imply (2.95c) and (2.96b). A simpler derivation of (2.96b) follows directly from (2.96a) and (2.25).

2.13 Let ω_{p1} be the plasma frequency corresponding to the suprathermal distribution, and let the "one-dimensional" axis be the 3-axis. The contribution is

$$-\delta_{i3}\delta_{j3}\frac{\omega_{p1}^2}{\omega^2}\frac{v_\phi^2}{\{(v_s-v_\phi)^2+\bar{v}^2\}^2}\left[(v_s-v_\phi)^2-\bar{v}^2+i\bar{v}(v_s-v_\phi)\right].$$

3.1 Assume that there are no ion sound waves for $k\lambda_{\mathrm{De}}>1$, and that for $k\lambda_{\mathrm{De}}<1$ they have $\omega=kv_s$. Then one finds

$$P=\int\frac{d^3\mathbf{k}}{(2\pi)^3}\frac{4\pi e^2k^2v_s^4}{\omega_{pi}^2}\delta(kv_s-\mathbf{k}\cdot\mathbf{v})$$
$$=\begin{cases}\dfrac{e^2\omega_{pi}^2}{4v} & \text{for } v>v_s\\ 0 & \text{for } v<v_s.\end{cases}$$

3.2 The extraneous currents are

$$\mathbf{j}(\mathbf{k},\omega)=-i\omega\mathbf{d}2\pi\,\delta(\omega-\mathbf{k}\cdot\mathbf{v})$$

and

$$\mathbf{j}(\mathbf{k}, \omega) = ic\mathbf{k} \times \mathbf{m} 2\pi \delta(\omega - \mathbf{k} \cdot \mathbf{v}).$$

The powers radiated are

$$P = \frac{d^2}{vc^2} \int_{v > c/n(\omega)} d\omega\, \omega^3 \{1 - \cos^2 \theta_0 \cos^2 \chi_0 - \tfrac{1}{2} \sin^2 \theta_0 \sin^2 \chi_0\}$$

and

$$P = \frac{m^2}{vc^2} \int_{v > c/n(\omega)} d\omega\, \omega^3 n^2(\omega) \{1 - \cos^2 \theta_0 \cos^2 \chi_0 - \tfrac{1}{2} \sin^2 \theta_0 \sin^2 \chi_0\},$$

with $\cos \theta_0 := c/vn(\omega)$ and $\cos \chi_0 := \mathbf{v} \cdot \mathbf{d}/vd$ or $\mathbf{v} \cdot \mathbf{m}/vm$.

3.4 The power is entirely in the extraordinary mode:

$$P = \frac{q^2 v}{c^2} \int_{v^2 > c^2/\varepsilon_\perp(\omega)} \left(1 - \frac{c^2}{v^2 \varepsilon_\perp(\omega)}\right)\left(1 + \frac{c^2}{v^2} \frac{\varepsilon_\perp(\omega) - \varepsilon_\parallel(\omega)}{\varepsilon_\perp^2(\omega) \varepsilon_\parallel^2(\omega)}\right)$$

3.7 The generalization of (3.81) is $P^\sigma(\omega) = R_E^\sigma P(\omega)$ with $P(\omega)$ given by (3.81) and with

$$R_E^{\mathrm{o}} \approx \frac{1}{2}\left\{1 - \frac{2\Omega_e}{\omega} |\cos \theta|\right\}, \quad R_E^{x} \approx \frac{1}{2}\left\{1 + \frac{2\Omega_e}{\omega} |\cos \theta|\right\}.$$

3.10(*c*)

$$\mathbf{j}(\mathbf{k}, \omega) = 2\pi q R \omega_0 \sum_{n=1}^{\infty} \left[\left\{iJ_n'(ne), \frac{(1 - e^2)^{1/2}}{e} J_n(ne), 0\right\} \delta(\omega - n\omega_0) + \left\{-iJ_n'(ne), \frac{(1 - e^2)^{1/2}}{e} J_n(ne), 0\right\} \delta(\omega + n\omega_0)\right]$$

(d)

$$P = \frac{q^2 R^2 \omega_0^4}{2c^3} \sum_{n > n_0} \left(1 - \frac{\omega_p^2}{n^2 \omega_0^2}\right)^{1/2} \left\{n^2 J_n'^2(ne) + \frac{(1 - e^2)}{e^2} n^2 J_n^2(ne)\right\},$$

where n_0 is the smallest integer greater than ω_p/ω_0.

(e) $P_{2\omega_0}/P_{\omega_0} = 2e^2$.

4.1 (a)

$$\eta(s, \omega, \theta) = \frac{n(\omega) e^2 \Omega_e^2}{2\pi c} \frac{(s\beta_\perp/2)^{2s} \{n(\omega) \sin \theta\}^{2s-2}}{\{(s-1)!\}^2} \times (1 + \cos^2 \theta)\, \delta(\omega\{1 - n(\omega)\beta_\parallel \cos \theta\} - s\Omega_e)$$

(b) $$\bar{\eta}(s, \omega, \theta) := \frac{n_1}{(2\pi)^{3/2}\beta_0^3} 2\pi \int_{-\infty}^{\infty} \mathrm{d}\beta_\parallel \int_0^\infty \mathrm{d}\beta_\perp \beta_\perp \eta(s, \omega, \theta) \exp\left[-\frac{(\beta_\perp{}^2+\beta_\parallel^2)}{2\beta_0^2}\right]$$

$$= \frac{n_1 e^2 \Omega_e (1+\cos^2\theta)}{(2\pi)^{3/2}\beta_0 c|\cos\theta|} \frac{(s^2\beta_0^2/2)^s (n(\omega)\sin\theta)^{2s-2}}{(s-1)!}$$

$$\times \exp\left[-\frac{(\omega - s\Omega_e)^2}{2\{n(\omega)\omega\beta_0\cos\theta\}^2}\right]$$

(c) A non-relativistic Maxwellian distribution implies a finite probability of finding a particle with arbitrarily large $\beta_\perp$. There exist (non-physical) particles for which the assumption $n(\omega)\beta_\perp \sin\theta \ll 1$ breaks down, and such particles appear to radiate an arbitrarily large power.

4.2 From (4.28), $T=0$ requires $z=0$, and then (4.124) implies $\cos\theta = n(\omega)\beta\cos\alpha$, or for $\xi \gg 1$ (cf. (4.68)), $\theta = \alpha + \frac{1}{2}\xi^{-2}\cot\alpha$.

4.3 The result (4.102) is reproduced but with

$$J^{11}(a) = \frac{5}{48}\left(\frac{15\sqrt{3}}{8}\right)^{-(a-1)/2}, \quad J^{22}(a) = 7J^{11}(a).$$

4.4 (P4-3) follows from (4.113) and (4.115).

4.5 The generalization of (3.55) for gyromagnetic emission is

$$P^l(\mathbf{k}) = \frac{4\pi^2 q^2 \{\omega^l(k)\}^4}{k^2\omega_p^2} \sum_{s=-\infty}^{\infty} J_s^2\left(\frac{k_\perp v_\perp}{\Omega}\right) \delta(\omega^l(k) - s\Omega - k_\parallel v_\parallel).$$

The generalization of (3.56) is

$$P^l = \frac{q^2\omega_p^2}{2\pi|v_\parallel|} \sum_{s=-\infty}^{\infty} \int_0^{k(s)} \frac{dk_\perp k_\perp J_s^2(k_\perp v_\perp/\Omega_e)}{k_\perp^2 + (s\Omega_e - \omega_p)^2/v_\parallel^2}$$

with

$$k(s) := \left\{\lambda_{\mathrm{De}}^{-2} - \left(\frac{s\Omega_e - \omega_p}{v_\parallel}\right)^2\right\}^{1/2}.$$

4.6 The probability is defined in (5.30), and (4.16) for $k_\perp v_\perp \ll \Omega$ gives

$$\mathbf{V}(0, \mathbf{p}, \mathbf{k}) \approx (0, 0, v_\parallel),$$

$$\mathbf{V}(\pm 1, \mathbf{p}, \mathbf{k}) \approx \tfrac{1}{2} v_\perp (1, -i\epsilon s, 0),$$

from which one finds

$$|\mathbf{e}^{\sigma *}(\mathbf{k})\cdot\mathbf{V}(s,\mathbf{p},\mathbf{k})|^2\approx\begin{cases}\frac{1}{4}v_\perp^2|e_1^\sigma(\mathbf{k})+i\epsilon s e_2^\sigma(\mathbf{k})|^2 & \text{for } s=\pm 1\\ v_\parallel^2|e_3^\sigma(\mathbf{k})|^2 & \text{for } s=0.\end{cases}$$

The wave properties to be inserted in (5.30) are

Alfvén mode

$$\omega=kv_A|\cos\theta|,\ \mathbf{e}=(1,0,0),\ R_E=v_A^2/2c^2;$$

Magnetoacoustic mode

$$\omega=kv_A,\quad \mathbf{e}=(0,i,0),\quad R_E=v_A^2/2c^2;$$

whistler mode

$$\omega=\frac{k^2c^2}{\omega_p^2}\Omega_e|\cos\theta|,\quad \mathbf{e}=\frac{(1,i|\cos\theta|,0)}{(1+\cos^2\theta)^{1/2}},\quad R_E=\frac{\omega\Omega_e}{\omega_p^2}\frac{1+\cos^2\theta}{|\cos\theta|}.$$

4.7 In (4.39) *et seq.* set $\beta_\parallel=0$, $\beta_\perp=\beta n(\omega)$ and $\Omega=\beta c/R$, carry through the derivation to (4.48) and then evaluate $P=P^{11}+P^{22}$.

5.4 The following terms are added to the RH sides of (5.36) and (5.37) respectively:

$$\frac{1}{2}\int d^3\mathbf{p}\left\{\hbar\mathbf{k}\cdot\frac{\partial}{\partial\mathbf{p}}w^\sigma(\mathbf{p},\mathbf{k})\right\}\hbar\mathbf{k}\cdot\frac{\partial f(\mathbf{p})}{\partial\mathbf{p}}N^\sigma(\mathbf{k}),$$

$$\frac{1}{2}\int\frac{d^3\mathbf{k}}{(2\pi)^3}\left(\hbar\mathbf{k}\cdot\frac{\partial}{\partial\mathbf{p}}\right)^2\{w^\sigma(\mathbf{p},\mathbf{k})f(\mathbf{p})\}.$$

5.5 Use (5.68) with α the angle between $\mathbf{p}$ and the axis of symmetry, and with

$$\Delta\alpha=\left(\hbar\mathbf{k}\cdot\frac{\partial}{\partial\mathbf{p}}\right)\arctan\frac{(p_x^2+p_y^2)^{1/2}}{p_z}$$

$$=\frac{\hbar k}{p}\sin\alpha\,\{\cos\alpha\sin\theta\cos\phi-\cos\theta\},$$

and with $d^3\mathbf{k}=k^2\,dk\,d\cos\theta\,d\phi$ in (5.68).

5.8

$$\Phi(\alpha,t)=\sum_{n=0}^{\infty}\Phi_n(0)P_n(\cos\alpha)\exp\{-n(n+1)\nu_0 t\}.$$

6.1 Only $s=0,\ \pm 1$ contribute, and the approximation in the answer to problem (4.6) applies. After summing over $s=0, \pm 1$, one needs to establish the result (for electrons, i.e. $\epsilon=-1$, and magnetoionic waves)

$$R_E^\sigma(\mathbf{k})\left\{|e_3^\sigma(\mathbf{k})|^2+\frac{\omega^2}{2(\omega-\Omega_e)^2}|e_1^\sigma(\mathbf{k})-ie_2^\sigma(\mathbf{k})|^2+\frac{\omega^2}{2(\omega+\Omega_e)^2}|e_1^\sigma(\mathbf{k})+ie_2^\sigma(\mathbf{k})|^2\right\}$$

$$=\frac{R_E^\sigma(\mathbf{k})e_i^\sigma(\mathbf{k})e_j^{\sigma *}(\mathbf{k})}{\omega_p^2}\frac{\partial}{\partial\omega}\{\omega\varepsilon_{ij}(\omega)-\omega\delta_{ij}\}$$

$$=R_p^\sigma(\mathbf{k})/\omega_p^2,$$

where (2.92) has been used.

6.2 (a) One has

$$\langle j_i j_j^*\rangle(\mathbf{k},\omega)=\int \mathrm{d}^3\mathbf{r}\,\mathrm{d}t\exp\left[-i(\mathbf{k}\cdot\mathbf{r}-\omega t)\right]\int \mathrm{d}^3\xi\,\mathrm{d}\tau\int \mathrm{d}^3\mathbf{p}f(\mathbf{p})$$
$$\times q^2v_i(t)v_j(t)\delta^3(\mathbf{r}-\mathbf{r}(t))\delta^3(\mathbf{r}+\xi-\mathbf{r}(t+\tau)).$$

For $\mathbf{r}(t)=\mathbf{v}t$ the result (P6-2) follows trivially, and for magnetized particles (P6-2) follows after expanding in Bessel functions, e.g. using (4.12).

(b) The result follows directly from the solution of the wave equation

$$E_i(\mathbf{k},\omega)=-\frac{4\pi i}{\omega}\frac{\lambda_{ij}(\mathbf{k},\omega)j_j(\mathbf{k},\omega)}{\Lambda(\mathbf{k},\omega)}.$$

6.3(a) Use

$$\langle j_i(\mathbf{k},\omega)j_j^*(\mathbf{k}',\omega')\rangle=\langle j_ij_j^*\rangle(\mathbf{k},\omega)(2\pi)^4\delta^3(\mathbf{k}-\mathbf{k}')\delta(\omega-\omega')$$

for $\mathbf{k}'=\mathbf{k}$ and $\omega-\omega'$ to find

$$\langle j_i(\mathbf{k},\omega)j_j^*(\mathbf{k},\omega)\rangle=VT\langle j_ij_j^*\rangle(\mathbf{k},\omega),$$

and hence, directly from (3.18),

$$\frac{P^\sigma(\mathbf{k})}{V}=4\pi R_E^\sigma(\mathbf{k})e_i^{\sigma *}(\mathbf{k})e_j^\sigma(\mathbf{k})\langle j_ij_j^*\rangle(\mathbf{k},\omega^\sigma(\mathbf{k})).$$

The result (P6-4) follows by balancing this emission with absorption.

6.10

$$\frac{\mathrm{d}N^{\alpha\beta}(\mathbf{k})}{\mathrm{d}t}=\int\frac{\mathrm{d}^3\mathbf{k}'}{(2\pi)^3}\int \mathrm{d}^3\mathbf{p}w^{\alpha\beta\gamma\delta\alpha'\beta'}(\mathbf{p},\mathbf{k},\mathbf{k}')$$
$$\times\left[f(\mathbf{p})\{N^{\alpha'\beta'}(\mathbf{k}')\delta^{\gamma\delta}-N^{\gamma\delta}(\mathbf{k})\delta^{\alpha'\beta'}\}+\hbar(\mathbf{k}-\mathbf{k}')\cdot\frac{\partial f(\mathbf{p})}{\partial\mathbf{p}}N^{\gamma\delta}(\mathbf{k})N^{\alpha'\beta'}(\mathbf{k}')\right],$$

$$w^{\alpha\beta\gamma\delta\alpha'\beta'}(\mathbf{p}, \mathbf{k}, \mathbf{k}') := \tfrac{1}{2}\{w^{\alpha\gamma\alpha'\beta'}(\mathbf{p}, \mathbf{k}, \mathbf{k}')\delta^{\beta\delta} + w^{\delta\beta\alpha'\beta'}(\mathbf{p}, \mathbf{k}, \mathbf{k}')\delta^{\alpha\gamma}\};$$

$$\frac{\mathrm{d}N_A(\mathbf{k})}{\mathrm{d}t} = \int \frac{\mathrm{d}^3\mathbf{k}'}{(2\pi)^3} \int \mathrm{d}^3\mathbf{p}\, w_{ABA'}(\mathbf{p}, \mathbf{k}, \mathbf{k}')$$

$$\times \left[f(\mathbf{p})\,\{N_{A'}(\mathbf{k}')\Sigma_B - N_B(\mathbf{k})\Sigma_{A'}\} + \hbar(\mathbf{k}-\mathbf{k}') \cdot \frac{\partial f(\mathbf{p})}{\partial \mathbf{p}}\; N_{A'}(\mathbf{k}')N_B(\mathbf{k}) \right],$$

where the indices $\alpha\beta$, $\gamma\delta$ and $\alpha'\beta'$ are translated into A, B and A′, respectively, as in (6.50), and where Σ_A denotes that one is to sum the index A (from 1 to 4 or, equivalently, over I, Q, U, V).

REFERENCES

Akhieser, A. I., Akhieser, I. A., Polovin, R. V., Sitenkov, A. G., and Stepanov, K. N.: 1967, *Collective Oscillations in a Plasma*, M.I.T. Press (Cambridge, Mass.).

Alfvén, H.: 1942, *Arkiv. Mat. Astron. Fysik.* **29B** (2).

Alfvén, H. and Herlofson, N.: *Phys. Rev.* **78**, 616.

Allen, C. W.: 1973, *Astrophysical Quantities*, 3rd edition, The Athlone Press (London).

Appleton, E. V.: 1932, *J. Inst. Elect. Engrs.* **71**, 642.

Arons, J.: 1972: *Astrophys. J.* **177**, 395.

Åström, E.: 1950, *Arkiv. Fysik.* **2**, 443.

Baade, W. and Minkowski, R.: 1954, *Astrophys. J.* **119**, 206.

Batygin, V. V. and Toptygin, I. N.: 1964, *Problems in Electrodynamics*, Academic Press (New York).

Bekefi, G.: 1966, *Radiation Processes in Plasmas*, Wiley (New York).

Berstein, I. B.: 1958, *Phys. Rev.* **109**, 10.

Blandford, R. D.: 1972, *Astron. Astrophys.* **20**, 135.

Blandford, R. D. and Rees, M. J.: 1972, *Astrophys. Lett.* **10**, 77.

Blandford, R. D. and Scarlemann, E. T.: 1975, *Mon. Not. Roy. Astron. Soc.* **174**, 59.

Blumenthal, G. R. and Gould, R. J.: 1970, *Rev. Mod. Phys.* **42**, 237.

Boischot, A.: 1957, *C.R. Acad. Sci. Paris* **244**, 1326.

Bolt, E. and Serlimitsos, P.: 1969, *Astrophys. J.* **157**, 557.

Bolton, J. G.: 1948, *Nature* **162**, 141.

Bolton, J. G. and Stanley, G. J.: 1948, *Nature* **161**, 312.

Bolton, J. G., Stanley, G. J., and Slee, O. B.: 1949, *Nature* **164**, 101.

Bonometto, S. and Scrascia, L.: 1974, *Astron Astrophys.* **32**, 115.

Braginskii, S. I.: 1965, in M. A. Leontovich (ed.), *Rev. Plasma Phys.* **1**, 205. Consultants Bureau (New York).

Brown, J. C.: 1971, *Solar Phys.* **18**, 489.

Brown, J. C.: 1975, in S. R. Kane (ed.). *Solar gamma-, X- and EUV radiation*, IAU Symp. No. 68, D. Reidel (Dordrecht, Holland), p. 245.

Brown, L. S. and Kibble, T. W. B.: 1964, *Phys. Rev.* **133**, A705.

Brown, R. L.: 1970, *Astrophys. J.* **159**, L187.

Budden, K. G.: 1961, *Radio Waves in the Ionosphere*, Cambridge Univ. Press.

Burke, B. F. and Franklin, K. L.: 1955, *J. Geophys. Res.* **60**, 213.

Canuto, V. and Ventura, J.: 1972, *Astrophys. Space Sci.* **18**, 104.
Cavaliere, A., Morrison, P., and Pacini, F.: 1970, *Astrophys. J.* **162**, L133.
Cheng, F. T. and Fung, P. C. W.: 1977, *Astrophys. Space Sci.* **49**, 427.
Chiuderi, C. and Veltri, P.: 1974, *Astron. Astrophys.* **30**, 265.
Colgate, S. A.: 1967, *Astrophys. J.* **150**, 163.
Colgate, S. A., Colvin, J. D., and Petschek, A. G.: 1975, *Astrophys. J.* **197**, L105.
Cronyn, W. M.: 1966, *Astrophys. J.* **144**, 834.
Dawson, J. M.: 1968, *Adv. Plasma Phys.* **1**, 1.
de Hoffman, F. and Teller, E.: 1950, *Phys. Rev.* **80**, 692.
Denisse, J. F. and Delcroix, J. L.: 1963, *Plasma Waves*, Interscience (New York).
Dewar, R. L.: 1973, *Phys. Fluids* **16**, 1102.
Dewar, R. L.: 1976, *J. Phys. A.* **9**, 2043.
Dreicer, H.: 1964, *Phys. Fluids* **7**, 735.
Drummond, W. E. and Pines, D.: 1962, *Nucl. Fusion Suppl. Part 3*, 1049.
Drummond, W. E. and Rosenbluth, M. N.: 1960, *Phys. Fluids* **3**, 45.
Einstein, A.: 1917, *Phys. Zeit.* **18**, 121.
Énomé, S.: 1969, *Publ. Astron. Soc. Japan* **21**, 367.
Epstein, R. I.: 1973, *Astrophys. J.* **183**, 593.
Epstein, R. I. and Feldman, P. A.: 1967, *Astrophys. J.* **150**, L109.
Epstein, R. I. and Petrosian, V.: 1973, *Astrophys. J.* **146**, 686.
Ewan, H. I. and Purcell, E. M.: 1951, *Nature* **168**, 356.
Fried, B. D. and Conte, S. D.: 1961, *The Plasma Dispersion Function*, Academic Press (New York).
Fukai, J. and Harris, E. G.: 1972, *J. Plasma Phys.* **7**, 313.
Fung, P. C. W.: 1969, *Astrophys. Space Sci.* **5**, 448.
Gailitis, A. K. and Tsytovich, W. N.: 1964, *Zh. Eksp. Teor. Fiz.* **46**, 1726; *Soviet Phys. JETP* **19**, 1165.
Gaunt, J. A.: 1930, *Phil. Trans. Roy. Soc. London* **A229**, 163.
Ginzburg, V. L.: 1939, *J. Phys. (USSR)* **2**, 441.
Ginzburg, V. L.: 1946, *Dokl. Acad. Nauk. SSSR* **52**, 487.
Ginzburg, V. L.: 1951, *Dokl. Acad. Nauk. SSSR* **76**, 377.
Ginzburg, V. L.: 1953, *Uspekhi Fiz. Nauk.* **51**, 343.
Ginzburg, V. L.: 1957, *Uspekhi Fiz. Nauk.* **62**, 37.
Ginzburg, V. L.: 1967, in C. De Witt (ed.), *High Energy Astrophysics*, Gordon and Breach (New York), Vol. I, p. 17.
Ginzburg, V. L. and Syrovatskii, S. I.: 1964, *The Origin of Cosmic Rays*, Pergamon (Oxford).
Ginzburg, V. L. and Syrovatskii, S. I.: 1965, *Ann. Rev. Astron. Astrophys.* **3**, 297.
Ginzburg, V. L. and Zheleznyakov, V. V. 1958, *Astron. Zh.* **35**, 694; *Soviet Astron. AJ*, **2**, 653.
Ginzburg, V. L. and Zheleznyakov, V. V.: 1965, *Phil. Mag.* **11**, 197.
Glauber, R. J.: 1963, *Phys. Rev.* **131**, 2766.
Goldreich, P., McCray, R. and Rees, M. J.: 1968, *Nature* **217**, 931.
Goldstein, H.: 1959, *Classical Mechanics*, Addison-Wesley
Gunn, J. E. and Ostriker, J. P.: 1971, *Astrophys. J.* **165**, 523.
Gurevich, L. E. and Panov, A. N.: 1976, *Zh. Esksp. Teor. Fiz.* **70**, 61; *Soviet Phys. JETP* **43**, 32.
Hamilton, W. R.: 1931, *Mathematical Papers, Vol. I, Geometric Optics*, Cambridge Univ. Press.
Harris, E. G.: 1969, *Adv. Plasma Phys.* **3**, 157.
Hartree, D. R.: 1931, *Proc. Camb. Phil. Soc.* **27**, 143.
Haselgrove, J.: 1955, in *The Physics of the Ionosphere*, Physical Society Conference, Cambridge Univ. Press. p. 355.

Heitler, W.: 1954, *The Quantum Theory of Radiation*, Oxford Univ. Press.
Helliwell, R. A.: 1967, *J. Geophys. Res.* **72**, 4773.
Herlofson, N.: 1950, *Nature* **165**, 1020.
Hewish, A., Bell, S. J., Pilkington, J. D. H., Scott, P. F., and Collins, R. A.: 1968, *Nature* **217**, 709.
Hey, J. S.: 1946, *Nature* **157**, 47.
Hey, J. S., Parsons, S. J., and Phillips, J. W.: 1946, *Nature*, 158, 234.
Hornby, J. M. and Williams, P. J. S.: 1966, *Mon. Not. Roy. Astron. Soc.* **131**, 237.
Ivanenko, D. and Sokolov, A. A.: 1948, *Dokl. Acad. Nauk. SSSR* **59**, 1551.
Jansky, K. G.: 1933, *Proc. I.R.E.* **21**, 1387.
Jansky, K. G.: 1935, *Proc. I.R.E.* **23**, 1158.
Jelley, J. V.: 1958, *Čerenkov Radiation and Its Applications*, Pergamon Press (Oxford).
Jones, F. C.: 1971, *Astrophys. J.* **169**, 503.
Jones, T. W. and O'Dell, S. L.: 1977, *Astrophys. J.* **214**, 522.
Kai, K.: 1965, *Publ. Astron. Soc. Japan* **17**, 3.
Kaplan, S. A. and Tsytovich, V. N.: 1969, *Astron. Zh.* **46**, 199; *Soviet Astron. AJ* **13**, 150.
Kaplan, S. A. and Tsytovich, V. N.: 1973, *Plasma Astrophysics*, Pergamon Press (Oxford).
Kawabata, K.: 1964, *Publ. Astron. Soc. Japan* **16**, 30.
Kiepenheuer, K. O.: 1950, *Phys. Rev.* **79**, 738.
Klauder, J. R. and Sudarshan E. C. G.: 1968, *Fundamentals of Quantum Optics*, W. A. Benjamin Inc. (New York).
Kompaneets, A. S.: 1956, *Zh. Eksp. Teor. Fiz.* **31**, *Soviet Phys. JETP* **4**, 230.
Krall, N. A, and Smith, D. F.: 1974, *Astrophys. J.* **199**, 500.
Landau, L. D.: 1946, *J. Phys. (U.S.S.R.)* **10**, 25.
Legg, M. P. C. and Westfold, K. C.: 1968, *Astrophys. J.* **154**, 499.
le Roux, R.: 1961, *Ann. Astrophys.* **24**, 71.
Levich, E. V. and Sunyaev, R. A.: 1970, *Radiofizika* **13**, 1873; *Radiophysics and Quantum Electronics* **13**, 1445.
Lindhard, D. J.: 1954, *Mat. Fys. Medd. Dan. Vid. Selsk.* **28**, No. 8.
Lorentz, H. A.: 1905, *Verlag. Asterdam Acad.* **14**, 518.
Manageney, A. and Veltri, P.: 1976, *Astron. Astrophys.* **47**, 165 & 181.
Martyn, D. F.: 1946a, *Nature* **158**, 632.
Martyn, D. F.: 1946b, *Nature* **158**, 308.
Martyn, D. F.: 1947, *Nature* **159**, 26.
McCray, R.: 1966, *Science* **154**, 1320.
Melrose, D. B.: 1968a, *Astrophys. Space Sci.* **2**, 171.
Melrose, D. B.: 1971a, *Astrophys. Space Sci.* **10**, 186.
Melrose, D. B.: 1971b, *Astrophys. Space Sci.* **10**, 197.
Melrose, D. B.: 1971c, *Astrophys. Space Sci.* **12**, 172.
Melrose, D. B.: 1971d, *Astrophys. Space Sci.* **13**, 56.
Melrose, D. B.: 1972a, *Astrophys. Space Sci.* **18**, 267.
Melrose, D. B.: 1972b, *Nuovo Cim.* **7A**, 669.
Melrose, D. B.: 1974a, *Astrophys. Space Sci.* **29**, 211.
Melrose, D. B.: 1974b, *Plasma Phys.* **16**, 845.
Melrose, D. B. and Sy, W. N.: 1972, *Aust. J. Phys.* **25**, 387.
Mikhailovskii, A. B.: 1967, in M. A. Leontovich (ed.), *Rev. Plasma Phys.* **3**, 159. (Consultants Bureau, New York.)
O'Dell, S. L. and Sartori, L.: 1970, *Astrophys. J.* **161**, L37.
Oort, J. H. and Walraven, Th.: 1956, *Bull. Astron. Soc. Netherlands* **12**, 285.

Oster, L.: 1960, *Phys. Rev.* **119**, 1444.
Oster, L.: 1961, *Rev. Mod. Phys.* **33**, 525.
Oster, L.: 1963, *Astrophys. J.* **138**, 761.
Oster, L.: 1968a, *Nature*, **217** 148.
Oster, L.: 1968b, *Astrophys. Space Sci.* **2**, 438.
Oster, L. and Altschuler, M. D.: 1971, *Astrophys. Space Sci.* **10**, 431.
Osterbrock, D. E.: 1961, *Astrophys. J.* **134**, 347.
Ostriker, J. P. and Gunn, J. E.: 1969, *Astrophys. J.* **157**, 1395.
O'Sullivan, R. A. and Derfler, H.: 1973, *Phys. Rev.* **8A**, 2645.
Pacholczyk, A. G.: 1973, *Mon. Not. Roy. Astron. Soc.* **163**, 29P.
Pacholczyk, A. G. and Swihart, T. L.: 1970, *Astrophys. J.* **161**, 415.
Payne-Scott, R., Yabsley, D. E., and Bolton, J. G.: 1947, *Nature* **160**, 256.
Pines, D. and Schrieffer, J. R.: 1962, *Phys. Rev.* **125**, 805.
Prentice, A. J. R.: 1967, *Plasma Phys.* **9**, 433.
Ramaty, R.: 1969, *Astrophys. J.* **158**, 753.
Razin, V. A.: 1960, *Radiofizika*, **3**, 584.
Reber, G.: 1940, *Astrophys. J.* **91**, 621.
Rees, M. J.: 1971, *Nature* **229**, 312.
Sazonov, V. N.: 1969, *Zh. Eskp. Teor. Fiz.* **56**, 1075; *Soviet Phys. JETP* **29**, 578.
Sazonov, V. N.: 1972, *Astrophys. Space Sci.* **19**, 25.
Sazonov, V. N. and Tsytovich, V. N.: 1968, *Radiofizika* **11**, 1287; *Radiophysics and Quantum Electronics* **11**, 731.
Scheuer, P. A. G.: 1960, *Mon. Not. Roy. Astron. Soc.* **120**, 231.
Scheuer, P. A. G.: 1965, in I. Robinson, A. Schild and E. L. Schucking (eds.), *Quasi-Stellar Sources and Gravitational Collapse*, Univ. Chicago Press, p. 373.
Scheuer, P. A. G.: 1968, *Astrophys. J.* **151**, L139.
Schott, G. A.: 1912, *Electromagnetic Radiation*, Cambridge Univ. Press.
Schwinger, J.: 1949, *Phys. Rev.* **75**, 1912.
Shkarofsky, I. P.: 1966a, *Phys. Fluids* **9**, 561.
Shkarofsky, I. P.: 1966b, *Phys. Fluids* **9**, 570.
Shklovskii, I. S.: 1953, *Dokl. Acad. Nauk. SSSR* **90**, 983.
Silin, V. P.: 1960, *Zh. Eksp. Teor. Fiz.* **38**, 1577; *Soviet Phys. JETP* **11**, 1136.
Silin, V. P.: 1961, *Zh. Eksp. Teor. Fiz.* **40**, 616; *Soviet Phys. JETP* **13**, 430.
Sitenko, A. G. and Stepanov, K. N.: 1956, *Zh. Eksp. Teor. Fiz.* **31**, 642; *Soviet Phys. JETP*, 512 (1957).
Sitenko, A. G.: 1967, *Electromagnetic Fluctuations in Plasma*, Academic Press (New York).
Sivukhin, P. V.: 1966, in M. A. Leontovich (ed.), *Rev. Plasma Phys.* **4**, 93 (Consultants Bureau, New York).
Smerd, S. F. and Westfold, K. C.: 1949, *Phil. Mag.* **40**, 831.
Sokolov, A. A. and Ternov, I. M.: 1968, *Synchrotron Radiation*, Academic-Verlag (Berlin).
Southworth, G. C.: 1945, *J. Franklin Inst.* **239**, 285.
Stewart, P.: 1972, *Astron. Astrophys.* **32**, 13.
Stewart, P.: 1974, *Astrophys. Space Sci.* **18**, 377.
Stix, T. H.: 1962, *The Theory of Plasma Waves*, McGraw-Hill (New York).
Sunyaev, R. A.: 1970, *Astrophys. Lett.* **7**, 19.
Sunyaev, R. A.: 1971, *Astron. Zh.* **48**, 244; *Soviet Astron. AJ* **15**, 190.
Svetozarova, G. I. and Tsytovich, V. N.: 1962, *Radiofizika* **5**, 658.
Synge, J. L.: 1970, *Quart. Appl. Math.* **28**, 437.
Tamm, I. E. and Frank, I. M.: 1937, *Dokl. Akad. Nauk. SSSR* **14**, 109.

Tarnstrom, G. L.: 1966, *Astron. Astrophys.* **49**, 31.
Tonks, L. and Langmuir, I.: 1929, *Phys. Rev.* **33**, 195.
Trubnikov, B. A.: 1958, dissertation, Moscow Institute of Engineering and Physics (English translation: USA EC Tech. Information Service, AEC-tr-4073).
Tsytovich, V. N.: 1951, *Vestn. Mosk. Univ. Phys. No. 11*, 27.
Tsytovich, V. N.: 1961, *Zh. Eksp. Teor. Fiz.* **40**, 1775; *Soviet Phys. JETP* **13**, 1249.
Tsytovich, V. N.: 1962, *Zh. Eksp. Teor. Fiz.* **42**, 803; *Soviet Phys. JETP* **15**, 561.
Tsytovich, V. N. and Chikhachev, A. S.: *Astron. Zh.* **46**, 486; *Soviet Astron. Aj.* **13**, 385.
Twiss, R. Q.: 1958, *Aust. J. Phys.* **11**, 564.
Twiss, R. Q.: 1963, *Phil. Mag.* **8**, 1249.
Unno, W.: 1956, *Publ. Astron. Soc. Japan* **8**, 108.
van de Hulst, H. C.: 1945, *Ned Tijdschr. Natuurk.* **11**, 201.
van de Hulst, H. C.: 1951, in *Problems of Cosmical Aerodynamics*, Central Air Documents, Dayton Ohio, Chapter 6.
Vedenov, A. A., Velikov, E. P. and Sagdeev, R. Z.: 1962, *Nucl. Fusion Suppl. Part 2*, 465.
Vlasov, A. A.: 1938, *Zh. Eksp. Teor. Fiz.* **8**, 29).
Vlasov, A. A.: 1945, *J. Phys. (U.S.S.R.)* **9**, 25.
Walsh, D.: 1970, *Astrophys. J.* **159**, 733.
Watson, G. N.: 1944, *A Treatise on the Theory of Bessel Functions*, Cambridge Univ. Press.
Westfold, K. C.: 1959, *Astrophys. J.* **130**, 241.
Westfold, K. C.: 1974, in Gordon Newkirk Jr. (ed.), *Coronal Disturbances*, IAU Symp. No. 57, D. Reidel (Dordrecht-Holland) p. 295.
Weymann, R.: 1965, *Phys. Fluids* **8**, 2112.
Wild, J. P.: 1950a, *Aust. J. Sci. Res.* **A3**, 399.
Wild, J. P.: 1950b, *Aust. J. Sci. Res.* **A3**, 541.
Wild, J. P. and Hill, E. R.: 1971, *Aust. J. Phys.* **24**, 43.
Wild, J. P. and McCready, L. L.: 1950, *Aust. J. Sci. Res.* **A3**, 387.
Wild, J. P., Murray, J. D., and Rowe, W. C.: 1954, *Aust. J. Phys.* **7**, 439.
Wild, J. P., Sheridan, K. V., and Trent, G. H.: 1959, in R. N. Bracewell (ed.), *Paris Symposium Radio Astronomy*, Stanford Univ. Press (Stanford).
Wild, J. P., Smerd, S. F., and Weiss, A. A.: 1963, *Ann. Rev. Astron. Astrophys.* **1**, 291.
Windsor, R. A. and Kellogg, P. J.: 1974, *Astrophys. J.* **190**, 167.
Woltjer, L.: 1966, *Astrophys. J.* **146**, 597.
Wooley, R. v. d. R.: 1947, *Suppl. Aust. J. Sci. Res.* **10**, 1.
Yukon, S. P.: 1968, *Astrophys. Lett.* **2**, 181.
Zeldovich, Ya. B. and Levich, E. V.: 1970, *Zh. Eksp. Teor. Fiz. Pis. Red.* **11**, 57; *Soviet Phys. JETP Letters* **11**, 35.
Zheleznyakov, V. V.: 1966, *Zh. Eksp. Teo. Fiz.* **51**, 570; *Soviet Phys. JETP* **24**, 381.
Zheleznyakov, V. V.: 1967, *Astrophys. J.* **148**, 849.
Zheleznyakov, V. V.: 1969, *Astrophys. J.* **155**, 1129.
Zheleznyakov, V. V.: 1971, *Astrophys. J.* **168**, 281.
Zheleznyakov, V. V. and Suvorov, E. V.: 1972, *Astrophys. Space Sci.* **15**, 24.
Zheleznyakov, V. V., Suvorov, E. V., and Shaposhnikov, V. E.: 1974, *Astron. Zh.* **51**, 243; *Soviet Astron. AJ* **18**, 142.

LIST OF FREQUENTLY USED SYMBOLS

a spectral index 4
$\mathbf{a}$ unit vector 43
$a_{ij}(\mathbf{k}, \omega; \mathbf{k}', \omega'; \mathbf{v})$ 91
$\mathbf{A}$ vector potential 149

b impact parameter 80
$\mathbf{b}$ unit vector along $\mathbf{B}$
$\mathbf{B}$ magnetic induction

c speed of light *in vacuo* 18
c_s sound speed 15

$\mathbf{d}$ electric dipole moment 38
d_{ij} electric quadrupole moment 39
$d\lambda^{(n)}$ n-fold convolution integral 29
D cold plasma parameter 53
$\mathbf{D}$ electric induction 38
$D_{ij}(\mathbf{k})$, $D_{\alpha\beta}(\mathbf{k})$ diffusion coefficients 162–163

e elemental charge 18
$-e$ electronic charge
$\mathbf{e}_1$, $\mathbf{e}^\alpha$ unit vectors 36, 106
$\mathbf{e}^\sigma(\mathbf{k})$ polarization vector 42
$\mathbf{E}$ electric field strength
E^α transverse component of $\mathbf{E}$ 200
$\mathbf{E}^\sigma(\mathbf{k})$ electric amplitude for waves 64

f cyclic frequency $(\omega/2\pi)$ 3
f_B cyclic electron gyrofrequency $(\Omega_e/2\pi)$ 19
f_p cyclic plasma frequency $(\omega_p/2\pi)$ 19
$\mathbf{F}^\sigma(\mathbf{k})$ energy flux in waves 46

$g_{ij}(\mathbf{k}, \omega; \mathbf{v})$ 72
G gravitational constant 18
$G(T_e, \omega)$ Gaunt factor 84

$\hbar$ Planck's constant divided by 2π 18
$\mathbf{H}$ magnetic field strength 38
$H_\nu^{(1)}(z)$, $H_\nu^{(2)}(z)$ Hankel functions 232

I specific intensity 3
$I^{\alpha\beta}$ specific intensity in the Jones calculus 196
Im imaginary part
$I_\nu(z)$ modified Bessel function 232

$\mathbf{j}$ current density
$J_s(z)$ Bessel function 232

$\mathbf{k}$ wave vector 17
K normalization coefficient 122
$K_\sigma(\mathbf{k})$ longitudinal component of polarization vector 43
$K_\nu(z)$ Macdonald function 233

m_e electronic mass 18
m_i ionic mass
m_p proton mass 18
m_α mass of species α

m magnetic dipole moment 69
M magnetization 38
$M\odot$, solar mass 18

$n, n_\sigma(\omega, \theta)$ refractive index (kc/ω) 42
n simple harmonic quantum number 148
n_α number density for species α
$N(\varepsilon)$ energy spectrum 122
$N^\sigma(\mathbf{k})$ occupation number for photons 146

$\mathbf{p}$ momentum
P pressure 15
P cold plasma parameter 53
$\mathbf{P}$ polarization 38
$P_l(x)$ Legendre polynomial 178
$P^\sigma(\mathbf{k})$ power radiated 67
$dP^\sigma(\mathbf{k})/dV$ power per unit volume 152
$\mathscr{P}$ Cauchy principal value 66

q_α charge on species α
q_{ij} electric quadrupole moment 39
Q Stokes parameter 198
$Q^\sigma(\mathbf{k})$ rate of energy generation 47

r_c degree of circular polarization 122, 198
r_e classical radius of the electron 18
r_l degree of linear polarization 110, 198
$r^{\alpha\beta}, r_V, r_U, r_Q$ Faraday rotation coefficients 197, 200
R radius of gyration 100
$R_\pm$ cold plasma parameters 53
$R_E, R_E{}^\sigma(\mathbf{k})$ ratio of electric to total energy 47–48
$R\odot$, solar radius 18

s harmonic number 98
s distance along a ray path 188
S cold plasma parameter 53
S_A Stokes vector 199

t time
$\mathbf{t}$ unit vector 43

T_b brightness temperature 3
T_α temperature for species α
$T_\sigma(\mathbf{k})$ axial ratio of polarization ellipse 43

$u^{\sigma\sigma'\sigma''}(\mathbf{k}, \mathbf{k}', \mathbf{k}'')$ probability of coalesence or decay 173
U Stokes parameter 198
$U^\sigma(\mathbf{k})$ energy radiated 67
U_s streaming speed 58–59

$\mathbf{v}$ velocity
v_A Alfvén speed 17
$v_g{}^\sigma(\mathbf{k})$ group velocity 47
V normalization volume 27
V Stokes parameter 198
V_α thermal speed for species $\alpha((T_\alpha/m_\alpha)^{1/2})$

$w^\sigma(\mathbf{p}, \mathbf{k})$ probability of Cerenkov emission 156
$w^\sigma(s, p, \mathbf{k})$ probability of gyromagnetic emission 57
$w^{\sigma\sigma'}(\mathbf{p}, \mathbf{k}, \mathbf{k}')$ scattering probability 93
$W^\sigma(\mathbf{k}), W_E{}^\sigma(k), W_K{}^\sigma(\mathbf{k}), W_M{}^\sigma(\mathbf{k}), W_P{}^\sigma(\mathbf{k})W^\sigma(\mathbf{k})$ energy in waves 45–46

x Cartesian coordinate
x_1, x_2, x_3 Cartesian coordinates
x_σ 103
X magnetoionic parameter $(\omega_p{}^2/\omega^2)$ 55

y Cartesian coordinate
y_α 51
y_σ 104
Y magnetoionic parameter (Ω_e/ω) 55

z Cartesian coordinate
z argument of Bessel functions 101
z_σ 104
Z_i number of elemental charges on an ion 21

α spectral index 4
α fine structure constant ($\approx 1/137$) 87
α pitch angle 100
$\alpha^{\alpha\beta}$ spontaneous emission coefficient in the Jones calculus 196–199
α_A spontaneous emission coefficient in the Mueller calculus

β plasma beta 22
β (v/c) 100

γ adiabatic index 15
γ Lorentz factor 70
$\gamma^{\alpha\beta}(\mathbf{k})$ absorption coefficient for waves in mode σ 47
$\gamma^{\alpha\beta}(\omega, \theta)$ (temporal) absorption coefficient in the Jones calculus 214

Γ Euler's constant 84
$\Gamma(x)$ gamma function 118
$\Gamma_{ij}(\mathbf{k}, \omega)$ 60

$\delta(x), \delta^3(\mathbf{r})$ Dirac δ-function 25
δ_{ij} Kronecker delta 31

ε particle energy ($=\gamma mc^2$)
ε_{ijk} permutation symbol 31
$\varepsilon_{ij}(\mathbf{k}, \omega)$ dielectric tensor 34
$\{\varepsilon^a_{ij}(\mathbf{k}, \omega)\varepsilon^h_{ij}(\mathbf{k}, \omega)$ anti-hermitian and hermitian parts of $\varepsilon_{ij}(\mathbf{k}, \omega)$ 35
$\varepsilon^\sigma(\mathbf{k}, \omega)$ 48
ε_k kinetic energy ($=\varepsilon - mc^2$)
$\varepsilon(\mathbf{p})$ particle energy ($=(m^2c^4 + p^2c^2)^{1/2}$) 242
ϵ_α sign of charge for species $\alpha(=q_\alpha/|q_\alpha|)$ 40

η mass density 15
$\eta^\sigma(s, \omega, 0)$ emissivity at sth harmonic in mode σ 103
$d\eta/dV$ volume emissivity 122

θ wave angle (arc cos $\boldsymbol{\kappa} \cdot \mathbf{b}$)
$\theta(t)$ step function 66

$\boldsymbol{\kappa}$ unit vector along $\mathbf{k}$ 23

λ_D Debye length 12
$\lambda_{D\alpha}$ Debye length for species $\alpha(V_\alpha/\omega_{p\alpha})$
$\lambda_{ij}(\mathbf{k}, \omega)$ 44

$\Lambda_{ij}(\mathbf{k}, \omega)$ 34
$\Lambda(\mathbf{k}, \omega)$ 41

μ spatial damping coefficient 47
$\mu^{\alpha\beta}$ (spatial) absorption coefficient in the Jones calculus 197
μ_{AB} (spatial) absorption coefficient in the Mueller calculus 199

ν nonlinearity parameter 89
$\nu_0, \nu(\omega)$ collision frequency 19, 182

ξ modified Lorentz factor, 114

$\rho, \rho(\mathbf{r}, t)\ \rho(\mathbf{k}, \omega)$ charge density

σ label for wave mode
σ_T Thomas cross-section 18
$\sigma_{ij}(\mathbf{k}, \omega)$ conductivity tensor 34

τ normalization time 64

ϕ electrostatic potential 11
ϕ azimuthal angle 76
$\phi(\alpha)$ pitch angle distribution 113
$\phi(y)$ plasma dispersion function 50

χ angle denoting plane of polarization 198
χ polar angle 76

ω angular frequency 17
ω_p plasma frequency 13
$\omega_{p\alpha}$ plasma frequency for species α 12
$\omega^\sigma(\mathbf{k})$ frequency for waves in mode σ 42

Ω gyrofrequency 100
Ω_α gyrofrequency for species α 19

INDEX